Engineering Empires

Also by Ben Marsden

WATT'S PERFECT ENGINE: Steam and the Age of Invention (2002)

Also by Crosbie Smith

ENERGY AND EMPIRE: A Biographical Study of Lord Kelvin
(with Norton Wise, 1989)

THE SCIENCE OF ENERGY: A Cultural History of Energy Physics in Victorian Britain
(1998)

Engineering Empires

A Cultural History of Technology in Nineteenth-Century Britain

Ben Marsden

and

Crosbie Smith

First published 2005 by
PALGRAVE MACMILLAN
Houndmills, Basingstoke, Hampshire RG21 6XS and
175 Fifth Avenue, New York, N.Y. 10010
Companies and representatives throughout the world

PALGRAVE MACMILLAN is the global academic imprint of the Palgrave Macmillan division of St. Martin's Press, LLC and of Palgrave Macmillan Ltd. Macmillan® is a registered trademark in the United States, United Kingdom and other countries. Palgrave is a registered trademark in the European Union and other countries.

ISBN 0–230–50704–2
ISBN 978–0–230–50704–3

This book is printed on paper suitable for recycling and made from fully managed and sustained forest sources.

A catalogue record for this book is available from the British Library.

Library of Congress Cataloging-in-Publication Data
Marsden, Ben.
   Engineering Empires : a cultural history of technology in
   nineteenth-century Britain / Ben Marsden and Crosbie Smith.
      p.   cm.
   Includes bibliographical references and index.
   ISBN 0–333–77278–4
   1. Technology—Social aspects—Great Britain.   I. Smith, Crosbie.
   II. Title.

   T14.5.M37 2004
   306.4′6′094109034—dc22
                                                        2004052595

10   9   8   7   6   5   4   3   2   1
14   13   12   11   10   09   08   07   06   05

Transferred to Digital Printing in 2007

Contents

List of Figures

Preface

Some years ago senior colleagues in the School of History at the University of Kent suggested to the authors that they might contribute a compact history of technology to an 'Introductions to History' series. From their own teaching in the history of science, technology and medicine at undergraduate and postgraduate levels, the authors increasingly recognised the lack of a modern scholarly introduction to major issues in the cultural history of science and technology in Britain and the British Empire between 1760 and 1914.

Even an important survey like Donald Cardwell's *Fontana History of Technology* (1994) fails to engage significantly with a series of recent historiographical innovations. Cardwell's narrative is not alone in retaining strong elements of 'whig history'. Such history is the history of the winning side, valuing the past only where it matches, or approaches, the present, and all but ignoring the 'failures', 'dead-ends' or paths not taken, except where they stand as salutary reminders of the extent of human folly, nurtured by arrogance or fashion (as famously articulated in Butterfield 1931). Whig history of technology treats as unproblematic the categories of 'progress' and 'improvement'. Such categories then structure a narrative populated with heroic 'pioneers' and prophets, praised for being 'ahead of their time' as they presciently 'laid the foundations' for the 'advanced' forms of technological achievement 'we know today'.

Within such narratives, individual technologies rarely gain significance by virtue of their relationships, or interconnections, with others; the absence of female practitioners and the masculine gendering of technologies goes unremarked; the history of technology becomes a history of success; historical debate reduces to disputes over questions of priority in invention or implementation; more often than not, technologies are construed as merely 'science applied'; practical and material detail takes priority over cultural meaning.

More generally, historians of empire have followed Daniel Headrick's *The Tools of Empire* (1981) on the relationships between technology and colonial history. In this approach, 'eager imperialists', located in the outposts of empire, lacked 'the industry to manufacture the tools of conquest.... they needed British technology'. The suppliers were 'the creators of the tools of empire – people like...the Lairds [iron shipbuilders of Birkenhead], the arms manufacturers – [who provisioned] the empire with the equipment that the peripheral imperialists required'. Thus many of the technological innovations 'that proved useful to the imperialists...first had an impact in the two decades from 1860 to 1880'. In this approach, therefore, 'technology helps explain events' which previously relied on political, diplomatic, and business motives invoked by imperial historians. These 'technological explanations' claim that innovations such as the steamship 'lower the cost, in both financial and human terms, of penetrating, conquering, and exploiting new territories. ... It is because the flow of new technologies in the nineteenth century made imperialism so cheap that it reached the threshold of acceptance among the peoples and governments of Europe' (Headrick 1981: 204–10; see also, for critique of Headrick, Smith *et al.* 2003b: 383).

Forbes Munro's *Maritime Enterprise and Empire. Sir William Mackinnon and his Business Network, 1823–1893* (2003) follows this 'tools of empire' model with its language of 'needs', 'developments', 'applications', and 'impacts'. Mackinnon was the founder of the British India Steam Navigation Company, which came to rule the Indian Ocean trades in the second half of the nineteenth century. Munro represents him as one of a handful of 'pioneering shipowners who pushed forward the transition from sail to steam' serving, in particular, the needs of Empire. The model is causal: 'the technological lead which the pioneering development of the iron- and steel-hulled steamship, together with successive innovations in marine engines, propulsion systems, hull-design and on-board and dockside cargo-handling machinery, gave to British shipowners and managers' was a principal factor underlying 'the growth of the mercantile marine' and, arguably, 'the growth and development of the Empire' (Munro 2003: 8, 498 [on Headrick], 485–88 [on steamship technology]).

It is rare for such approaches to admit that the new technologies were not simply 'givens' but rather were historically highly problematic and culturally contingent. A central aim of our book, then, is to highlight the cultural contingencies which shaped the varied technologies of empire in the long nineteenth century. To take just one example, we show how in the 1860s Alfred Holt, former apprentice on the Liverpool & Manchester Railway and deeply imbued with the values of his Unitarian religious culture, *chose*, unlike Isambard Kingdom Brunel and other contemporaries, economy of fuel over size, speed, and luxury in designing high-pressure compound engines for long-distance ocean steamships (ch. 3).

The work of historians of technology and other historians concerned with technology is thus ripe for revision, not least because, from the 1980s, the discipline of cultural history of science and technology has witnessed any number of historiographic reconstructions: the elaboration of a 'technological systems' approach by Thomas P. Hughes, especially in his *Networks of Power* (1983); the introduction of gender issues, notably in Ruth Schwartz Cowan's *More Work for Mother* (1983); and re-evaluations of the politics and meanings of technological success and failure, by John Staudenmaier (1989b) and others. But their substantive applications in the history of technology are still largely confined to programmatic research statements, of which the most seminal is Wiebe E. Bijker, Thomas P. Hughes and Trevor Pinch's *The Social Construction of Technological Systems* (1987), compilations of key articles, most usefully Donald MacKenzie and Judy Wajcman's valuable *The Social Shaping of Technology* (1985; 1999), and periodical literature, particularly within the pages of *Technology and Culture*, the journal of the Society for the History of Technology.

Engineering Empires makes these and other insights accessible to a wider audience. Key themes include exploration and mapping, motive power, steamships, railways and telegraphy. By demonstrating, in historical context, the complex and changing interactions between science and technology in the decades of the imperial expansion from 1760 to 1914, *Engineering Empires* seeks to combat charges of insularity in the history of technology, as a discipline, and to offer something both new and provocative.

For their support in preparing this book we thank Luciana O'flaherty, our patient editor at Palgrave; our students in history and philosophy of science and in cultural history at the universities of Kent, Leeds and Aberdeen; our colleagues, past and present, at the University of Aberdeen, especially Rainer Brömer, Peter McCaffery, Allan Stewart, David Ditchburn, Seth Kunin, Leigh Clayton, John Reid and Elizabeth Hallam, and at the University of Kent, especially Ian Higginson, Phillip Wolstenholme, Anne Scott, Alex Dolby, Charlotte Sleigh and David Turley. We also particularly thank Jon Agar, Will Ashworth, Yakup Bektas, Ana Carneiro, Gillian Cookson, Colin Divall, Aileen Fyfe, Graeme Gooday, Amalia Hatjievgeniadu, Colin Hempstead, Frank A.J.L. James, Eoin Magennis, James Massender, David Miller, Simon Schaffer, Claire Taylor, Jon Topham and Steve Woodhouse. Crosbie Smith remains as ever grateful to Norton Wise. Librarians and archivists at the universities of Aberdeen, Cambridge, Kent, Glasgow and Leeds, University College London, the British Library, the Institution of Civil Engineers, Liverpool Central Library (Record Office), the Merseyside Maritime Museum Archives, and the National Library of Scotland offered invaluable assistance. The authors thank the School of Divinity, History and Philosophy, University of Aberdeen, for a grant towards illustrations and indexing.

Ben Marsden's work on engineering science, academic engineering, and the cultural history of technology began with the support of the ESRC (under the guidance of Crosbie Smith) and has been extended with the financial assistance of the British Academy and the Royal Society of London. He aired some of the material on Brunel at Brunel University and at the Royal Institution Centre for the History of Science and Technology. Jon Turney and Simon Flynn at Icon Books indulged his interest in Watt and his thanks go to them. He dedicates this book to his parents, Pat Marsden and Dennis Marsden.

Crosbie Smith acknowledges the generous support of the British Academy and the Leverhulme Trust which made possible work on the cultural history of technology in wider contexts (notably through studies of Henry Adams) and the Arts & Humanities Research Board's Large Grant Scheme which is currently funding an in-depth cultural history of the ocean steamship in the Victorian period. He dedicates this book to his mother, Doreen, and late father, Jack.

Ben Marsden   Crosbie Smith
University of Aberdeen   University of Kent

Introduction: Technology, Science and Culture in the Long Nineteenth Century

> It may be safely averred that railways, steamships, and telegraphs
> are combinedly our most powerful weapon in the cause of Inter-
> Imperial Commerce.
>
> – *Charles Bright affirms the mutually supporting role of the*
> *three nineteenth-century technological systems fundamental*
> *to the commerce of the British Empire (1911)*[1]

Disciplining technology in history

Engineers are empire-builders: active agents of political and economic
empire, they have worked to build and expand personal and business
empires of material technology founded on and sustained by durable networks of trust and expertise. It is our aim in this book to re-examine, from
within, the cultural construction of the large-scale technologies of empire.
Beginning with an analysis of collective adventures in exploration, mapping
and measurement, we consider technologies of power (especially steam), the
recruitment and refinement of these powers in steamships and in railways,
and the mechanisms of communication (especially electrical telegraphy) by
which those powers, and their applications, were surveyed and controlled.

All of these studies are located in the 'long nineteenth century' between
about 1760 and the beginning of the First World War. All are rooted in and
reflect the diverse cultural contexts of Britain and its empire, although for
the sake of comparison we have made forays into the contrasting environments of France and North America. It is our wish, throughout, to draw
from three historical genres which are often, and in our view wrongly, seen
as distinct by practitioners and audiences alike. They are the history of
technology, the history of science and cultural history. But technological
histories have taken diverse forms for diverse audiences.

Popular histories of technology, for example, are most seductive, para-
doxically, where they express familiar themes: the inexorable march of material
technological progress; the individual triumph over adversity and the forces of

1

conservatism; and the moralised life of the engineering 'visionary', outside – and yet ahead of – his (always his) time. Such works, the unconscious heirs to Victorian narrators of progress in a 'seen universe' simultaneously moral and practical, peddle the rose-tinted retrospect, cataloguing the archaeology of a 'golden age' of British engineering and empire, now sadly gone; the quaint idiosyncrasies of local cultural heritage, polished brass and varnished wood, lovingly preserved in an age of globalisation, standardisation and synthetic materials; and the immediacy of 'comprehensible', craft-based technologies so unlike the black boxes of modern electronic consumables. Such narratives – populated by the 'gung ho' imperialist, the simple untutored craftsman, the evermore gargantuan steamship, the railway minutely detailed by route and locomotive design, the electric telegraph as Victorian 'forerunner' of the Internet – might well be visually pleasing and stylistically straightforward, adorning coffee tables and within easy reach on the bedside cabinet. But they comfort rather than challenge.

Economic and business histories of our chosen technologies and their promoters, have, on the contrary, focused on quantifiable matters, on macroscopic questions and on 'impacts' (especially financial ones): how many steam engines produced, by what companies, with what royalty payments – and with what measurable impacts on industry and society? How many tons of shipping constructed, for what cargoes? How many miles of track laid down and at what cost to the canal interests? How many words 'telegraphed', with what speeds and with what returns to shareholders? Although such considerations, and parallel questions of relative economic performance,[2] cannot and should not be ignored, it is not our prime concern to chart numerically the economic trajectories of firms, regions or nations wedded (by production, consumption or exploitation) to specific technologies perceived as 'givens'. Our interests, rather, lie in questions that are qualitative, local and circumstantial; they encompass the genesis, transformation and even the demise of technologies which in some cases had little direct 'economic impact'.

Our interest in this 'shaping' of technologies implies a wish to move beyond what is, perhaps unfairly, dismissed as the 'antiquarian' in the history of technology. Works in this genre, although often valuable as sources of reference and of the suggestive example, dwell on detail for its own sake;[3] they attend primarily to the 'internal' dynamics, material construction and design particularity rather than to the broader, 'external' meanings or patterns of use associated with a technological artefact: the precise dimensions of a steam cylinder; the pitch of a screw propeller; the gaudy livery of a railway locomotive; the distinctive tap of the telegraph key and so on. In this study our concern will be to relate such local particularities to larger events and trends: the shorthand for this is 'context' and especially 'social context'.

A mature 'history of technology', seeks, then, to recognise the strengths but also the deficiencies of other genres (popular, economic, antiquarian) where they treat technology in history; but above all it seeks to admit social context, in all its richness and flux, as a framework for elucidation and explanation. Thus we understand an efficient steam engine as banishing waste construed as morally reprehensible; an electric telegraph promoted as the solution to problems of public order in an urban society; a plainly fitted-out steamship reflecting the ordered moral economy of its Presbyterian backers; a transcontinental railway literally building and binding a nation.

Yet there is a seeming paradox here. Part of that social context must be concerned with language, texts and words; yet the very word 'technology' is of recent origin. As historians of science know well, it was William Whewell who coined the term 'scientist' in the 1830s but not for decades did it become common parlance. Contemporaries like Michael Faraday preferred the address 'philosopher', with all its connotations of disinterested systematic, but not necessarily collective speculation into the causes of natural phenomena.[4] Likewise, Edinburgh professor George Wilson, at his lectern in 1855, needed to give prospective visitors to his new industrial museum a lengthy and laboured answer to the question 'what is technology?'[5] Campaigns by Whewell, Wilson and many others serve to remind us that, through the period of this study, we must be sensitive to the terms used by our historical agents ('actors' categories') and to the nuances of those terms.[6] That is, we must aim to deploy their language rather than adopting, or imposing, modern terms which fail to capture the richness of their contemporary experience, fail to appreciate their agendas for change – or worse, impose distinctions which were not theirs.

One such distinction, in fact, is to imagine a 'natural' separation between 'science' and 'technology' (rather as we now do between 'science' and 'religion'). Merely using the words 'technology' and 'science' carries the assumption that there are distinct well-defined entities of 'science' and 'technology' that have persisted through history; it raises questions, probably unresolvable ones, about the 'relationship' between them (in some respects analogous to the questions concerning the 'relationship' between science and religion).[7] Amongst professional scholars, the 'history of technology' and the 'history of science' have most often been seen as related and yet distinct: 'mirror-image' disciplines, perhaps, with different objects and methods, echoing what one author has memorably described as the 'mirror-image twins' of 'science' and 'technology' themselves.[8]

This disciplinary distinction has been convenient and valuable, especially in the United States, in encouraging authors to take the history of technology (and indeed of engineering) as seriously as the history of science has been taken. But making and taking an essentialist definition of 'technology' (whether Wilson's or anybody else's) cannot easily be squared with the constantly changing categories of the historical actors. Boundaries there may have been between science and civil engineering (as it emerged in Britain in the late eighteenth century), or science and the 'practical arts' (perhaps the closest approximation to technology before the alliance of science and engineering in 'engineering science' from the mid nineteenth century).[9] One of our concerns is to avoid present-centred boundaries and rather, as historians, to watch those contests play out in practical environments on all scales of endeavour. The 'craftsman' James Watt insisted on his status as 'philosopher' (or man of science); steamship promoters worked intimately with men close to academic natural philosophy; railways were 'experiments' to their nineteenth-century promoters; the chief agents of the telegraph advertised it as a practical element of experimental electrical science. The issues, then, for our historical actors were the shifting and contested boundaries of science and the practical arts.

Technology and culture

The kinds of history sketched above – popular, economic, antiquarian, or 'essentialist' histories of technology – tend to put beyond debate the questions that, on the contrary, are central to us. Cultural history, on the other hand, offers both an alternative and an opportunity. Although this expanding field is complex and contested, a workable definition of cultural history is the study of the construction (or production) and the dissemination (or reproduction) of meanings in varying historical and cultural settings.[10] Cultural historians, therefore, have many and large concerns: with power; with practice, process and contingent experience; with space and location; with the private and the public; with the distinctions between high (elite, learned) and low (popular, folk) cultures; with the emergence – or co-production – of elite and popular cultures;[11] with modes of representation, constitution and dissemination of knowledge, in oral, literary or visual cultures;[12] and with patterns of memory and commemoration. In particular, cultural history is concerned with the ways in which material artefacts have been used to convey such meanings, especially (but not exclusively) in exhibitions, museums, or other orchestrated displays.[13] The fact that the history of science, likewise, focuses increasingly on questions of practice and of material things suggests, therefore, a pragmatic convergence of cultural history, the history of science and the history of technology – since the last cannot but be a history of the physical, as much as the social or intellectual, products of engineering endeavour.[14]

In this study we want to place particular emphasis on a set of keywords adopted from the practice of cultural history but adapted, also, to our specific needs. We too will be concerned with technology as culture (rather than divorced from it), and with technological endeavour as fluid practice (rather than solid product). In admitting historical contingencies, we cannot but be open to the sometimes surprising attributions of success and failure which historical actors have applied to our core technologies. Questions of representation loom large here, and they lead us to re-examine the special relationships, in the worlds of engineering, between private and public and between high and low cultures. In a dramatic sense, we are intimately concerned with questions of power: but not merely with power as an isolated physical (or even economic) quantity. As engineers built empires, they jostled for cultural authority, managing, as they must, entirely new networks of 'trust', 'trustworthiness' and 'confidence' on a grand scale; they looked to assemble, to make coherent, and to make robust their 'systems' of technologies spanning vast geographical ranges; and, yet, 'experiments' in engineering, designed to trump competitors with the offspring of a fecund marriage of science and practical art, destabilised developing systems – and begged questions about the right relationship between technology and science.

We accept, then, that the study of technology in history is part of the study of *culture* more generally. That modern students will construe technology and culture as interlinked is evidenced by the leading journal in the field, *Technology and Culture*. One can argue about the advantages of asserting (more often tacitly accepting) a clear-cut distinction between 'technology' and 'culture', especially when that distinction appears to confer sense on two questions making one prior to the other: 'does technology produce culture' or 'does culture produce technology'? If it is unhelpful to see technology as a 'product' of culture, or vice versa, we might instead prefer to see 'technology' and 'culture' in simultaneous reciprocal transformation – each involved in the other's production and each conferring meaning on the other.[15]

When we ask how the history of technology is cultural history, we are asking to what extent technologies have been shaped by, and, simultaneously, to what extent they have shaped, the cultures they in part constitute.[16] That mutual shaping is not simply a matter of economics. A useful notion here is that of 'interpretative flexibility': different social groups having interpreted, or given meaning to, a technology differently according to their own local needs and demands, as famously shown in Pinch and Bijker's study of how the 'safety bicycle' achieved its stable form amidst a plethora of rivals, in form and function, from the Penny Farthing on. That study showed that the eventual stabilisation of a technology in one particular form, and with one particular meaning, is as much a social event as a material one.[17]

Our ambition, then, is to study the *practices* involved in the projection, creation, expansion and stabilisation of technologies – rather than focusing on technological 'end products', or disembodied ideas and theories, at the stage in which debates and contests over legitimacy have ceased. We accept, therefore, historical *contingency* rather than assuming the inevitable success of certain projects or technologies, especially those subsequently found to have been 'successful', in some sense, in the long term. It is not our job to fabricate stories implying the inevitable success of certain technologies, nor to produce narratives implying the pre-ordained failure of others, but rather to examine the contingent circumstances that led to the exclusion of viable technological possibilities (subsequently dubbed 'failures') – and thus to suggest that events might well have unfolded otherwise.

We are interested, furthermore, in technological process as it was construed at the time. We recognise, of course, that change in technology was not unguided or blind: clearly engineers and technological 'projectors' had intentions and ambitions; and they were opportunistic, sometimes wildly so. But we do not abuse hindsight to judge the 'rightness' or 'wrongness' of those ambitious intentions or imagined uses. We recognise that historical actors frequently hedged their bets, advocating multiple uses for their technologies, only to find that one alone, or none of them, were in fact ultimately taken up. Thus a 'useful thing' or everyday artefact designed with one end in mind might have been deployed in other quite different arenas.[18] But we accept that if their prophecies were ultimately unfulfilled, we have no right, as historians, to discard their statements from the debates of the day.

Whether or not a technology succeeds, or fails, is as much to do with the social as it is to do with any supposed inherent material worth. Classic studies illustrating this point include Ruth Schwartz Cowan's work on the domestic technology of the humble refrigerator: why did the gas fridge fail, and the electric fridge succeed?[19] There have also been detailed studies of the electric vehicle, a technology which might be seen as a failure relative to the internal combustion vehicle.[20] If historians have until recently been squeamish about technological failure, there has been an upsurge of interest, not least promoted by the realisation that engineers have been all too aware of the ubiquity of failure and of the lessons to be learnt from it.[21] Indeed, even Samuel Smiles, accused of taking too much notice of 'men who have succeeded in life... and too little of the multitude of men who have failed', gave a considered response to the question: 'Why should not Failure... have its Plutarch as well as Success?'[22]

For nineteenth-century examples of technological failure, we need not look far. Making compasses for the iron-built steamships of empire *reliable* was no straightforward proposition.[23] For every successful Watt engine with separate condenser, there were untold rivals, vaunted to supersede steam in all its industrial applications, yet in the final analysis, unable to realise the advantages promised on paper or indicated in elaborate demonstrations.[24] Railway innovators trumpeted the atmospheric system as a cheap successor to the expensive locomotive – yet this was a technology which expired with barely a gasp, never to succeed as a commercial concern.[25] Most remarkable of all, perhaps, was the electric telegraph (ch. 5) which, especially in its submarine manifestations, lurched from one deep-sea disaster to the next high-profile failure. And yet it would be central to a transformed information culture of the 1860s.

The fact that all these technologies could, at least briefly, simultaneously support interpretations at opposite poles – destined for 'success' *and* doomed to 'failure' – indicates that technological advocates were adept at attaching less crude, and indeed more ideologically driven, *representations* to their offspring: the transoceanic steamer embodying imperial power, the railway as annihilator of space and time, or the Atlantic telegraph as peacemaker, bonding Old and New Worlds. The Victorians created such representations in art and music, but above all they used the vast and changing possibilities of print. The media adopted for the dissemination of those representations alerts us, again, to questions of private versus public, and elite versus popular: our period saw the creation of elite institutions for specialised branches of engineering; yet it also witnessed the creation, through new printing technologies and through railway distribution, of vast popular audiences for coded news about science and the practical arts.

Engineers had a particular problem, more so, perhaps, than men of science, when it came to releasing, at the right moment and to the right audience, sanitised accounts of finished technological products. It was by no means easy to conceal the mess of technological process when the product was an innovative steam-powered mill, an 'experimental' railway or a great steamship ready to be launched. Both a Boulton and a Brunel were keenly aware of the rewards to be reaped from theatrical presentation; sober innovators, like Watt or Holt (chs 2, 3) wanted their works to speak for themselves. Part of the 'representation' of technologies involved cultures of *display* in which marketing and carefully staged demonstration, often in public, went hand in hand. Historians of science have noted how the context of display, or spectacle, and the constitution of the audience for such display, has raised epistemic questions.[26] To display was vital, but to manage that display well was equally important. Yet to be seen to manage was to be seen to manipulate – and to raise questions concerning credibility and trustworthiness.

Our engineers then, like their contemporaries in science, grappled with issues of trust as they sought to consolidate power, as they sought to develop and expand systems (whether scientific or technological), and as they sought to balance the promise of the experimental innovation (risky, yet of high kudos for the technological projector) with the reliability of the established routine. Steven Shapin observes that 'the role of trust and authority in the constitution and maintenance of systems of valued *knowledge* has been practically invisible'. Since the seventeenth century, he argues, scientific practitioners and apologists have represented trust and authority as standing against the very idea of scientific knowledge: 'Knowledge is supposed to be the product of a sovereign individual confronting the world; reliance on the views of others produces error.' Typically, then, 'the plausibility of a [knowledge] claim and the trustworthiness of a claimant can appear as independent variables, which, when summed, factored, or compared together, yield a reliable judgement of credibility'. In contrast, Shapin shows that schemes of plausibility, carrying the appearance of being independent of trust, were 'themselves built up by crediting the relations of trusted sources'.[27]

Shapin's insights, developed for the seventeenth century, seem especially appropriate to our period when the 'trustworthiness' of engineering projects was often far from self-evident to contemporary publics. We challenge conventional assumptions that first, 'trust in technology' is independent of those human actors who communicate that trust to others and second, that 'a morally consequential sense of trust' or trust in promised actions is *only* relevant to personal and business relations. As Shapin affirms, both these forms of trust (involving *communication* and *action* respectively) serve as 'systems of expectation about the world', as engendering states of belief, and are 'implicated in schemes of co-ordinated *action*'.[28] Communications of promised actions, especially with respect to re-engineering the iron steamer (ch. 3), constitute some of the issues that we shall attempt to explore in later chapters. In these explorations we indicate how historical actors might be afforded trustworthy status by virtue of appropriate behaviour (perhaps the behaviour of the 'gentleman' – although not *all* aspects of gentlemanly action served to induce trust in the engineering context); and that questioning the status of the engineer was one way in which the trustworthiness of the technology he advocated might be eroded.

That questions concerning trustworthiness, through 'confidence' and 'character', like questions of failure, are not simply the concerns of modern historians is fully demonstrated in our case studies; but for the moment, consider the remarks of Charles Babbage on the intimate relations of confidence, character and capital in the manufactures, and trade, of Britain and its empire in 1832. According to Babbage, 'character' generated 'confidence'; moreover, the 'character' of a firm correlated with 'capital' since a firm of great capital had more to lose if it were to act with low character. As an old trading nation, England had an advantage over its national rivals because of the 'well-grounded confidence' in the character of its merchants. As a result of that confidence fewer written contracts were required, there was less need for classes of 'middle-men' to assess and vouch for the quality of goods – and a freer market obtained.[29]

Most of our engineers were concerned with the extension of patterns of trust over ever-wider geographical areas in order to make technologies reliable. In fact, one of the most powerful approaches in recent history and sociology of technology has been to think in terms of the deliberate evolution of 'networks' or the design and advocacy of engineering 'systems'. The exemplar of this approach is Thomas Hughes's inspiring study of the growth of *Networks of Power* (1983) – here, electrical power distribution – in the late nineteenth and early twentieth centuries. Hughes's point was to see engineers working within an environment that was simultaneously social, political, legal and material; ambitions to expand their large technological systems could only be realised if they could solve not merely the material technical problems but also problems in the social and political arenas. More recently, Hughes's approaches have been hybridised with social constructivist methods to beget a vigorous programme under the banner of the 'social construction of technological systems'.[30] Stated simply, we gain interrogative purchase and narrative power, as historians, by considering technologies as complex systems of heterogeneous elements, given collective meaning by and within their social milieus.

In this study, we frequently track our historical actors working deliberately as system builders, responding to multifarious encounters, tackling and not always surmounting obstacles and resistances (although often grasping the media agenda to subdue the voices of counter-cultures). We see how economic geologists defined and imposed their preferred systems on masses of data in ventures little short of global colonisation; how heat-engineers co-ordinated the production and assembly of specialist engine parts, and recruited allies for international marketing of at-first fragile products; how shipping magnates networked ports at home and abroad with ships made robust and captains and crews made trustworthy; how railway engineers, like Brunel, went beyond the mere connection of town and country to weave webs which integrated rail, road and ocean steamer; how submarine electrical telegraphers spliced Old and New Worlds, building international systems of communication that linked existing national systems. In order to see that the historical actors themselves were actively aware of their building of systems, and even systems of systems, we need only recall Charles Bright's claims, quoted in the epigraph above, that rail, steamship, and telegraph combined in a *single* weapon of greater imperial power than any component technology.

Effective techniques by which systems have been built, stabilised and maintained include the imposition (or attempted imposition) of *standards*, whether technical, practical or social; and the related development of cultures of measurement.[31] We discuss a rise of 'chartism' or a generalist attempt to reduce the messiness of natural and technological variety to the graph, given meaning with respect to an absolute standard (ch. 1). We examine the marketing of steam that drove engineers' attempts to reduce mechanical complexity to a generally accepted and understood measure of power (ch. 2). Measures of position – underpinned by campaigns to export standards of longitude – underpinned maritime endeavour; controversies over standards of gauge for British railway track cut to the heart of system builders' ambitions and powers; and a commitment to electrical standards defined the new professionalism of the electrical engineer, spawned from cultures of electrical telegraphy (chs 1, 4, 5).

Our accounts contain numerous examples of attempts to standardise or, better, to make routine, aspects of the skills necessary to give technologies stability and expansive power – often through an interplay of disciplined human agent and the specialist instrumentation that embodied skill: making ships' compasses work, requiring protracted negotiations concerning man and machine; training up a new race of steam engineers or, by way of avoiding human error and fatigue, introducing self-acting instruments to allow the new engines constant self-monitoring; solving problems of management aboard the new, sometimes vast, steamships; fixing time for the smooth operation of railways; distributing instruments easily and reliably read to facilitate the spread of a global telegraphic system. Such programmes of standardisation lay at the heart of engineering empires.

In creative tension with the introduction of the routine (with its emphasis on reliability, predictability and rationalisation) was the notion of engineering innovation especially through *experiment* in technological culture, often, although not always, associated with the ambitions of the masculine engineering hero, and often, although again not always, the focus of vigorous debate in public arenas. Standardisation was a means of consolidation and also of exclusion (e.g. excluding rival operators from railway tracks); but it could be viewed – indeed was viewed by innovative engineers, including Brunel – as engendering stagnation in engineering and the practical arts (ch. 4). How could the arts progress when there was so great a push to rebuild a world according to a standard template of rules and conventions? The very robustness of standardised systems (e.g. a standard rather than broad railway gauge), by materialising the social and the technological created practical momentum – or inertia. It conditioned if not determined technological choices, and probably reduced the viability of engineering experiments. Technological constraints pleased a certain kind of sponsor – but they were surely constraints on radical action.

In our case studies, we see, then, just such a tension where engineers grappled with the notion of the engineering experiment, be it an experimental heat engine, a proposed new marine design dignified as 'experimental', the first railways represented explicitly as transport experiments, the land telegraphs as emerging from a context of electrical experimentation, or the transatlantic submarine lines dubbed the most vast electrical experiment ever. Such 'experiments' made cultures of technology 'scientific' through discourse and more; experiment emphasised innovation and augured progress; to some extent it provided discursive security against – and thus legitimated – failure; but the rush to experiment did not secure the trust and support of sponsors except in rare cases. Tradition and practical experience promised reliability; experiment appealed to a language of science for science's support; yet our case studies show the fragility of technologies ostensibly built on science.[32]

The case studies that follow can be read as illustrating key aspects of technological cultures; they are suggestive in the sense that techniques and approaches deployed to study them can be readily 'transferred' to other case studies and historical locations. More than that, however, they have a bearing on the larger historiography of empires in the nineteenth century. Conventional historical accounts of the British Empire focus on high politics, military coercion, religious or cultural impulse and imperative; another form of histories of empire recognises the inadequacy of such accounts without the requisite tools. But even those have tended to take the tools as 'givens'. We will argue, as we have suggested above, that any technology needs to be understood not as a given but as both shaped by, and shaping, culture. Our study, then, attempts to penetrate the black boxes of 'steam-power', 'steamship', 'railroad' and 'telegraph' to see each of those 'tools of empire', not simply as a fixed product with a given role, but also as a dynamic system, formed according to the contexts of exhibitions, experiments, standardisation and so on, for varied and specific ends including empire – but not limited to it. Thus our technologies were not only vital to the formation, maintenance and shaping of empire; they were also, themselves, shaped within contexts including but not restricted to political or military empire. Moreover, they were shaped by active agents concerned to build not only their own personal and business empires but also informed of the possibilities of British imperialism.

1
'Objects of national importance': Exploration, Mapping and Measurement

I have already adverted to what the influence of the [British] Association [for the Advancement of Science] may effect, in causing the spaces yet vacant on the map, in the British possessions in India and Canada, to be filled. But beyond all comparison, the most important service of this kind, which this or any other country could render this branch of science [terrestrial magnetism] would be filling the void still existing in the southern hemisphere, and particularly in the vicinity of those parts of that hemisphere which are of principal magnetic interest. This can only be accomplished by a naval voyage; for which it is natural that other countries should look to England.... Viewed in itself and in its various relations, the magnetism of the earth cannot be counted less than one of the most important branches of the physical history of the planet we inhabit; and we may feel quite assured, that the completion of our knowledge of its distribution on the surface of the earth would be regarded by our contemporaries and by posterity as a fitting enterprise of a maritime people, and a worthy achievement of a nation which has ever sought to rank foremost in every arduous and honourable undertaking.

> – *Roderick Murchison addresses the British Association for the Advancement of Science on the national importance of mapping the earth's magnetism (1838)*[1]

Leading geologist and gentleman of science Roderick Murchison's words expressed a mapping imperative widely shared by European cultural elites since at least the eighteenth century. From the point of view of an imperial power, 'spaces yet vacant on the map' – especially within existing British territories – meant that 'possession' was incomplete, that control, if any, still resided with other forces, and that 'civilisation' – in the form of Western 'rationality' and 'discipline' – had yet to be introduced to that region. For

Murchison, the British Association for the Advancement of Science (BAAS) –
then still in its infancy – could and should serve as a principal agent for
securing such British power, both on land and sea, and above all in 'filling
the void still existing in the southern hemisphere'.

In the second half of the eighteenth century, British audiences had read
the three voyages of Captain James Cook as destroying myths concerning
an undiscovered Southern Continent. Cook's practices of accurate charting –
made possible through the skilled use of navigational instruments – imposed
geometrical order on the coasts and seas of a region long regarded by his
countrymen as a vast space yet vacant on any map. Such quests for accurate,
'scientific' knowledge bore witness to the major cultural shifts in European
society which historians have identified – at least for convenience – by the
terms 'Scientific Revolution' (seventeenth century) and 'Enlightenment'
(eighteenth century). In Britain the *new* sets of cultural values were located
in both Court patronage and, more widely, in a gentlemanly culture which
found expression in the Whig party and its adherents, in metropolitan learned
societies, and, increasingly, in provincial literary and philosophical societies.[2]

Traditionally identified with 'heroic' Europeans such as Copernicus,
Galileo and Newton, the 'Scientific Revolution' was more about the promotion of a fresh set of cultural values. Experiment, reliably undertaken
and properly communicated by men of status, and trust in observation of
the Book of Nature replaced trust in the ancient wisdom. In Britain, specific
institutions embodied these values. Established under King Charles II's
patronage during the period of the restoration of the monarchy following
the social, political and epistemological instabilities of the English Civil
War, both the Royal Society of London (founded 1660) and the Royal
Observatory at Greenwich (1675) acquired enduring cultural authority
through their command of architectural and social space in the commercially
thriving metropolis.[3]

The Restoration Court expected in return to gain prestige from the new
discoveries in natural history, natural philosophy and astronomy which would
accrue from trustworthy methods of acquiring knowledge. Francis Bacon,
writing his *New Atlantis* in the earlier part of the seventeenth century, had
held out a vision of the fruits of this new knowledge, of the benefits to
mankind, and had coined the maxim that 'knowledge itself is power'. By
the end of the seventeenth century, the gentlemen of the 'Whig Revolution'
in English politics had embraced the new sciences as their own. In particular,
they consolidated a moderate Anglican theology at the core of Church and
State and regarded a knowledge of God's creation as a foundation of social
and political order, a bastion against any return to the disorder of the earlier
Civil War and the religious enthusiasm of much of the sixteenth and
seventeenth centuries throughout Europe.[4]

Famously asked whether he lived in an enlightened age, the German
philosopher Immanuel Kant replied that it was not so; rather, he was living

in an 'Age of Enlightenment'. Impossible to define, the Enlightenment itself seemed to contemporaries such as Kant a historical process characterised by a proliferation of new sciences which had taken their cue from the mathematical and experimental sciences of the previous century. Above all, it was an age of exploration and measurement made tangible in maps and charts.[5]

Among modern analysts, Michel Foucault identified the relations, embedded in space, between knowledge and power. He claimed that space expressed in a geopolitical language of 'region', 'domain', or 'territory', provided the key to understanding the possible relations between power and knowledge: 'Once knowledge can be analysed in terms of region, domain, implantation, displacement, transposition, one is able to capture the process by which knowledge functions as a form of power and disseminates the effects of power.' Politico-strategic terms such as 'domain', moreover, are indications of 'how the military and the administration actually come to inscribe themselves both on a material soil and within forms of discourse'.[6]

Jacques Revel's work on the national territory of France offers an excellent case study of knowledge of the territory in relation to political power and control. Compared to Foucault's abstract spatial discourses, Revel's account is securely embodied in the political history of France. By 1300, France had become associated with a specific geographical area to be invoked as 'a garden of perfection', an 'intangible territory' and 'a homeland in defense of which its inhabitants would soon learn to die'.[7] But control over the national domain was far from permanent and assured. Over the next five centuries, therefore, the public authorities sought to acquire knowledge of, establish control over, and lend homogeneity to the diverse regions within the national borders.

With respect to statistical surveys, Revel draws attention to two distinctive approaches widely deployed during the last century of the Ancien Régime which ended with the French Revolution of 1789. The first approach recorded data (such as population, yields and prices) *over time* in order to gauge developments and tendencies. The State often promoted such surveys on a national scale. In contrast, the second approach reflected nature as closely as possible, was descriptive and inclusive of all aspects of a locality, and sought to reconstruct the relations among these variables. Spatial in character, the particular 'natural history' of a region could be juxtaposed with the natural history of other regions to produce a collage. Here the promoters were often Enlightenment figures, acting in the public interest and claiming unique 'field experience': 'a spontaneous network of travelers, geographers, economists, agronomists, doctors, low-ranking administrators, and notables of the area'.[8]

In the immediate aftermath of the Revolution, the new regime utilised both approaches. On the one hand censuses provided the Republic with statistical information on the nation's material and human resources for rapid decision-making in a time of crisis. On the other hand the ideology of the new order

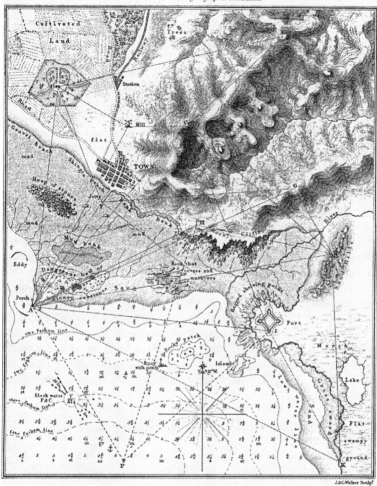

Figure 1.1   In *A Manual of Scientific Enquiry; Prepared for the Use of Officers in Her Majesty's Navy; and Travellers in General* (1851) and edited by Sir John Herschel, Captain F.W. Beechey instructed readers in the techniques of chart-making and surveying in order that ' "golden opportunities" of acquiring a knowledge of distant ports' would not be lost.

Source: J.F.W. Herschel (ed.), *A Manual of Scientific Enquiry; Prepared for the Use of Officers in Her Majesty's Navy; and Travellers in General*. Second edition. London: Murray, 1851, facing p. 1. (Courtesy of Aberdeen University Library.)

aspired to unify the diversity of the people throughout the territory. Descriptive surveys, 'systematic and even encyclopedic' in character, were thus 'centrally initiated and locally implemented'. Advocates of such public surveys sought to encourage the transition from regional diversity to national unity. With the increasingly centralised and authoritarian regime, however, local co-operation soon yielded to state power and privilege. Ken Alder has also shown the ways in which the revolutionary regime introduced the metric system and thereby attempted to replace local custom and tradition by uniformity.[9]

Turning to another representation of the territory in France, Revel notes that maps combined practical and symbolic value. For their makers and users they served as a form of power and 'the preferred means of transmitting political intentions'. Cartography thus 'became inseparable from the affirmation of monarchic power; delineation of the territory was first and foremost the king's business'. Indeed, the king need no longer travel in order to survey his territory. It was rather the role of official surveyors, engineers and royal geographers to provide 'a new visual support for the monarchy's ambitions' in war or peace.[10]

'Enlightenment' carried the practices of surveying and map-making into the very processes of making empires, especially maritime empires (Figure 1.1).[11] Central to such projects in Britain and in France, astronomical observatories located in Greenwich and Paris had as their primary object of national importance the accurate determination of longitude. State-sponsored authorities such as Britain's 'Astronomer Royal' – whose title complemented such national symbols as the 'Ark Royal', the name often given to the Royal Navy's most prestigious warship – now advanced their own reputations as expert determiners of longitude. Trusted to provide the respective – and rival – British and French fleets, both naval and merchant, with the knowledge to navigate safely across hostile oceans, these astronomers championed the Enlightenment preoccupation with exploration, measurement and mapping.

'The whole civilised world has heard of Greenwich'

By the third quarter of the nineteenth century the Royal Observatory at Greenwich regulated standardised and centralised time throughout the country and served the goals of commercial and imperial expansion. Through an expanding network of underwater electric telegraph lines largely controlled by Britain (ch. 5), Greenwich time could be transmitted to most of the British Empire by the end of the century. And in 1884 Greenwich Mean Time became the internationally accepted basis of the world's time zones.[12]

The Royal Observatory at Greenwich did not mark the first meridian because of some 'natural' characteristic. At the International Meridian Conference held in Washington (1884) to fix on an agreed first meridian for time and longitude, Sandford Fleming, the British delegate representing Canada and engineer-in-chief of the Canadian Pacific Railway (chs 4, 5), presented a table showing that 72 per cent of the world's commercial shipping used Greenwich as the first meridian. The remaining 28 per cent, Fleming showed, was divided among ten different first meridians. Indeed, around half the world's shipping flew the Red Ensign and by convention used charts which assumed zero degrees of longitude to pass through Greenwich. Whatever the attractions of a theoretical meridian independent of nationality and empire, practice pointed to Greenwich. As a result, the Conference voted overwhelmingly for Greenwich, with France and Brazil abstaining and only San Domingo expressing opposition.[13]

Joseph Conrad's novel *The Secret Agent* (1907) vividly illustrates the almost-sacred significance attached to the Observatory during the later years of the Victorian Empire. The author introduces, through the character of a foreign 'diplomat' named Vladimir, a sinister plot to destroy the Greenwich Observatory, and thus symbolically to attack the first meridian. Deeply angered by British toleration of all kinds of revolutionaries plotting the overthrow of Continental autocracies, Vladimir plans a terrorist outrage designed to provoke the British middle classes to demand repressive action against their dangerous guests and accordingly informs his 'secret agent':

> The fetish of today is neither royalty nor religion. Therefore the palace and the church should be left alone.... The sacrosanct fetish of today is science... [The middle classes] believe that in some mysterious way science is at the source of their material prosperity...The attack must have all the shocking senselessness of gratuitous blasphemy...What do you think of having a go at astronomy?...The whole civilised world has heard of Greenwich...the blowing up of the first meridian is bound to raise a howl of execration...Go for the first meridian.[14]

The Royal Observatory's authority in the later nineteenth century as the great regulator of the nation's time – and thus of its trade, transport, commerce and industry at home and across the world – had been neither inevitable nor self-evident. Instead, through a series of contingent processes, the Observatory became one of the most potent symbols of Britain's maritime wealth and power. Back in 1673 a Committee, appointed by Charles II and including among its members Lord Brouncker (President of the Royal Society of London and Controller of the Navy) and Robert Hooke (curator to the Royal Society), was investigating the merits of proposals for measuring longitude at sea. The following year the King appointed a Royal Commission with Sir Christopher Wren (the King's Surveyor-General) added to the membership of the existing Committee, to look at another longitude claim. Meanwhile, the Royal Society planned an observatory at Chelsea, well upriver from London Bridge. Sir Jonas Moore, who with Samuel Pepys and others had founded the Royal Mathematical School within Christ's Hospital to provide training in navigation for boys in the King's service, offered to fund the new observatory. Moore was also patron of the young astronomer John Flamsteed whom he sought to make observer at the Chelsea site.[15] Flamsteed would become first Astronomer Royal. Persuaded (in Flamsteed's later account) of the urgency of astronomical data for longitude determination, the King acknowledged that he 'did not want his ship-owners and sailors to be deprived of any help the Heavens could supply, whereby navigation could be made safer'. A new state-funded observatory had the royal seal of approval.[16]

Wren, who designed the buildings, chose the royal park at Greenwich as the site for the new Observatory, its elevated position making it clearly visible from the winding River Thames with an ever-changing stock of commercial shipping.[17] But accurate determination of longitude *at sea* remained a matter of major concern both to ocean voyagers and to the nation, especially in the wake of disastrous losses of ships and men. Responding to petitions from navy captains, commanders of merchant ships and London merchants, Parliament passed a Longitude Act in 1714 which offered 'a Publick Reward for such Person or Persons as shall Discover the Longitude at Sea'. The Act empowered the Commissioners, known subsequently as the Board of Longitude, to award a sum of up to £20,000. Before the award could be paid, however, the Bill insisted that any proposed method 'shall be tried and found Practicable and Useful at Sea'. This trial, described as 'the Experiment', would take the form of an ocean voyage from the British Isles to a port in the West Indies. The latter destination would be decided by the powerful Commissioners who included the Lord High Admiral, the Master of Trinity House (responsible for lighthouses in England), the President of the Royal Society, the Astronomer Royal and professors of mathematics at Oxford and Cambridge Universities.[18]

By the mid-eighteenth century astronomers, cartographers and navigators were familiar with mathematical methods for longitude calculation. The annual *Nautical Almanac*, first produced in 1766 from data supplied by the Royal Observatory, provided accurate tables of moon and star positions to enable mathematically skilled captains (mostly naval officers) to determine longitude at sea. Finding a system independent of such skilled practitioners lay behind the prize money on offer. The award eventually went, over time, to the instrument maker John Harrison for his work on chronometers which would keep accurate time in a variety of sea-going environments (ch. 2).[19] More generally, the existence of a flourishing trade, centred on London, in the manufacture of reliable navigational and scientific instruments in the eighteenth century did much to promote the construction of a new generation of compact chronometers which could be carried easily aboard ship, where they would be protected in the most privileged space, the captain's accommodation.[20]

The chronometer method of determining longitude, legitimised by Cook on his second Pacific voyage, rendered the calculations practical and accessible to deep-sea merchant ship captains by the nineteenth century. Determining local noon by the sun's position, and relying upon a knowledge of Greenwich Mean Time from the chronometer, navigators obtained a time difference in which each hour equated to fifteen degrees of longitude east or west of the first meridian (24 hours equating to the 360 degrees of longitude around the earth). Far from rendering mariners independent of Greenwich, however, the new system depended not only on trust in the reliability of the instrument, but also upon its being set accurately to Greenwich time at the outset of the voyage.[21]

During the final years of John Pond's tenure as Astronomer Royal in the 1830s, the Lords Commissioners of the Admiralty (who controlled the Observatory) announced:

> that a ball will henceforward be dropped, every day, from the top of a pole on the Eastern Turret of the Royal Observatory at Greenwich, at the moment of one o'clock P.M. mean solar time. By observing the first instant of its downward movement, all vessels in the adjacent reaches of the river as well as in most of the docks, will thereby have an opportunity of regulating and rating their chronometers.[22]

The installation, constructed by London engineers Maudslay Sons & Field, further enhanced the authority of Greenwich, its imposing range of Royal Navy buildings leading the eye of the river user gently upwards to the Observatory itself. But with no direct communication beyond the range of the visible, ships elsewhere had to make similar checks with local observatories in ports such as Liverpool and Glasgow.[23]

The appointment of the Cambridge mathematician George Biddell Airy as Astronomer Royal in 1835 brought a new disciplinary regime to Greenwich. Airy's sense of his own worth was legendary. In his *Autobiography* he wrote of his graduation in the University's Senate House as Senior Wrangler in the Mathematics Tripos (1823): 'I, as Senior Wrangler, was led up first to receive the degree, and rarely has the Senate House rung with such applause as then filled it. For many minutes, after I was brought in front of the Vice-Chancellor, it was impossible to proceed with the ceremony on account of the uproar.'[24]

With Airy at the helm, the Observatory became a veritable factory, complete with a division of labour in which different groups of employees were set to work making observations, recordings and tortuous calculations for very little reward. As E.P. Thompson argued in a classic study (1967), the new culture of time discipline in Britain marked a radical shift from traditional task-orientation such as harvesting (where the objective was to complete the task rather than to measure the time taken) to timed labour (where time meant money), that is, a shift from a rural economy to industrial capitalism. Airy's own life too was highly disciplined, every incoming and outgoing letter being assigned its slot in a grand filing system preserved in Cambridge to this day.[25]

The office and person of the Astronomer Royal himself fully represented the Royal Observatory's growing authority in the Victorian period. The construction of networks of land telegraph lines in the 1840s and underwater telegraph lines in the following decades (ch. 5) made possible the near-instantaneous transmission of Greenwich Time far beyond the River Thames, simultaneously extending the power of the Observatory and its Director.[26] By 1847 the Secretary of the Liverpool & Manchester Railway, Henry Booth (chs 3, 4) wanted the Government to take the initiative with an Act of Parliament authorising uniformity of time throughout the country. Privately sceptical of his chances of winning support from the Astronomer Royal, Booth published his direct appeal to Parliament and people in the form of a pamphlet putting the case for an end to the chaos arising from different 'local times' (traditionally based on readings from sundials), especially for railway timetabling (ch. 4).[27] Not everyone rallied to the 'progressive' cause. In 1851, for example, *Chambers's Edinburgh Journal* published a satirical piece entitled 'Railway-time Aggression' making the case for 'real' rather than 'railway' time. It concluded with the exhortation to 'rally around Old Time with the determination to agitate, and if needs be, to resist this arbitrary aggression. Let our rallying cry be: "The Sun or the Railway!"'[28]

Arrangements with the local South Eastern Railway nevertheless meant that Airy did link his Observatory to telegraph networks through a line laid from Greenwich to the nearest railway station at Lewisham in 1852. Time signals were therefore communicated from an astronomically regulated electric clock in the Observatory, via automatic electro-magnetic mechanisms (which obviated unreliable human agents) and networks of land telegraph lines, to key public sites such as the new Houses of Parliament, the Royal Exchange, the Electric Telegraph Company offices in the Strand, and at Deal overlooking the Downs anchorage at the south-east tip of Kent that many ships used as their departure or arrival point for long-distance voyages. At the last two sites, electro-magnetic mechanisms automatically released prominent time-balls. The network of telegraphic time signals also extended, for astronomical purposes, across the channel to link Greenwich and Paris Observatories.[29]

In the nineteenth century, its promoters made the Royal Observatory at Greenwich stand as the symbolic and actual regulator of time for disciplining the useful work of the nation, of the empire and ultimately of the 'civilised' – that is, British imperial – world. An intricate *system* of electric telegraphs and time signals represented Greenwich Mean Time as exerting control over business, railways and everyday life in capitalist society. Airy could therefore reflect in 1853: 'I cannot but feel a satisfaction in thinking that the Royal Observatory is thus quietly contributing to the punctuality of business through a large portion of this busy country.' By 1865 he could affirm that he had always considered it 'the very proper duty of the National Observatory to promote... the dissemination of a knowledge of accurate time which is now really a matter of very great importance... We can, on an English railway, always obtain correct time, but not so on a French or German railway, where the clocks are often found considerably in error.'[30]

Greenwich Time was no longer simply time for the local setting in Greenwich. By 1872 all British Post Offices had to operate by Greenwich Time. Eight years later, it was 'legal time' throughout the land and by 1884 it had become the standard for the world's time.[31] The disciplined practice of navigation, moreover, deployed Greenwich time to ensure that British ships, especially steamships, could be trusted to transform widely scattered, isolated colonies and outposts into something like a coherent imperial system. The Victorian culture of mechanised, astronomical time had, in the eyes of its projectors, taken on a universal, objective character in the shadow of which personal and local cultures of time seemed as powerless as Conrad's fictional anarchists and their plot against the first meridian. But Conrad's fiction – most notably *Heart of Darkness* (1899) – provides dramatic reminders of the very *limited* extent of the authority and power of western systems of discipline and control, including time, on most regions of the nineteenth-century world.[32]

The British Asses and the rise of 'chartism'

In the late 1770s the Royal Society came under the presidential rule of Sir Joseph Banks. Banks's autocratic regime lasted until his death in 1820 and provoked a younger scientific generation to establish new specialist scientific societies in London (notably the Geological Society in 1807 and the Astro-nomical Society in 1820) to challenge the monopoly of the Royal Society.[33] The new generation also worked to change the Royal Society away from an aristocratic leadership, but encountered strong opposition. Charles Babbage's publication of his *Reflections on the Decline of Science in England, and on Some of its Causes* (1830), indicting the Royal Society's ruling coterie, only deepened the gulf between radicals like Babbage and the old guard, many of whom did not practise science themselves. Following the failure soon after to secure election of the Cambridge-educated John Herschel, representing scientifically active Fellows of the Royal Society, the reformers acted elsewhere for change. Appropriately, perhaps, the Royal Society would become known as the scientific equivalent of the House of Lords, to its critics based less on merit than on social connection and status.[34]

The British Association for the Advancement of Science (BAAS or British Asses) was the scientific reformers' eventual response to the Royal Society's failure to reform itself from within. Launched in 1831 at York (seen as the symbolic geographical centre of the British Isles and lacking the prejudices associated with the metropolis), the BAAS emerged in a decade marked by the Great Reform Bill (1832). With the Whig party in Government, the politics of progress confronted Old Corruption in the form of state monopolies and traditional institutions such as the East India Company (ch. 3). Consistent with Whig ideology, BAAS promoters stressed values of political breadth and religious tolerance, though in practice most of the core 'gentlemen of science' were liberal Anglicans in religion and Whigs or moderate conservatives in politics. There was one token Quaker (the chemist John Dalton) but there were almost certainly no Catholics, Jews or high Tories. Although membership was wide, an inner core carefully managed the BAAS. This 'high priesthood' included the geologists William Buckland, Adam Sedgwick and Roderick Murchison, the Cambridge polymath William Whewell, the astronomer Airy, and the Scottish natural philosopher James David Forbes. A high proportion had Oxford and Cambridge University connections, either through undergraduate education or through academic status.[35]

Although far from being a 'democratic' institution, the BAAS seemed (in the eyes of its promoters) the scientific equivalent of the House of Commons, itself scarcely a democratic institution on account of the very small proportion of the population permitted to cast votes in General Elections.[36] The BAAS met annually in a different provincial city or town and thus, over a period of time, circulated around the Kingdom with meetings in Oxford, Cambridge, Edinburgh, Dublin, Liverpool, and even Cork in Ireland. The Association chose mainly academic centres, provincial capitals and seaports, but its professed liking for industry took it to Birmingham in 1839 and, somewhat reluctantly, to the restless manufacturing city of Manchester in 1842. In several cases – Liverpool, Manchester, Glasgow and Newcastle – gentrified Literary and Philosophical Societies, legacies of Enlightenment cultures of 'improvement' through scientific knowledge, played leading roles in attracting the BAAS to a specific location by highlighting local traditions, scientific heroes, or technological prowess.[37]

The very hierarchical structure of BAAS meetings – and of the published *Reports* – institutionalised its centralised and non-democratic character. The Association famously divided into sections, with Section A ('Mathematics and Physics') taking precedence. Here the premier 'lions' of the BAAS such as Airy, Whewell and Herschel would grace the meetings with their authoritative presence (Figure 1.2). Section B ('Chemistry'), now aspiring to high rank among the quantitative sciences, followed. Section C ('Geology') also occupied an elevated place on account of the prestige and influence of gentlemanly geologists such as Murchison, Adam Sedgwick, and Charles Lyell, many of whom played a major part in shaping the BAAS. At the lowest end of the hierarchy, Section G ('Mechanical Science') occupied a subservient position, professing to 'apply' the laws and theories of Section A while simultaneously using that strategy to distance itself from the mechanical arts of the practical engineer. Founded at the Bristol meeting (1836), Section G, though supported by very distinguished engineers such as the naval architect John Scott Russell, struggled to survive in the mid-1840s.[38]

The very first BAAS Report emphasised the benefits for science of 'combining the Philosophical Societies dispersed through the provinces of the empire in a general co-operative union in order to facilitate the collection of geographically distributed data prior to analysis'.[39] Widely distributed

23

Figure 1.2 Gentlemen of science: the 'Mathematical Section' (Section A) of the BAAS congregates at Southampton in 1846.

Source: Illustrated London News 9 (1846), p. 184. (Courtesy of Aberdeen University Library.)

local scientific societies throughout Britain and the empire would operate under the unifying umbrella of the BAAS, whose gentlemen of science would bring to bear on the data thus collected their specialist analytical skills or mathematical expertise. The 'benefits' were seen to combine the 'advancement of knowledge' with the advancement of the socio-cultural status of 'science', that is, making scientific knowledge not merely helpful but essential to the prosperity and power of the modern state of Britain and the British Empire. The pursuit of knowledge for its own sake would enhance national prestige through scientific 'discovery' or 'invention' while the promise of 'improvements' to the human condition would channel state support for scientific endeavour on a grand, empire-wide scale. Individual 'scientists' (Whewell's new term coined in 1833) would of course enjoy considerable career advancement within this new imperial scientific culture.[40]

The collection of quantitative data on a global scale characterised BAAS scientific practice from the early years. Typically, BAAS gentlemen of science represented their data graphically, in the form of maps or charts. The astronomer John Herschel, whose authority in such matters derived from his work on mapping the Southern skies from the Cape of Good Hope as well as from his famous father William, remarked on the obsession with 'chartism' in this period, referring not to the threat to social order from disaffected and displaced artisans or 'Chartists' but to the rising fashion for chart-making in every science from astronomy to political economy. 'Chartism', the orderly mapping not only of land and sea but of terrestrial magnetism, tidal flows, meteorological phenomena and of the health and diseases of the body politic, became a defining characteristic of British science in a period threatened by increasing political chaos. Indeed, while we now remember the voyage of HMS *Beagle* as inseparable from the life and work of Charles Darwin, the voyage was primarily one of charting and surveying the coasts and channels of southern South America for the Admiralty.[41]

Coinciding with the foundation of the BAAS, British natural philosophers increasingly insisted on the inadequacy of older qualitative traditions of scientific investigation. Quantification, rather than verbal description, characterised the new order of physical science.[42] Nowhere was this significance of quantitative investigation more forcefully expressed than in Herschel's *Preliminary Discourse on the Study of Natural Philosophy* (1830). 'In all cases which admit of numeration or measurement', he wrote, 'it is of the utmost consequence to obtain precise numerical statements, whether in the measure of time, space or quantity of any kind. To omit this, is, in the first instance, to expose ourselves to illusions of sense, which may lead to the grossest errors.' As 'the very soul of science', he stressed, numerical precision affords 'the only criterion, or at least the best, of the truth of theories and the correctness of experiments'. Chemistry, for example, had as a result – presumably of the labours of Antoine Lavoisier and his followers – become, in 'the most pre-eminent way, a science of quantity', leaving behind the 'mistakes and confusions' of earlier qualitative chemistry. At the same time, Herschel emphasised the need for convenient standards of the quantities measured: standards common to all enquirers and without change over time.[43] The gentlemen of science now promoted precise measurement and the sound advancement of scientific knowledge as wholly interdependent.

While the Admiralty under its hydrographer, Admiral Beaufort, undertook the production of some 1600 accurate navigational charts of seas and coasts worldwide, BAAS gentlemen produced charts of tides, of weather, and of terrestrial magnetism. Such 'objects of national importance' became eligible for government funding. The gentlemen of science represented them as vital to the future security, well-being and prestige of empire. With the capacity to determine longitude accurately aboard ship, gentlemen of science like Murchison saw no limit to exploring geographically the remaining portions of the globe and to charting with precision every coast and island around the world. Indeed, the physical construction of lighthouses and other navigational marks at the behest of agencies such as Trinity House, closely allied to the Admiralty, ensured that Britain literally made her mark in space from the coasts of Kent to the coasts of China. Michael Faraday, more usually associated with the genteel lecture theatre of the Royal Institution in London's Albemarle Street, played a key scientific and advisory role with respect to the operation of English lighthouses.[44]

The BAAS allocated its first research grant (£200) in 1833 'to the *discussion* of observations of the Tides, and the formation of Tide Tables, under the superintendence of Mr Baily, Mr Lubbock, Rev. G. Peacock, and Rev. Whewell', the last two being well known as Cambridge gentlemen of science. The meeting further recommended that 'the Association should endeavour to procure the establishment of systematic Tidal Observations along the coasts of Great Britain and Ireland'.[45] Already in 1832 Beaufort had written to John Lubbock that 'no one can be impressed with a stronger conviction than myself of the urgent necessity of acquiring proper data for the construction of our Tide Tables, that I considered it to be a national object, and that the Government sh^d. take it in hand when they found a person qualified like you'. At the meeting of the BAAS in Edinburgh (1834), the junior secretary and new professor of natural philosophy in the University of Edinburgh, Forbes, reported with satisfaction that when objects of 'national importance presented themselves, the Association has fulfilled its pledge of stimulating Government to the aid of science ... the observations recommended by the Committee on Tides have been undertaken, by order of the Lords of the Admiralty, at above 500 [coastguard] stations on the coast of Britain'. Over the next few years the BAAS allocated almost £1000 to William Whewell and over £550 to his one-time pupil, Lubbock, for tidal investigations.

Whewell's 'tidology' stood as a new ideal for British scientific practice. At a high theoretical level, it belonged to the realm of astronomy, queen of the sciences, whom the Cambridge mathematical elite served with loyalty and devotion. At a practical level, lesser mortals had to observe, measure, and record such data as the high priests required. Results appeared in visual form, co-tidal lines joining all places with a common time of high water and providing a firm geometrical basis for calculation and thus for compiling accurate tide tables so essential to the efficient and safe commerce of an island nation and empire. Long after Whewell, the BAAS's Tidal Committee monitored its tide gauges in the ports and estuaries of empire, aided by mechanised tide predictors designed to facilitate the production of tide tables vital for the safe navigation of every ship.[46]

Meteorology was equally a subject of national importance. Prompted by his friend and advisor Whewell, Forbes prepared a lengthy 'Report on Meteorology' for the Oxford meeting of the BAAS (1832). Recognising the immature and scattered state of the science, he nevertheless believed that 'Five or six Registers in Great Britain and Ireland, carried on by learned Societies or by Government, would afford [supply] the great normal quantities required for establishing the numerical data for the climate of this kingdom with regard to the great elements of temperature, pressure and humidity in relation to that of other parts of the globe.' Of the many branches of physics involved in meteorology, the most important was heat:

No one department [of Meteorology] is exempt from its influence; no one substance in nature seems independent of the action of this subtile [sic] element. Impalpable though it be, yet since we possess such accurate means of investigating many of its laws, it is surprising how very imperfect are the notions entertained by mankind at large, and even by the scientific world, as to the importance of the part which it assumes in the economy of nature. To attempt to study Meteorology without it, is like trying to read a cipher without previously mastering the key.[47]

Forbes's agenda gradually took practical shape. The electrician William Snow Harris, Whewell and others worked on self-registering anemometers, instruments to record the properties of wind over time and without the need for human observers. Herschel and an assistant produced graphical representations, projections of atmospheric pressure waves based on barometric measurements. Working within Plymouth Dockyard, Snow Harris received BAAS funding to record hourly temperature measurements. By 1840 he had some 120,000 recordings from thermometers, barometers and hygrometers. When the BAAS acquired the former Royal Observatory at Kew in 1842, self-recording instruments for meteorology, atmospheric electricity and terrestrial magnetism provided a stable core of activity for the otherwise peripatetic organisation.[48] Gentlemen of science thus promoted meteorology, like tidology, not as a matter of pure science, but of immediate concern to a maritime nation.

Magnetic crusaders and the magnetism of ships

Neither tidology nor meteorology, however, attained anything approaching the rank of terrestrial magnetism on the scale of objects of national importance. Recognising the subject as one of paramount significance, William Whewell dignified the campaign to map the earth's magnetism and especially the regions close to the earth's magnetic poles by the title of 'the Magnetic Crusade'. Articulated by its projectors as necessary to the improvement of navigation through a more complete knowledge of the earth's magnetic field, this exemplar *par excellence* of 'chartism' not only acquired high national importance but was one of the most expensive state-funded scientific expeditions of the century.[49]

The story of the Magnetic Crusade is an amalgamation of personal ambitions, institutional and group politics, rivalries and collaboration, high politics and national interest. The aims included an empire-wide chain of fixed observatories set up to measure the changing characteristics of the earth's magnetic field. For example, navigators and natural philosophers had long known that a compass needle points to magnetic north. Magnetic north, however, does not stay constant relative to 'true north' as given by the position of the Pole Star, the difference between magnetic north and true north being the 'magnetic variation' and appearing to follow a path which over time revolves around the North Pole. One objective therefore was to map accurately this magnetic variation in order that Admiralty charts could carry up-to-date information. But the most spectacular goal required the despatch of ships to investigate terrestrial magnetism in the southern seas around the Antarctic.[50]

In his address to the Newcastle meeting of the BAAS (1838), the geologist Roderick Murchison (one of the general secretaries) initiated the latest and ultimately successful appeal for an expedition to make magnetical observations in the Antarctic seas. Unlike earlier attempts to lobby the Government, this part of the campaign co-ordinated the influence in high places of individual gentlemen of science – notably Herschel and Whewell – together with the combined authority of BAAS and Royal Society. Murchison's remarks, quoted as the epigraph to this chapter, identified mapping with imperial power. In his own fieldwork as a gentleman geologist in the 1830s, he had extended his Silurian system of rock strata into vast new territories around the globe, including Czarist Russia.[51] Now through Admiralty charts and magnetic observatories, he predicted that Britain had an even grander opportunity to assert global maritime power through an integrated campaign of navy, army and 'scientists'.

Murchison also made clear BAAS policy of taking 'special care' to ask 'nothing of Government but *what belongs to the national interests or honour to effect,* and *what cannot but be effected by national means'.* The BAAS, he asserted, had a crucial role in promoting key scientific projects to legislators too busy and thus unable to discriminate among the numerous projects offered by mere individuals.[52] The new scientific elite, the gentlemen of science, would henceforth be the arbiters of objects of national importance in matters scientific. The BAAS had made itself a significant part of the Establishment in early Victorian Britain.

Orchestrated by Major Edward Sabine, Fellow of the Royal Society (himself a target of Charles Babbage's biting critique of British science in his *Decline of Science*), the magnetic campaign resulted in James Clark Ross's Antarctic expedition (1839–41) at a cost of over £100,000 to the Admiralty, and the establishment of a network of empire-wide fixed magnetic observatories equipped with precision instruments. The latter were largely funded by the East India Company, the War office, the Admiralty and (in India) the Rajah of Trivandrum and the King of Oude. Sabine effectively controlled his own magnetic empire and arranged for the processing of data at Woolwich, mainly in the form of charts representing the magnetic character of the earth and enabling magnetic parameters to be obtained at a given time and place.[53] A young Glasgow-born lawyer and recent Cambridge graduate, Archibald Smith, conducted much of the complex mathematical analysis (required to take into account the effects of the magnetic character of the ships on the delicate measuring instruments). Smith later provided a more general handbook to facilitate the analysis of ships' magnetism in order that readings from their magnetic compasses could be corrected from tables of magnetic deviation at each point of the compass.[54]

From the 1830s construction of iron-hulled ships, most notably by Lairds of Birkenhead on the Mersey, posed problems for navigation. Many Laird vessels were primarily intended to serve the causes of empire-builders on the rivers of Asia and Africa, especially since steam-powered, iron-hulled riverboats demonstrated immense versatility through their ability to conquer strong currents, tow heavy barges, manoeuvre independently of wind and weather, and display all the weapons of conquest in the most inaccessible waterways.[55] But without the means of ascertaining and if possible neutralising their magnetic qualities, these craft seemed unpromising rivals to traditional sea-going wooden ships.

In July 1838 the Admiralty assigned Airy full powers to carry out magnetic experiments on the iron steamer *Rainbow* at Deptford, just upriver from Greenwich. By identifying the various disturbing magnetic effects on the compass (resulting in an error of some 50 degrees), Airy introduced fixed magnets and iron correctors to cancel out the disturbances. In his report after two voyages to Antwerp, the captain informed Airy that 'I am perfectly satisfied as to the correctness of the compasses, and feel quite certain they will continue so. I took particular notice from land to land from our departure and found the bearings by compass to be exact.'[56]

Such optimism, as we shall now see, was not borne out in practice. Just four years after the *Rainbow* experiments, two staunch projectors of iron vessels on Merseyside identified problems which threatened the whole future of iron shipbuilding. With interests in the iron trade, Thomas Jevons had constructed some of the earliest iron vessels on Merseyside, while the future of John Laird's Birkenhead shipbuilding yard rested heavily on the success or failure of the iron ship which had to be made credible (ch. 3). But in the early 1840s that credibility was difficult to construct, as Jevons told Laird:

> It having been represented to me by several individuals that the progress of Iron Shipbuilding is greatly retarded by the difficulty of Insurances upon them and their Cargoes, I venture to request you to meet a few friends to the cause of Iron Ships at my Counting House New Quay at one o'clock on Thursday the 8th December to discuss the matter and to devise some means by which the prejudice at present existing against that class of vessel may be removed.[57]

In her study of the conflict over the magnetism of iron ships between the Astronomer Royal and the former whaling captain and evangelical Anglican preacher, William Scoresby, Alison Winter has raised the fundamental historical issue of public disagreements about what kind of person could be trusted to resolve such problems. At stake were matters of credibility with key sections of the public – captains, ship-owners, merchants and shippers, passengers, engineers, underwriters, scientific bodies and other significant cultural groups within the seascape of Victorian Britain. Winter shows just how central religious and moral status were to an individual's public credibility: Scoresby for example secured credibility with middle-class audiences by embodying in his religious and scientific life 'what a trustworthy religious, scientific, and nautical figure should be'.[58]

It was at the BAAS meeting in Liverpool in 1854 that Scoresby's attack on the Astronomer Royal's 'scientific' methods of compass correction on board iron ships reached maximum intensity after the loss of the iron sailing ship *Tayleur,* and around 290 lives, on the Irish coast earlier that year while on passage from Liverpool to Australia.[59] Scoresby's criticisms were unwelcome to an Association that prided itself on projecting a public image of harmony in the scientific community, ruled over by high priests like Airy. Indeed, the BAAS President that year, the Earl of Harrowby, had concluded his address by calling Liverpool 'the missionary of science' by virtue of 'her ships and commercial agencies, by her enterprising spirit and her connexion with every soil and climate... the importer of the raw material of facts and observations, – the exporter of the manufactured results arising from their scientific discussion'.[60] Now Scoresby questioned the trustworthiness of those scientific practitioners – centred on Airy's authority as Astronomer Royal and member of the BAAS elite – upon which that missionary endeavour had been assumed to depend.

Scoresby's Section G audience received his presentation, grounded on the authority of a shipmaster's direct experience, as a very serious threat to the future prospects of the iron ship. He began by complaining that his views on the problems of iron ships in relation to the compass had either not been received or had simply been denied by 'the principal body of scientific men engaged in the consideration of compass adjustment and compass action in iron ships' with the exception of Archibald Smith, 'a gentleman eminent as a mathematician'. He also complained that his presentation to Section A had been cut short through lack of time, implying that the BAAS lions might have tried to suppress his views. Section G, however, provided a different kind of audience. It had attracted a large public concerned with maritime affairs (ch. 3) and was more likely to respond to the emotional issues surrounding the heavy loss of life in the *Tayleur* tragedy. Certainly the *Liverpool Mercury* published a very detailed report of his presentation. Perhaps wisely, Professor Airy, as *The Times* reported on 30 September under the headline 'A Philosopher in a Coal-pit', had migrated to the deepest coal mine on Tyneside to prepare for pendulum experiments to measure the density of the earth.[61]

Scoresby based his credibility on personal experience, most notably the practical experience of ocean storms and waves. His Evangelical perspective interpreted nature, especially the sea, as expressing the infinite power of Providence. These displays of Divine omnipotence demonstrated the immeasurable superiority of the ways of God over the ways of fallen and fallible man, compelling man to acknowledge the need for total trust in God rather than in human science.[62] As a manifestation of God's power, the ocean contained, in the reported words of fellow captain Andrew Henderson, 'elements whose violence and fury so often overwhelm theories and wreck mechanical science' (ch. 3). Such views scarcely resonated with those of the BAAS leadership, especially in Section A.[63]

In contrast, liberal Presbyterians and Unitarians had no reservations about (in Scott Russell's words) 'consulting' and even 'cross-examining' nature through experiment with a view to 'imitating Providence' through the harmony of human artefacts with nature's mechanisms.[64] Although Scoresby displayed experiments to demonstrate to his audiences the way in which the magnetic character of iron could be altered by severe shocks, his aim was not to engineer a system of compass correction but rather to show the fallibility of such an endeavour. He therefore placed the emphasis on the moral duty of the redeemed Christian by offering the following advice (which drew applause from the audience) to captains of ships:

> A watchful look-out do not cease
> Nor observation ever miss;
> Be of compass changes wary;
> Running near to danger chary;

On duty personally attend;
And above all on Providence divine depend.[65]

Scoresby's views drew polite but telling criticism from Liverpool marine engineer and shipbuilder John Grantham (again reported in the *Liverpool Mercury*) who immediately expressed his own 'apprehension' at the 'able remarks made by Dr Scoresby'. As already communicated to Section A, these remarks, he had heard, had had 'a most injurious tendency' on the Liverpool Exchange:

> The statements made by their honoured friend had partaken of that degree of partiality which was likely to endanger a principle of the highest degree of importance to this country. It was now so essential to the well-being of our mercantile marine, that if proved that iron ships could not be navigated with safety, it would strike at one of the greatest improvements of modern times, especially in cases where steam-boats were employed. *The matter was very much more dangerous in the hands of Dr Scoresby, for his well-known experience, ability, and talent, to deliver his sentiments gave such weight to all he said as to render them doubly dangerous* ... Dr Scoresby said that his paper was cut short in section A when he was coming to a remedy. He [Grantham] 'had not heard him give a distinct remedy that day.[66]

Grantham then seized the opportunity presented by the media to offer a robust defence of iron ships against wooden ones. High-profile accidents to iron ships, he claimed, masked 'the total amount of life lost in wooden ships' which was very much greater in proportion to their number. He therefore urged Liverpool's 'mercantile community ... not to allow themselves to be influenced by particular and striking cases ... – [and] to look at the mass of iron ships that had been built and safely worked for years and years ...'[67]

Scoresby certainly had his supporters within the Liverpool shipping communities. John Thomas Towson, 'scientific examiner of masters and mates' and a member of the Liverpool Literary & Philosophical Society since 1851, rose to say 'that he had equally with Mr Grantham had the opportunity of observing the workings of compasses in iron ships, and he had never found that after going a long voyage they were in a proper state of adjustment'. Affirming that 'No one believed more than himself in the importance of the progress of iron ships to the mercantile progress of the country', he nevertheless considered it impossible at present to obtain a correct compass in an iron ship.[68]

Such disagreements over the causes and solutions to the problem of compass deviation in iron ships were symptomatic of tensions within the BAAS where groups and individuals competed for scientific authority. Opponents of the gentlemanly coterie could represent the elite as having little or no experience of the ocean, for example, and therefore lacking that trustworthiness grounded on practical skills rather than theoretical and cloistered knowledge. Likewise, men with such seafaring experience were highly sceptical of 'experimental' ships of science, the products of Section A, such as Scott Russell's *Leviathan* (later the *Great Eastern*) (ch. 3). Equally, rising middle-class gentlemen such as the Manchester experimental philosopher James Prescott Joule could feel nervous enough to 'decline sailing in an iron ship' and to record his observation of 'the blundering arrangements for launching the *Leviathan*'.[69]

The theme of engineering projects as experiments – ships, railways, and telegraphs – recurs throughout our chapters. On the one hand, protagonists could readily invoke the authority of 'science' and its practitioners as a way of legitimising a new project for the benefit of investors and wider publics in the form of potential users of the new technology. A project could thus be promoted as 'scientific' and 'progressive'. On the other hand, however, the risks of untested 'experimental' schemes – challenging tried and tested practices and traditions – were high and could threaten public trust in an entire technology (chs 3–5). Section G in particular peddled an ideology of 'mechanical science' as science – especially the sciences of Section A – applied in experimental engineering. It was this kind of ideology, exemplified in this case by Airy, that Scoresby confronted as presumptuous and arrogant.

Airy's personal arrogance isolated him from even the Magnetic Crusade and its leading figures, most notably Sabine and Archibald Smith. From the mid-1840s, the recently appointed Glasgow professor of natural philosophy (physics) William Thomson corresponded with Smith on the magnetism of iron ships and its effect on compasses. Smith's methods won the trust of the Glasgow professor. They also received endorsement by the Admiralty and ultimately the rather grudging award from them of a gold watch. In the 1870s, however, and especially after Smith's death in 1872, Thomson appropriated the subject of ships' compasses as his own. Using his 120-ton schooner yacht *Lalla Rookh* as a floating laboratory for the trial and testing of navigational instruments, he worked out ideal theoretical conditions for a magnetic compass and practical techniques for realising them. The result took material form in Thomson's patent compass.[70]

This instrument was actually much more than a simple compass. It was a self-contained technological system with delicately arranged components, each with a specific purpose. The light-weight card itself (marking the points of the compass) was fixed to a set of parallel magnetic needles arranged below the card in such a way as to produce steadiness in a heavy sea without the need for the dampening effects of a liquid medium. Two large soft-iron spheres, fixed on either side of the compass, neutralised the effects of the ship's 'induced' or 'transient' magnetism. A series of bar magnets, whose positions were adjustable, compensated for the 'permanent' magnetism of the ship. The mahogany binnacle, accessible only to the captain and expert compass adjusters, neatly housed all these magnets. Although his system seemed like a re-invention of Airy's methods of correction, Thomson had avoided Airy's scientific elitism whereby masters were deemed too stupid to understand, let alone adjust, a scientifically fixed set of magnets. Thomson's system recognised instead a need for frequent adjustments to take account of any changes in the ship's magnetic character (e.g. after dry-docking and prior to a long outward voyage).[71]

Early in 1876 Thomson sent a prototype to Airy who remarked tersely: 'It won't do.' Dismissing Airy's negative verdict with the equally scornful words 'So much for the Astronomer Royal's opinion',[72] Thomson quickly took out a patent and in the same year fitted his first commercial compass to a cross-channel steamer owned locally by G. & J. Burns. The Burns family was intimately allied to the Cunard Steamship Company (ch. 3) and it was not long before Thomson's firm was fitting, testing and approving his patent compass as standard navigational instrumentation aboard the Cunard liners. Other imperial lines – P&O, British India, the Castle Line, the Union Steamship Company, the Glen Line, the Shire Line, the Clan Line, the White Star Line and many others – followed. Persuasion often depended on winning over senior captains in each fleet, although on occasions both captains and owners would complain that the system was too scientific or, worse still, that Sir William was experimenting with an owner's precious ships. The reputation of Clyde shipbuilders counted for much in giving the compass a trustworthy pedigree (ch. 3). London and Liverpool agents too played a vital role in the battles to win over the elite vessels of the merchant service. The Admiralty, however, resisted Thomson's marketing efforts until 1889.[73]

Thomson set up a Glasgow factory for production of these and other instruments including the contemporaneous sounding machine for measuring depths while steaming at speed. And unlike Smith's reliance on the Admiralty's award of a gold watch, he rose in social and economic status to be the first scientist to become a member of the House of Lords, as Baron Kelvin of Largs in 1892. Production of the compass rose from around 150 in 1880 to a peak of over 500 in each of the years 1892 and 1893. By the time of Kelvin's death in 1907, the Glasgow firm had supplied no fewer than 10,000 compasses to the world's merchant fleets and fighting navies. But by the 1900s the Royal Navy abandoned the Kelvin dry-card patent in favour of new types of liquid compasses which offered greater stability at sea and especially when heavy guns were being fired.[74]

The brethren of the hammer

'Geology, in the magnitude and sublimity of the objects of which it treats, undoubtedly ranks, in the scale of the sciences, next to astronomy', wrote John Herschel in 1830. Like astronomy, he noted, the progress of geology depended upon the continual accumulation of data derived from observation, but, unlike astronomy, the practitioners of geology had scarcely begun that accumulation. As we have seen, the BAAS also placed geology, if not quite next to astronomy, high in its ranking of the Sections.[75]

Roy Porter and Martin Rudwick have analysed the social structure of geological practices in the early nineteenth century. At one level, provincial learned societies located in mining regions provided a forum for shared concerns with issues of practical geology. The Newcastle-upon-Tyne Literary and Philosophical Society (founded in 1793) brought together diverse members of the community associated with Britain's most important coalfield. With practical geology and mine safety dominating their activities, members ranged across religious, professional and class divisions: Unitarians and Anglicans, mine owners, tradesmen, medical practitioners and coal-viewers. Similarly, the Royal Geological Society of Cornwall (founded in 1814) offered common ground to members from the local aristocracy downwards. Deep and capital-intensive copper and tin mines demanding a variety of expertise (from steam power engineering to mining technologies) meant that economic and practical concerns subdued disturbing forces arising from sectarian and political disputes.[76]

At a second level, industrialisation and urban growth from the later years of the eighteenth century provided a market for new skills such as those of land surveyors, canal builders, coal prospectors and viewers, drainage experts, mineral assayers, quarrymen and civil engineers. These practical men in some cases sought to emulate the learned professions (law, medicine and the church) and enhanced their credibility through the formation of new societies with related publications, thus differentiating themselves from traditional craftsmen whose skills were passed from one generation to the next by means of apprenticeships. These 'practical professionals' played a leading role in the making of nineteenth-century geological science. On the one hand, they focused on rock strata rather than on questions of fossils *per se*, origins of the earth, or landforms. William 'Strata' Smith was one of many such professional men who used their practical surveying skills to detail rock strata. On the other hand, these practical professionals emphasised detailed, meticulous observation in opposition to speculative theorising. What differentiated them, however, from the third level, the 'gentlemen geologists', was their primary aim of enhancing their marketable skills.[77]

The Geological Society of London included among its early members bankers, publishers and businessmen. These exclusively male representatives of the liberal professions and gentlemen of leisure paid lip-service to useful knowledge, but had little interest in a direct practical and personal pay-off from the study of geological science. But they were no dilettantes. Gentlemanly specialisation increasingly characterised the Society's meetings. With a spatial arrangement resembling the House of Commons, 'front-bench' elite specialists dominated the proceedings under the authority of the President as 'speaker'. Sometimes, indeed, controversial issues would lead to members sitting on opposite sides of the room. Unlike the aristocratic Royal Society, moreover, 'back-benchers' took part in heated discussion following presentation of papers.[78]

The dynamic Geological Society with its growing body of specialists inspired by its example a whole series of new metropolitan scientific societies (including the Astronomical in 1820 and the Geographical in 1830) which, peopled with ambitious scientific specialists, challenged the ageing Fellows of the authoritarian Royal Society. Contemporaries continued to regard the Geological Society as the most lively and innovative learned scientific society of the day. Charles Babbage, always the severest critic of the Royal Society, wrote in 1830 that the Geological 'possesses all the freshness, the vigour, and the ardour of youth in the pursuit of a youthful science, and has succeeded in a most difficult experiment, that of having an oral discussion on the subject of each paper read at its meetings'.[79] By the 1830s, the Society had spawned a list of powerful and distinguished gentlemanly geologists, each ready to defend their individual reputations but united by common practices and ready to reshape British scientific territory.[80]

Fieldwork formed the leading feature of gentlemanly geology. The field sciences meshed neatly with 'manly' pursuits (akin to hunting and shooting) as well as with a 'romantic' urge to travel and to witness natural wonders such as Mount Etna in Sicily or Fingal's Cave on the West of Scotland. The gentleman geologist socialised with and learnt from local gentry, paid the lower orders (quarrymen, for instance) or low-status (occasionally female) collectors for specimens of rocks or fossils, and engaged local artisans (such as boatmen) to assist in the pursuit of geological knowledge. Covering around 20 miles a day, geologists had no complex instrumentation such as telescopes to transport. Instead, their very basic tools included a geological hammer, compass, clinometer, some acid, a hand lens and a notebook.[81]

In its early years the Geological Society's primary concern was with the structural sequence of rocks. Spatial rather than temporal arrangement avoided not only speculative questions concerning the earth's origins and age but most theorising as well. At a local level, transverse sections (rather like a section of cliffs seen from seaward) showed both the vertical sequence and the tilt or dip of strata. Columnar sections (rather like a vertical descent down a mineshaft) displayed both sequence and thickness of strata. Correlations could then be made between localities, using 'physical markers' such as type of rock.[82]

If, for example, limestone were common to both localities, it could be used as a marker, allowing other strata to be mapped above or below. Sandstone might be found in some localities to be above the limestone and in others below, prompting a differentiation into 'old red sandstone' (below) and 'new red sandstone' (above), with coal deposits a possibility between the limestone and the new red sandstone. Fossils characteristic of specific rocks would be increasingly used as identifying markers. Geologists thus built up visual stratigraphical representations which soon came to be merged into a sequence that could be deployed in any region of the globe. At the same time, territorial disagreements over classification schemes often generated bitter priority battles between warring individuals and demonstrated the problems of attempts to impose order upon the earth's crust (Figure 1.3).[83]

The young science of geology nevertheless flourished alongside the mining, quarrying and engineering practices of industrialising Britain. Promoted and controlled initially by metropolitan gentlemen geologists, geology too became subject to the chartist fashion. As in astronomy and physics, gentlemen of geology such as the Cambridge professor of geology Adam Sedgwick and the mathematical coach William Hopkins promoted the value of accurate observation and measurement. Addressing the BAAS meeting in Cambridge (1833), Sedgwick pronounced that 'By science I understand the consideration of all subjects...capable of being reduced to measurement and calculation. All things comprehended under the categories of space, time, and number properly belong to our investigations.'[84]

Such strictures served both to exclude – or neutralise – potentially contentious sciences, including ethnology and agriculture, which carried controversial and divisive religious or political freight.[85] Indeed, accurate observation and measurement helped to define a Cambridge style of geological enquiry. Writing to the Yorkshire geologist and BAAS secretary, John Phillips, in 1836, Hopkins emphasised that he had little by way of suggestion for *general* observers in geological matters. What geology needed were '*accurate* observers': 'you will be doing a great service to geology if you can make it well understood that every accurate observation however limited must have its value, and every inaccurate one must be worse than useless'. But he feared that at present 'few geologists possess a good *geometrical eye*, and clear geometrical conceptions of the relations of space'.[86]

In 1832 the new Whig Government awarded a grant of £300 for colouring geologically the one-inch-to-the-mile Ordnance Survey maps of Devon. Three years later Henry de la Beche began receiving a state salary of £500 per annum thanks to the support of gentleman geologists William Buckland, Adam Sedgwick and Charles Lyell. Four years later De la Beche was ready to move across the Bristol Channel to begin surveying the coalfields of South Wales, his staff consisting of four full-time assistants and a palaeontologist, John Phillips. By 1845 the Geological Survey Act had been passed by Parliament, consolidating De la Beche's position, extending his power to include control of the Geological Survey of Ireland and adding considerably to his staff. The Act demonstrated the Government's commitment to the production of a comprehensive geological mapping of the British Isles.[87] As with studies of terrestrial magnetism, the tides, and meteorology in the 1830s and 1840s, the Geological Survey of Great Britain linked accurate mapping with objects of national importance.

THE PRESIDENT'S GEOLOGICAL LECTURE ON BOARD THE "DE SAUMAREZ" STEAMER, WHITECLIFF BAY, ISLE OF WIGHT.

Figure 1.3 A steamship provides the ideal vantage point for a discourse on the geology of the south coast of England by one of the lions of the British Association for the assembled gentleman – and ladies – of science in 1846.

Source: Illustrated London News 9 (1846), p. 185. (Courtesy of Aberdeen University Library.)

The Survey's national importance resided in its service to an expanding industrial and commercial empire increasingly powered by coal on land and sea (chs 2–4). Indeed, its early economic goals had taken it from the tin-mining of Cornwall to the very heart of the steam economy: the coal mines of South Wales, noted for the high calorific value of their product. De la Beche and Lyon Playfair issued their 'First Report on Coals Suited to the Steam Navy' in 1848 in response to the British Admiralty's desire to have access to coal on a worldwide scale as steam began to power ships of war.[88] By mid-nineteenth century, geologists wielded new powers as they mapped the nation's vast material storehouses and pronounced upon the future economic prospects for far-flung outposts of empire. No one fostered this trend better than De la Beche's successor as Director of the Geological Survey, Sir Roderick Murchison.

Murchison's ambitious drive to conquer empire, if not the world, through geological mapping derived from his own recent work as a leading gentle-man geologist. With a land-owning pedigree in the north-west Highlands of Scotland and an army career at the close of the Napoleonic Wars, Murchison pursued his personal geological campaigns with all the self-confidence of a Scottish aristocrat and the discipline of a British military man. In the 1830s Murchison identified along the Welsh borders the oldest known fossil-bearing rocks which he named 'Silurian' after the Silures, an ancient and war-like tribe of Britons who had once inhabited the region. With remark-able energy, the new 'King of Siluria' rapidly extended this classification to other parts of the country and around the globe, resulting in his monumental *The Silurian System* (1839).[89] The recognition of Silurian strata containing mainly marine invertebrates and devoid of terrestrial vegetation offered Murchison a powerful tool in the form of a baseline below which it was pointless to search for coal. He and fellow geologists could therefore market their expertise: for example, to advise against extravagant and speculative mining ventures in regions where coal might be needed but could not, according to the geologists, exist. Equally, geologists pronounced these ancient sedimentary rocks to be rich in metallic ores.[90]

In two campaigns during the early 1840s, Murchison conquered Russia. Geologising and mapping as he travelled vast distances, he extended the Silurian system from the Baltic to the White Sea and from Moscow to the Urals. Once again the Silurian system was both a matter of personal territory and a tool for evaluating and enhancing imperial wealth. Crucial to the latter, for Murchison, was coal as 'the motor and metre of all commercial nations'. Without coal, he asserted, 'no modern people can become great, either in manufactures or in the *naval art of war*'. And Russia appeared poorly endowed with coal. During and after these campaigns, Murchison also worked out a theory of exploitable gold deposits which he believed could only be found where Silurian rocks occurred. On this theory he predicted from the mid-1840s the presence of gold in Australia, a theory that appeared to be vindicated with the 'gold rush' of the early 1850s.[91]

Four years earlier the metropolitan-centred School of Mines and Museum of Practical Geology had been formally opened in the South Kensington district of West London. A supporter of a national school of mines (long a feature in Continental countries with deep mines) since the late 1830s, Murchison later referred to the institution as 'the first palace ever raised from the ground in Britain, which is entirely devoted to the advancement of science'. With his powerful Royal Geographical Society connections, Murchison instructed prospective explorers as to the collection of specimens for the Museum, while the School's laboratories provided facilities and techniques for the analysis of overseas coal samples on behalf of the Admiralty as well as analyses of minerals for the Foreign, Colonial and India Offices.[92]

Most importantly for Murchison's role as imperial geologist was to despatch the best graduates to geological surveys in the colonies or to take up positions as mining engineers, assayers, or metallurgists in parts of the world where British influence seemed to be extending onwards and ever-upwards. The Murchisonian geological empire extended not only to British-controlled territories in the Antipodes, Canada, India and Africa but also to those regions of the world sometimes designated as parts of the 'informal empire': the Middle East, South America, China and beyond. Famously, the German geologist Leopold von Buch quipped that 'God made the world, Sir Roderick arranged it.'[93]

'Day by day it becomes more evident that the Coal we happily possess in excellent quality and abundance is the mainspring of modern material civilization.' With these words, the Liverpool – later Manchester – political economist William Stanley Jevons opened the first chapter of his celebrated *The Coal Question* (1865), a work which would awaken anxieties as to the future of Britain's wealth at the very highest level of Government. 'As the source of fire, it is the source at once of mechanical motion and of chemical change. Accordingly it is the chief agent in almost every improvement or discovery in the arts which the present age brings forth.... coal, therefore, commands this age – it is the Age of Coal.'[94] Jevons's message, however, was a sombre one:

> [My aim is to] reflect upon the long series of changes in our industrial condition which must result from the gradual deepening of our coal mines and the increased price of fuel.... we cannot long progress as we are now doing. I give the usual scientific reasons for supposing that coal must confer mighty influence and advantages upon its rich possessor, and I show that we now use much more of this valuable aid than all other countries put together. But it is impossible we should long maintain so singular a position; not only must we meet some limit within our own country, but we must witness the coal produce of other countries approximating to our own, and ultimately passing it.[95]

At the foundation of Jevons's case lay the verdict of 'geologists of eminence'. These unnamed authorities, 'acquainted with the contents of our strata, and accustomed, in the study of their great science, to look over long periods of time with judgment and enlightenment, were long ago painfully struck by the essentially limited nature of our main wealth'. Jevons then cited the very recent opinion of the Tyneside industrialist Sir William Armstrong, in his capacity as President of the BAAS (1863), that added new weight to the subject while addressing the meeting of the Association at Newcastle, 'the very birthplace of the coal trade'. Estimating that the entire quantity of available coal in Britain would – at the present rate of increase of consumption – last just over 200 years, Armstrong argued that long before complete exhaustion 'other nations, and especially the United States of America, which possess coalfields thirty-seven times more extensive than ours, will then be working more accessible beds at a smaller cost, and will be able to displace the English coal from every market'.[96]

By the 1890s, these anxieties for the future of Britain's wealth and power had assumed a new form, especially in the guise of writings by social prophets on the other side of the Atlantic. In a popular book published in 1895 and entitled *On the Law of Civilization and Decay*, Brooks Adams (great-grandson of the second, and grandson of the sixth, United States President) argued that the rise and fall of historical empires could be explained in terms of the concentration and dispersal of natural energies, including those of coal, culture and race. Coinciding with the rapid rise of the United States as the world's most prosperous nation and with much speculation over the declining prospects for the British Empire, Brooks (and his brother Henry) appropriated Lord Kelvin's energy physics to claim that technological development permitted those concentrations of wealth and power that characterised the USA in the modern world. The foundations of modern economic power, and ultimately that of political power, rested on energy resources (especially coal) available for the production of useful work and hence for the creation of wealth.[97] It was a reminder of the fundamental roles that 'coal' had played in the making of the first industrial empire in the nineteenth century.

2
Power and Wealth: Reputations and Rivalries in Steam Culture

As is well known, iron and fire are the mainstay, the very life-blood of the mechanical arts. There is perhaps not one industrial establishment in England which does not depend on them for its existence and which does not make extensive use of them. If you were now to deprive England of her steam engines, you would deprive her of both coal and iron; you would cut off the sources of all her wealth, totally destroy her means of prosperity, and reduce this nation of huge power to insignificance. The destruction of her navy, which she regards as the main source of her wealth, would probably be less disastrous.
> – *Sadi Carnot, writing in Paris, assesses the significance of steam engines for the power and wealth of Britain (1824)*[1]

'Labour', asserted the Scottish moral philosopher Adam Smith in his *An Inquiry into the Nature and Causes of the Wealth of Nations* (1776), 'is the real measure of the exchangeable value of all commodities'. As such, labour was the key to understanding the wealth of nations since the total 'annual labour of every nation is the fund which originally supplies it with all the necessaries and conveniences of life which it annually consumes'. Moreover, the wealth of a nation was regulated on the one hand by 'the skill, dexterity, and judgment with which its labour is generally applied' (Smith's famous 'division of labour') and on the other by 'the proportion between those who are employed in useful labour, and that of those who are not so employed'. Reversing Thomas Hobbes's seventeenth-century dictum that 'wealth is power', Smith effectively made 'useful labour' or 'power' the basis of 'real' wealth for nation and empire.[2]

In the eighteenth century, and especially in the Scottish Enlightenment, 'labour' or 'work' divided into useful or productive labour and useless or unproductive work. For a political economist such as Adam Smith, the production of marketable commodities and the manufacture of marketable

41

goods belonged to the former class; the labour of servants working for aristocratic gentlemen fell under the latter category. Thus the former equated with the production of wealth while the latter was conducive to idleness. In Calvinist Scotland, these divisions reflected traditional Presbyterian beliefs in the moral value of work and in the sinful nature of idleness and waste.[3]

During this period, 'labour', 'work' and 'power' became increasingly associated with *engines* other than human and animal agents. Common water and windmills were the focus of extended debates over questions of efficiency: how might the power of wind or water best be harnessed by mechanical means? Answers to questions like these focused not simply on the ability to do a thing, but on doing it with the utmost economy. The engineer John Smeaton, a man well known to Watt, undertook one of the most famous analyses of types of watermill. Smeaton's concern was to establish whether the overshot wheel or the undershot wheel was most efficient. An ideal, or perfect, waterwheel could pump the same amount of water as that used to drive it; although no actual engine could do this, the ratio of water pumped (multiplied by height pumped) to water used up gave a simple measure of efficiency. Engineering measures of 'work done', understood as weight raised to a height, became part of the practice of such engineers in the eighteenth century.

While a very high proportion of non-human and non-animal power to drive the mills and factories of late eighteenth- and early nineteenth-century Britain was sourced from water and harnessed through waterwheels, promoters of heat engines quickly seized upon their superior versatility. Waterwheels were always fixed in space and limited in purpose. Steam engines pumped water from mines, hoisted goods at ports, and, in the new factories, provided the power for tools to pummel and reshape. Steam power applied to specialist tools added speed, dexterity or simply strength to human capacity. By the 1820s a foreign observer like Sadi Carnot, concerned with what he and his circle of political radicals perceived as a backwardness in French industry in the wake of the Napoleonic Wars, identified the special properties of such steam engines:

They seem destined to bring about a great revolution in the civilized world. The heat engine is already at work in the exploitation of our mines, for driving our ships, digging out our ports and rivers, forging steel, fashioning wood, milling grain, spinning and weaving cloth, transporting the heaviest loads ... It seems that one day it must become a universal source of power and in this respect supplant animals, water and wind. Over the first of these sources of power the heat engine has the advantage of economy; over the other two, the invaluable advantage that it can be employed and remain in uninterrupted use irrespective of either time or space.... This [moreover] is not merely a case of a powerful and convenient source of power, universally available and easily transportable, taking the place of machines already in use; where the heat engine is employed, it rapidly stimulates improvements in techniques and may even give birth to other techniques that are completely new.[4]

The value of stationary power rose in tandem with the increasing mechanisation of processes and the division of labour – the classic example being the late eighteenth-century mechanisation of the textile industry, with weaving machines, and then machines for spinning. Adam Smith linked a great increase in the quantity of work from the same number of people to the division of labour. He attributed the increase to three different circumstances. First, 'reducing every man's business to some one simple operation' aided 'the increase of dexterity in every particular workman'. Second, a division of labour, whereby the simple operation, carried out in one place, became the sole employment of a man's life, facilitated 'the saving of time which is commonly lost in passing from one species of work to another'. In contrast, the 'habit of sauntering and of indolent careless application... acquired by every country workman who is obliged to change his work and his tools every half hour, and to apply his hand in twenty different ways almost every day of his life...renders him almost always slothful and lazy... [T]his cause alone must always reduce considerably the quantity of work which he is capable of performing.' And third, there was 'the invention of a great number of machines which facilitate and abridge labour, and enable one man to do the work of many'.[5]

Adam Smith identified the origin and refinement of such mechanical inventions with different kinds of human agents. To begin with, many of the machines 'made use of in those manufactures in which labour is most subdivided, were originally the inventions of common workmen, who, being each of them employed in some very simple operation, naturally turned their thoughts towards finding out easier and readier methods of performing it'. As his principal example, he chose not one of the new mechanical techniques for speeding up or replacing human labour in manufacturing but rather one of the very 'engines' which provided a primary source of power:

> In the first fire-engines, a boy was constantly employed to open and shut alternately the communication between the boiler and the cylinder, according as the piston either ascended or descended. One of those boys, who loved to play with his companions, observed that, by tying a string from the handle of the valve, which opened this communication, to another part of the machine, the valve would open and shut without his assistance, and leave him at liberty to divert himself with his playfellows. One of the greatest improvements that has been made upon this machine, since it was first invented, was in this manner the discovery of a boy who wanted to save his own labour.[6]

Inventions, however, also derived from human agents engaged not in using machines but in making them. These 'makers of machines' formed a 'peculiar trade' and the 'fabrication of the instruments of trade [had become]...itself the object of a great number of very important manufactures'. Finally, still other 'improvements' had been made by 'those who are called philosophers or men of speculation, whose trade it is, not to do anything, but to observe every thing', a trade itself increasingly subject to the division of labour.

In an early draft of *The Wealth of Nations* Smith dignified the inventive 'artist' (artisan) with the epithet of 'real philosopher' regardless of 'nominal profession'. Thus it was 'a real philosopher only who could invent the fire-engine, and first form the idea of producing so great an effect by a power in nature which had never been thought of'. Thereafter, he noted, many 'inferior artists, employed in the fabric of this wonderful machine, may afterwards discover more happy methods of applying that power than those first made use of by its illustrious inventer [sic]'.[7]

Nowhere in *The Wealth of Nations* (or, indeed, in its draft) did Smith name the fire engine's 'illustrious inventer'. Given his recognition of an established tradition of working fire engines, we can be pretty certain that he was not according his erstwhile contemporary at Glasgow College, James Watt, that honour. The moral philosopher's interests in mathematics and scientific subjects kept him in close friendship with medical professors William Cullen and Joseph Black; his former student and the future Edinburgh natural philosophy professor John Robison was a key friend of Watt; and Smith dealt directly with Watt over issues of space in the cramped medieval College at the time of Watt's service as 'Mathematical Instrument Maker to the University' and especially to its science professors, and his first attempts to improve the economy of existing fire engines by adding a separate condenser. It is therefore more plausible that Smith would have regarded the then-humble Watt as one of those 'artists' who, as a 'real philosopher', had indeed discovered 'more happy methods' of applying the power of steam.[8]

Half a century later and from the perspective of a nation defeated by Britain, itself made powerful and wealthy by steam, Sadi Carnot was clear where credit for the 'creation' of the heat engine lay: 'Savery, Newcomen, Smeathon [Smeaton], the famous Watt, Woolf, Trevetick [Trevithick], and certain other English engineers are the true creators of the heat engine.' But Carnot's primary concern lay less with assigning credit to individuals than with the 'development' of power from nature's gifts of fuel. 'By providing us everywhere with fuel, nature has given us the means to produce at any time and in any place both heat and the motive power to which heat gives rise', he told his readers at the outset. 'It is the purpose of heat engines to develop this power and to harness it to our use.'[9]

Carnot's *Reflexions on the Motive Power of Fire* (1824) addressed, then, the question as to whether there was a limit to improvements to heat engines, 'a limit which, by the very nature of the things, cannot be in any way surpassed'. Carnot recognised Cornish pumping engines as the supreme examples of engine economy, yielding the greatest quantity of work (for a given amount of coal) of any steam engine yet constructed.[10] These engines used markedly higher pressures than previous steam engines. They also used a two-stage expansion whereby high-pressure steam first expanded in a small high-pressure cylinder before passing to a larger low-pressure cylinder to complete the expansion (later known as the compound engine). In formulating his theory of the motive power of heat engines, Carnot raised also the issue of whether 'there might be working substances preferable to steam for the development of the motive power of fire'.[11] This question was to haunt a later generation of Carnot's readers, especially a group of scientific engineers and natural philosophers in Britain, who sought to replace steam power with the motive power of air. Steam engine designs, originally established in the patents of James Watt during the last quarter of the eighteenth century and subsequently transformed with immense variety, were thus not seen, even in their heyday, as inevitable forms of motive power. And indeed, engines which harnessed rare substances, like carbonic acid gas, or ubiquitous natural forces, like electro-magnetism, threatened to topple steam engines in the decades after Carnot wrote.

Work and wealth: Watt's perfect steam engine

Born in 1736 in the old Clyde seaport and shipbuilding town of Greenock, James Watt began life in a maritime culture that combined the useful mathematics of navigation and surveying with an austere Scottish Presbyterianism that frowned upon wasteful and extravagant pursuits. The son of a merchant, shipbuilder and ship chandler, he went to Glasgow in 1753 or 1754 and to a city with a far greater concentration of skills, London, in 1755 to train as a mathematical instrument maker. Back in Greenock from 1756, he tried to establish himself as an instrument maker and general merchant but in 1757 he again relocated to Glasgow to take advantage of its wider opportunities. He continued to provide telescopes, compasses and especially quadrants for Clyde navigators, whilst looking enviously at larger markets around the ports of Liverpool, London and also Bristol, where his brother 'Jockey' acted as his agent. He became involved in an industry new to Glasgow in the 1760s as technical adviser to, and investor in, the Delftfield Pottery. In partnership with John Craig from 1759, James Watt, mathematical and musical instrument maker and retailer, provided work for at least a dozen journeymen and apprentices, whilst James Watt, 'merchant', developed a portfolio of commercial interests including, from 1763, a large shop selling everything from shoe buckles to corkscrews (and occasionally staffed by his wife Margaret).[12] Taken together, these ventures made up what Richard Hills has called 'Watt's little empire'.[13]

It was the environment of Glasgow College, as much as the city's commercial streets, that was crucial to Watt. His mother, Agnes *née* Muirhead was
related to George Muirhead, college professor of humanity. Both Henry Drew
and George Jarden supported the practical work of the College as 'laboratory
stewards'.[14] One of them – most probably Jarden – perhaps trained Watt in
Glasgow during 1754. Professor of natural philosophy Robert Dick helped to
arrange Watt's nine-month apprenticeship with John Morgan in London. It
was the College that provided Watt with work in the autumn of 1756 fixing
up astronomical instruments recently shipped from Kingston, Jamaica as a
bequest for a new observatory. When in 1757 he came to work permanently
as 'Mathematical Instrument Maker to the University' he entered, literally,
an institution which was ancient but not atrophied: its professors, especially
in natural philosophy and medicine, formed part of an emerging scientific
community responsive to Scottish agriculture, industry and commerce.
Dick's successor John Anderson, for example, gave accessible 'anti-toga'
(i.e. without the classical clothing of a gentlemanly education) lectures to
non-matriculated artisans, while the chemist Black supported Watt's
commercial ventures with loans.[15]

Watt expected to profit most from routine business: turning out instruments
like quadrants he could push down prices, increase sales and enhance
profits by standardising work, for himself and his assistants, with special
tools and using mass production lines.[16] With a workshop in the university
precincts, however, Watt, like Drew and Jarden, was expected to work ad
hoc, repairing and augmenting the stock of apparatus used by the science
professors in practical and theoretical class demonstrations. Anderson and
Black were amongst his customers for this financially unprofitable work.
But there were other benefits: this role brought Watt into contact with the
talented Glasgow student John Robison who, in 1759, encouraged Watt to
think about making money from steam-driven carriages.[17]

In August 1757 Robison had published a design for an improved
Newcomen atmospheric steam engine: he had become fascinated by the
prospects of machines designed to harness the great forces of nature as an
alternative to animal power. Windmills and watermills were commonplace
sources of power for pumping, grinding and for any tools requiring rotary
motion.[18] But from the seventeenth century, craftsmen and natural philosophers had begun to harness steam as a source of power. At the young Royal
Society of London, skilled curators like Robert Hooke and expatriate
Huguenot Denis Papin staged experiments with steam (notably in the latter's 'digester' – or pressure cooker) as indicators of Baconian utility and to
court institutional legitimacy. Papin recognised that a vessel filled with steam,
then quickly cooled, would create a vacuum, against which atmospheric
pressure would act – thus delivering power. In the first decade of the
eighteenth century he audaciously offered to provide the Society with a
'fire-engine' which would propel an 80-ton vessel.[19]

Thomas Savery, likewise, found it was easier to make claims than to realise them. He proposed to use steam to develop power in two ways: to create a vacuum (where cold water injected into a steam-filled receiver created a void to suck water from a lower level); and as a source of pressure (to push water upwards through a pipe). Although elite patrons like William III (1698) and the Royal Society's Fellows (1699) witnessed demonstrations of this apparatus, and in *The Miners' Friend* (1702) Savery boasted his machine was the pump that industry needed, the sporadically functioning engine was more common in country houses than in deep mines. But this was barely a problem for its inventor since a patent granted in 1698, and extended to 1733 in 1699 by Act of Parliament, gave him a monopoly on machines for 'raising water by the impellent force of fire'. The broad terms of the patent meant that competitors paid up on premiums or else worked with Savery.[20]

One such man was Hooke's friend the Dartmouth ironmonger Thomas Newcomen. He worked alongside Savery and also with the 'proprietors' of Savery's patent who schemed to profit from it.[21] By 1712 a full-sized engine to Newcomen's design was at work above the Earl of Dudley's colliery, in sight of Dudley Castle. Among the 'erectors' of Newcomen's engines was the London-based Newtonian natural philosopher J.T. Desaguliers. He described the engine in his monumental *A Course of Experimental Philosophy* (1734–44), a vast illustrated compendium of useful mechanical techniques cashing out a programme of practical Newtonian science.[22] In Cornwall, Newcomen engines dominated the tin- and copper-mines which extended deep under the sea, requiring constant pumping. The continued 'improvement' of the fuel consumption of the Newcomen engine by philosophical engineers like Smeaton, and the struggles to introduce such engines in diverse contexts, however, made it very much a 'technology on trial'.[23]

When Robison and Watt came to think about power in the late 1750s, the Glasgow region had no large-scale Newcomen engine. Glasgow College did, by 1760, have a model Newcomen engine. Retrospectively Robison claimed of Watt that 'Every thing became Science in his hands': repair work might have been unprofitable financially, but daily encounters with the science professors, and with men like Robison, brought other rewards.[24] Even with Robison away, Watt experimented with Papin's steam digester in 1761 or 1762; and during the winter of 1763–64 he responded to Anderson's request to repair the model Newcomen engine. The first full-sized engine in the Glasgow neighbourhood (at Shettleston) was erected between 1763 and 1764: Anderson had been attempting to have the Newcomen model fixed from 1760 but the local engine may have added urgency to Watt's task and interest amongst Anderson's students. Whatever the case, it was this job, just one in a series of similar tasks, that Watt and his contemporaries alike interpreted as the key event in the creation of the steam engine with separate condenser.[25]

When Watt came to the machine, he found that it very quickly used up the steam available to it from a small boiler, ceasing to function after only a few strokes. Watt's task was to get the model engine to use less steam for each stroke and thus to work for longer. The question was how to do this. Perhaps the model functioned poorly because it was small: the cylinder might condense the steam more rapidly than a full-scale engine (depending upon its surface area), or it might transmit heat to the atmosphere more quickly or for a longer period (depending on the thickness or material of the walls). After experimenting with the original, he constructed alternative models (with a wooden instead of brass cylinder designed to transmit heat outwards to the atmosphere less quickly; and with a larger cylinder); when he failed to solve the problem (wood could not stand the onslaught of steam), he returned to conventional materials, measured the heat lost in and through the cylinder walls, and (using that famous kettle) also quantified the large amount of water needed to condense a quantity of steam: when that amount surprised him, he turned to Black for advice.[26]

Watt came to believe that he could create what he called a 'perfect engine'.[27] Such an engine would use only a single cylinder of steam for each stroke, rather than the four required, according to Watt's calculations, by the Newcomen engine; and the vacuum after condensation would be perfect. Similar schemes to create theoretical standards in mechanics, or to 'perfect' mechanical devices can be found in Smeaton's studies of perfection in waterwheels, against which actual waterwheels might be gauged, and in the search for a perfect marine chronometer by John Harrison to solve the problem of longitude (ch. 1). Hills suggests that Smeaton's study, read to the Royal Society in 1759, was one of the many technical articles provided by Robison for Watt in an intense programme of study; further, Robison had travelled to the West Indies on the Navy's experimental trials of Harrison's chronometer, just before returning to Glasgow in 1761.[28] Writing in 1796 Watt claimed that he had 'laid it down as an axiom that to make a *perfect steam-engine*, it was necessary that the cylinder was always as hot as the steam that entered it, and that the steam should be cooled down below 100° [Fahrenheit] to exert its full powers'.[29] In such an engine, there would be no waste from heating and cooling the cylinder walls; and the steam would condense to make an almost perfect vacuum. The question was to find a way of meeting two conditions which appeared to be mutually exclusive.[30]

Whilst walking on Glasgow Green one Sunday in the spring of 1765, Watt hit upon an idea of remarkable simplicity. The Newcomen engine had one cylinder which was heated, then cooled, then heated again in a cycle of actions that consumed steam wastefully. Watt realised that since steam was elastic it would rush into an exhausted space. All he needed to do was to produce a vacuum in a separate vessel and open a communication between the steam and that vessel. He could do this using an air pump. A simple model, using an anatomical brass syringe as air pump, proved the viability of this scheme.[31] Thus, Watt would have two cylinders. The first, the working cylinder, would contain the steam. It would be kept permanently hot; the second, the 'separate condenser', would be permanently cold. It would be in this space that the steam from the first would be condensed in order to create a vacuum (Figure 2.1).[32] He needed an air pump to suck the steam from the hot cylinder to the cold one. With this arrangement, Watt removed one source of waste in the Newcomen engine. In the 1790s, Robison recounted how he first learnt of Watt's 'perfect engine':

> I came into Mr. Watt's parlour without Ceremony, and found him sitting before the fire, having lying on his knee a little Tin Cistern, which he was looking at, I entered into conversation on what we had been speaking of at last meeting, something about Steam. All the while, Mr. Watt kept looking at the fire, and laid down the Cistern at the foot of his Chair – at last he looked at me and said briskly 'You need not fash [trouble] yourself any more about that Man, I have now made an Engine that shall not waste a particle of Steam...' ...In short I had no doubt that Mr. Watt had really made a perfect steam-engine.[33]

There has been much discussion of the role of Joseph Black in Watt's invention. As a chemistry lecturer, Black developed two formulations (latent heat and specific heat) which could have given Watt a theoretical vocabulary for describing the transmission of heat in real and imagined heat engines, including the Newcomen model and his own steam engine. 'Latent heat' was the heat absorbed by a substance, for example a liquid at its boiling point, which did not lead to a change of temperature registered on the thermometer. Black was lecturing on the heat required to melt ice into water from 1762; in October 1764 he measured the heat needed to turn water at boiling point into steam – and even asked Watt to make more accurate experiments.[34] The 'heat capacity' was the heat absorbed by a substance as its temperature increased by a fixed amount on an established scale. This heat varied from substance to substance.

The conventional claim is that Watt could only have invented the separate condenser with Black's aid, and in particular with the theoretical formulation of 'latent heat'. Although Black, Robison and Watt were allies they gave divergent accounts of their roles, and of the role of theoretical science, in the creation of the separate condenser. In his chemistry lectures, edited and published by Robison in 1803, Black ventured to suggest that he had made a considerable contribution to the public good by suggesting to Watt his improvements; Robison dedicated the text to Black's 'most illustrious pupil' and insisted that he (i.e. Watt) owed his improvements to instructions and information gleaned from Black. Yet Watt carefully and, given his business concerns, plausibly explained that he had not been Black's pupil.[35] Indeed, he only turned to Black to give general explanations of specific phenomena

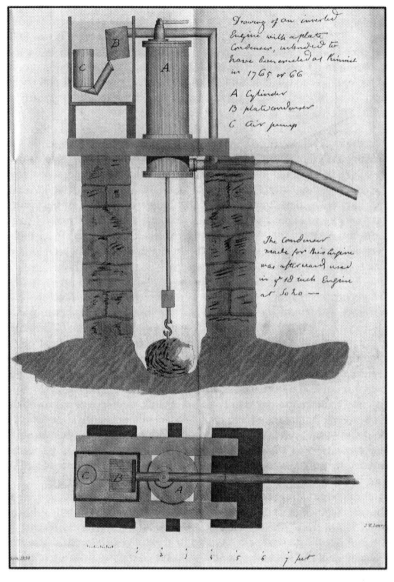

Figure 2.1 Sketch of an experimental 'inverted engine', showing Watt's separate condenser and air pump, 1765.

Source: J.P. Muirhead, *The Origin and Progress of the Mechanical Inventions of James Watt*. 3 vols. London: John Murray, 1854, p. iii, plate 1. (Courtesy of Aberdeen University Library.)

he had witnessed in experiments designed to understand the practicalities of the Newcomen model, and his experimental engines. As far as Watt was concerned, the steam engine with separate condenser was not a mere application of scientific ideas delivered to him by Black.[36]

Watt gave his own account when the optician David Brewster, as a successor to Desaguliers, came to edit John Robison's *System of Mechanical Philosophy* (1822), a tome which developed a history of steam power via Savery and Newcomen to Watt himself.[37] Watt there claimed to have measured experimentally a series of quantities: the capacity for heat of different substances (as compared with water); the bulk of steam (again as compared with water); the quantity of water evaporated in a boiler by a pound of coal; the elasticities of steam at temperatures above the boiling point of water; the quantity of steam required for each stroke of a Newcomen engine of given proportions; and the quantity of cold water required in each stroke to condense the steam in the cylinder of a Newcomen engine (an experiment which informed him that steam could heat six times its own weight of water to 212 degrees Fahrenheit). When he mentioned this fact to Black, the professor explained the doctrine of latent heat to him; but it was Watt who, independently, had come across a particular fact which supported Black's theory. What this fact alone indicated (without any theoretical explanation) was that there was a surprising reservoir of heat within steam; although Black helped Watt to understand this point, that explanation did not lead Watt to the idea of the separate condenser; instead, Watt was motivated towards that invention by considering the waste of cooling and heating the cylinder, about which the non-existence of latent heat would have made no difference. It is better, again as Watt himself indicated, to see Watt's experiments as part of an altogether bigger enterprise to understand 'back pressure' (the fact that the boiling point of water falls at lower pressure, so the temperature of the cylinder had to be low to get a good vacuum). Rather than seeing Watt's invention as a demonstration of the reliance of practical arts on scientific explanations, we should accept, in Watt's words, that Black suggested to him 'correct modes of reasoning, and of making experiments of which he set me the example'.[38]

'Not only to form my own Engines but also my own Engineers': making steam culture

Watt himself was keenly aware that to reform the Newcomen atmospheric engine as a large-scale steam power meant refashioning himself as a mechanical engineer, and then presenting himself as a steam engineer and an advocate for steam. This self-fashioning, by assembling and accumulating existing tropes of social presentation as well as skill, paralleled the fashioning of the new steam engine, which assembled existing technical material elements, associated practices and, indeed, meanings. The fashioning of the steam engine meant winning approval for it as a technology that was trustworthy, and reputable, in certain defined tasks. Initially, both the engineer and the engine were matters of promise: Watt and his allies worked to forge Watt's reputation as a philosophically informed steam engineer, even as he was developing, and (with Boulton) advocating, the engine as a reliable, and useful (in the specific contexts of mine and mill) practical machine.[39]

Watt realised, also, that he would need to breed – which is to say, to train – a race of engineers with the peculiar skills required for the new culture of steam engineering. Engineers collectively trained in the culture of the Newcomen engine were worse than novices: 'When I first began to construct my engine I found the workmen or Engineers accustomed to the erection of former engines so opinionated and obstinate that I had to discontinue employing them, and not only to *form my own Engines but also my own Engineers*.' Yet Watt himself decided, in the autumn of 1765, to acquire precisely those practical skills of design, erection and maintenance of common atmospheric engines.[40] This stage in the process of reforming himself as the archetype of the steam engineer would be taken in collaboration with John Roebuck.

Roebuck was introduced to Watt as a source of capital by Joseph Black, when, with Craig's death at the end of 1765, 'Watt's little empire' lost its chief financier and was in danger of collapse.[41] Roebuck was then building a fortune from the production of bleach for the Scottish linen industry. As the founder, in 1759 with Samuel Garbett, of the Carron Company, Roebuck had the resources of a large-scale foundry to develop Watt's condensing engine from a philosophical model to a practical reality. From January 1765, the Company was making cylinders and pipes for the Cornish engine market, and it continued to supply steam engine parts (including boiler plates), especially but not only in Scotland.[42] Roebuck also wanted efficient pumping engines for coal mines he leased near his Kinneil estate, in the hope of using coke, not charcoal, in iron smelting. He provided Watt with the chance to build a full-scale engine at Kinneil; and in addition he gave him that much-needed opportunity to erect 'common fire engines', or Newcomen engines.[43]

By experimenting on, fixing and making Newcomen engines in partnership with the canal surveyor Robert Mackell, Watt was learning about the opposition, and at the same time gaining a 'reputation' for his engine work and for himself as engine designer and constructor: in 1770 one satisfied customer expected to go 'mad with Joy' at the success of one of Watt's *Newcomen* engines, so superior was it to his neighbours'.[44] Ironically, experiments carried out on one of Roebuck's common engines confirmed to Watt (at least in a retrospective commentary) the 'great waste of fuel and steam' in large-scale Newcomen engines, just as with philosophical models; continuing to undermine the atmospheric engine as late as the 1810s, he complained of their badly made cylinders.[45] But the culture of the Newcomen engine was still vibrant. During the 1770s and 1780s John Smeaton continued to work systematically, making evolutionary alterations to atmospheric engines that would, eventually, create an engine twice as efficient as the Newcomen engines of the early eighteenth century. Such engines remained not merely as machines to be superseded by steam (and there were at least 223 working between 1733 and 1781) but as rivals.[46] In 1785, one disgruntled mine manager – motivated by wrangles over royalties, late delivery and poor quality castings – considered installing one of Smeaton's improved engines *after* he had built a Watt engine.[47] Here Watt had signally failed to sustain the reputation of the engine – leaving the door open for the opposition.

The experimental engine at Kinneil was sufficiently promising to persuade Roebuck to finance a patent for Watt's engine, on the understanding that he would take a two-thirds interest of the proceeds. This would become the famous patent of 5 January 1769 describing a 'New Method of Lessening the Consumption of Steam and Fuel in Fire Engines.' In framing the patent, Watt consulted colleagues and friends but not professional lawyers. The resulting text asserted the scientific principles underlying many potential fire engines with separate condenser – rather than specifying one, of a number, of actual practical machines. Thus the patent listed the principles of a steam jacket to keep the cylinder hot, the vital separate condenser and air pump, and other practical details designed to lessen steam consumption. A patent framed thus *after* Mansfield's judgement of 1778 might easily have been overturned, since from that time the patent 'specification' was expected to provide sufficient description that a competent artist might make or work the invention as well as the patentee. But in 1769 this was not explicitly the case. If Watt was stretching conventions, they were already elastic. The prolonged legal scrutiny of the patent from 1795 encouraged Watt to appraise himself fully of the law – and to argue for its reform.[48]

Nevertheless, it remained the case that with this patent approved, Watt and Roebuck had a legal monopoly on practically any machine using the separate condenser principle; and as a result, the opposition was open to prosecution and in theory immobilised. Roebuck provided Watt with the facilities and financial resources to develop his engine, which he continued to do in a clandestine fashion, notwithstanding the patent's protection. Roebuck's resources also allowed Watt to experiment with alternatives to the cylinder engine like the 'steam wheel', designed to deliver rotary power directly – something Newcomen engineers like Keane Fitzgerald were experimenting with from 1758.[49] Not least because Watt was exploring so many different avenues, it was only in the spring of 1770 that he succeeded in making the separate condenser work adequately well. But by June 1770 Roebuck was bankrupt, Watt's 'experiments on the engine' ceased, and as to the worth of the engine: as Watt told Boulton, 'None of his creditors value the engine at a farthing.'[50]

It was hardly surprising, then, that in September 1770 Watt reminded his friend William Small that the outcome of his engine experiments was 'uncertain'; he imagined himself 'growing gray' without being able to provide for wife and children. Moreover, other work, associated with the key transport revolution of his age (the canal) beckoned. Indeed, from about 1766, with proposals for a canal cutting across Scotland from the Forth, and Edinburgh, in the east to the Clyde, and Glasgow, in the west high on the agendas of Scottish merchants, James Watt had set up office as a land surveyor, now styling himself 'Engineer'. Surveying routes for a Forth and Clyde canal through coalfields to the east of Glasgow, asked to travel to London to support the necessary Bill, from 1770 working as part-time resident engineer on the Monkland Canal at a salary of £200, surveying the Strathmore Canal in the spring of 1770 – such activities offered quick money from contracts gained on the back of Watt's connections (who encouraged 'a man that lived among them rather than a stranger'). But they drew Watt away from his Glasgow shop, and now from the 'uncertain' steam engine experiments at Kinneil.[51]

Throughout this period, Watt's links with Scotland were loosening. As an apprentice he had studied in London; he travelled in England to buy stock, to drum up business, to fraternise with other instrument makers and, eventually, to lobby parliament. In this summer of 1767 Watt passed through Birmingham with a letter from John Roebuck. That letter provided the opportunity for Watt to meet two men, Erasmus Darwin and William Small, at the core of Birmingham's secretive but close-knit Lunar Society. In addition to the poet, evolutionary writer and medic Darwin and the Aberdeen-trained natural philosopher and medic William Small (the Society's driving force), amongst the members were the potter Josiah Wedgwood, the glass manufacturer James Keir and chemist Joseph Priestley, the clock-maker John Whitehurst, and the educationalist and philosopher of transport Richard Lovell Edgeworth.[52] As we noted earlier (ch. 1), the eighteenth century saw the establishment of numerous societies for science and technology, many of which publicised their activities and agendas. The Royal Society purported to speak out for science and, to a lesser extent, for the practical arts throughout Britain; but local 'literary and philosophical' societies and, eventually, specialist societies catered, respectively, for the needs of particular urban communities and for groups of interested parties who were dissatisfied with, or excluded from, the ethos of the gentlemanly amateur prevalent in a Royal Society widely viewed as in decline.[53] Watt himself had taken part in reading groups as a young man;[54] now he was welcomed into the mystery that was the Lunar Society.

Here was a group consisting of barely more than a dozen men of industry. Men, in the main, of considerable wealth, they offered support, investment, advice and alliances, all brokered at informal monthly congregations of self-defined 'fellow-schemers'.[55] In this 'clearing house' for information in natural philosophy and practical art, science and industry worked in tandem if, indeed, one could separate them in the projects pursued in chemistry (especially ceramics and dyeing), mechanics, instrument making and transport – especially canals.[56] Darwin and Wedgwood together lobbied for canals, like the Trent and Mersey and the 'Grand Trunk Canal', that would offer Wedgwood access to world markets; Priestley and Wedgwood traded chemical expertise on products and processes with specialist experimental apparatus, like the potter's standard pyrometer. In this forum, Watt too was a chemist, struggling with Keir to extract alkali from sea-salt. That scheme, begun in 1765 or 1766 with Black and Roebuck, looked, to his fellow Lunatics, at least as promising as the steam engine, effectively mothballed at Kinneil in 1770.[57] This was a forum in which personal aggrandisement and national benefit did not conflict.

Prominent within the Lunar Society, especially after Small's death in 1775, was Matthew Boulton. Boulton owned the Soho Works in Birmingham, an establishment he had transformed in the early 1760s from the buckle factory of his father to a state-of-the-art highly mechanised manufactory employing up to six hundred men in the production of fine steel 'toys' and hardware.[58] Boulton was avid for a reliable source of power immune, unlike water, to seasonal variation. He had a wide range of connections amongst local finan- ciers, politicians, town worthies and industrialists. Watt had been mightily impressed by Boulton's Soho Works when he was shown round, by Darwin and Small, in the summer of 1767, and Boulton, Small and Darwin were already curious about the possibilities of the steam engine.[59] By 1769 Boulton was trying to entice Watt to become his 'neighbour' in Birmingham, keen, he wrote, to act as 'midwife to ease you of your burthen, and to intro- duce your brat into the world'. Boulton had two stated reasons: 'love of you [Watt] and love of a money-getting ingenious project';[60] a third reason, perhaps, was the 'reputation' Watt was establishing as a fine builder of heat engines which added significantly to the credibility of a scheme barely off the ground at Kinneil.[61] In 1770 Watt was buoyant and the prospects of the canal business were excellent; he had effectively abandoned his own engine and his paid work on Newcomen engines diverted time and energy away from his 'brat'. By 1773 the canal business was drying up leaving Watt's Monkland salary unpaid and little indication of new work. In these circum- stances, Boulton's persistent offers to support Watt in Birmingham, and to resurrect the engine, looked enticing. Watt arrived in Birmingham in May 1774; the Kinneil engine was dismantled, packed, and transshipped to Birmingham also.[62]

Famously, the chief problem with the Kinneil engine had been the poor fit of piston and cylinder, Roebuck's Carron works notwithstanding. Through Boulton, Watt contacted John Wilkinson, brother-in-law of Joseph Priestley and iron founder. Based at Bersham in North Wales, Wilkinson had recently taken a boring mill for cannon-making and redeployed it to make extremely accurate cylinders for Newcomen engines.[63] Working sporadically with Roebuck, Watt had been unable to get a steam cylinder adequate to his special needs: he needed greater steam-tightness than a standard Newcomen engine. But once Wilkinson had created a cylinder good enough to meet Watt's stringent demands, Boulton and Watt had the basis for a company.

When Watt did the cost calculation, Boulton and Watt found that it made greater sense to obtain a new act of parliament extending the original patent than merely to start afresh. But taking the question to Westminster ran the risk of giving a voice to opponents of the bill – of which there were many, not simply amongst the mining interests who would rather efficient pumping engines were available without having to pay premiums to men like Boulton and Watt. If Boulton called upon connections like those who had assisted him when he established the Birmingham Assay Office, Watt, in a smaller way, could now solicit testimonials from Scottish worthies he had come across in his engineering duties. It was a winning combination – and led to the granting of a bill extending the life of the patent to the end of the century. From the beginning, the nation's ruling elite had given its assent and blessing to the new engine.[64]

Disseminating steam: exhibits, experts and exports

The ramifications were immense. Boulton and Watt entered into a formal partnership in 1775. With a patent extended to 1800, they worked in the well-founded expectation of a return on their investment. Much of their time between 1775 and 1785 would be spent marketing and distributing their engines. Boulton was the financier, organiser and salesman; Watt was primarily responsible for innovations in design, protected in additional patents of 1781, 1782 and 1784.[65]

In December 1769 Boulton had turned down an offer from Roebuck that he take a share in the property of Watt's invention merely 'as far as respects the countys of Warwick, Stafford, and Derby'. Boulton's plan had been 'to make for all the World'.[66] In order to build that business empire from 1775, Boulton and Watt worked outwards from those, often nearby, with whom they had relations of mutual confidence: the proprietors of the Broomfield Colliery at Tipton, site of Keir's chemical business, had a pumping engine by 1776; John Wilkinson had an engine providing the blast for his foundry soon after; another early engine served a distillery in Stratford – although that was allegedly sabotaged by an engineer made drunk at the behest of John Smeaton.[67]

That story, whether designed to destroy the reputation of an engine or an engineer, indicates the importance of effective marketing through public displays and the management of subsequent reports. Lunar Society members were familiar with such stagecraft: tourists had wondered at Wedgwood's 'Etruria' in Staffordshire, Keir's Chemical Works and Boulton's Soho Manufactory.[68] As Keir later explained:

The desire of visiting Soho became a fashion among the higher & opulent ranks, foreigners of distinction & all who could gain access to it...It was not without great difficulty & repeated instances of his friends that Mr B. could be persuaded to abridge or abandon this luxury of obliging. But however inconvenient it might be at the time to bestow so many hours in indulging the curiosity of others, he was not without a subsequent reward. He thus gained the acquaintance of most men distinguished for rank, influence & knowledge in the Kingdom, the good effects of which he felt in his several applications to Parliament.[69]

Boulton and Watt now displayed their new pumping engines to potential customers, erecting an engine at Soho as 'a specimen for the examination of mining speculators'.[70] Cornish miners were carefully courted since Cornwall was the most important concentrated market. There, deep mines needed pumps like those provided by the Newcomen engine; but the Cornish were avid for economical engines since they were in short supply of coal. Eventually reports would be published, giving comparisons of fuel consumed by the different, highly efficient (in contemporary terms) machines.[71] Boulton and Watt therefore offered their engines to Cornish mine-owners as improvements on the standard Newcomen engine, rather than the more efficient engines which Smeaton had worked on; they asked for payments based on money saved on fuel in comparison to the Newcomen engine, rather than, for example, simply selling on an engine (which would have been an unattractive capital outlay for the miners). Once again, the Watt engine was woven into a pre-existing fabric of Newcomen culture, even as it began to displace that culture.

In addition to this scheme of payment, Boulton and Watt offered regimes of construction ('erection'), maintenance (for the engines) and training (for the engineers). Companies would purchase the parts of the engines from recommended suppliers (Boulton and Watt often providing only the precision valves) and would then have them constructed, with Boulton and Watt's employees on hand to ensure that the machines ran smoothly, economically and productively. In the second half of the 1770s, Boulton and Watt often left Birmingham to oversee construction on site – leaving Lunar Society friend James Keir to manage Soho from 1778 (although he was unconvinced that the steam business would be a success).[72] Ultimately, the company produced handbooks indicating schedules of care, in an attempt to make routine what had originally been radical. Part of the programme of engineering 'formation' Watt had alluded to would be crystallised in the 1790s in an account of 'Points necessary to be known by a Steam Engineer' – a list which began with 'The Laws of Mechanics as a Science' and treated practical issues as essential, but secondary.[73]

From 1775 Boulton and Watt had an effective motive power with limited application: first pumping and later 'blowing' in blast furnaces. It was the businessman Boulton who predicted a rising demand for factory power and the more cautious Watt who responded by attempting to make the engine universal, not in space but in function. To make steam a commonplace in the textile industries, replacing water power in weaving and eventually spinning, Watt posed – and solved – a series of technical problems: to make an engine which was 'double-acting' so that both strokes delivered power (and also to economise on space and materials); and to make the engine generate rotary (circular) rather than merely reciprocal (back and forth) motion.

In solving these problems, during the 1780s, Watt created the 'sun and planet motion' as a response to an opportunistic patenting, by another engineer, of the crank, applied to the steam engine, to transform back and forth motion into rotary power. He invented the famous 'parallel motion', which was essential to allow both up and down strokes of the piston to develop power, in both a push and a pull: he was inordinately proud of this invention as a manifestation of pure geometry in actual mechanism (later integrated into a science of 'kinematics').[74] He was at least implicated in the broader use, in steam engines, of a 'governor' by which the engine could regulate its own rate of working (such 'self-acting' devices had been deployed already in mills). Arago later praised the governor (or 'regulator') as the 'true secret of the astonishing perfection of the manufacture of our epoch': with its consequent regular movement 'it can as easily embroider muslin as forge anchors'. For those who believed a gentle movement meant a loss of power, he replied: 'the apophthegm "Much noise and little work," is not only true in the moral world, – it is also an axiom in mechanics'.[75]

Furthermore, the 'indicator diagram' (developed with John Southern) aimed to provide the owner of early steam-driven textile mills, familiar with horse- and water-power, with a convincing, intelligible and permanent record of the power the new steam engines developed.[76] The key to the indicator diagram was a mechanism by which the engine recorded data without the overt intervention of human agency or intelligence. The indicator diagram thus helped to advertise the meaning of the steam engine as an economic factory power, and also a machine which was reliably self-reporting.

With these elements integrated into a reliable double-acting rotative engine, Boulton was keen to see the new power applied to make money (literally, in steam-driven coin presses) and to make bread. After trials with an 'experimental corn mill' at the Soho Manufactory, the partners invested in their deliberately famous Albion Flour Mill on the south bank of the Thames.[77] This project, floated in 1782 (once the rotative engine was patented) and granted its charter in 1784, was a showcase for steam-powered industry: capacious buildings by Samuel Wyatt; power from 50-horsepower Boulton & Watt rotative engines; extensive cast-iron construction and Watt's innovative parallel motion; each engine capable of driving six pairs of millstones simultaneously and with power to spare for machines for loading, lifting, sifting and dressing designed by the ambitious young millwright John Rennie brought down especially from Scotland (Figure 2.2). Watt said of the Mill that 'no edifice of the kind had been constructed with similar conveniences or powers'. The project integrated Boulton and Watt into elite metropolitan engineering and its situation in London added weight to Boulton's claim that it was 'a national object'.[78]

The Mill indicated dramatically the benefits and the dangers of reputation attendant on public display. The first trials of March 1786 drew in crowds of

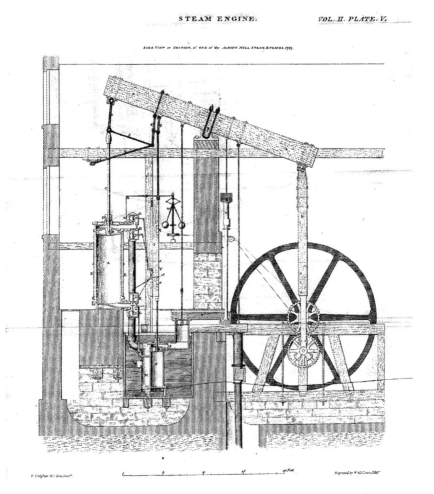

STEAM ENGINE.   *VOL. II. PLATE. V.*

Figure 2.2   The double-acting rotary steam engine with parallel motion and governor designed by Boulton and Watt for the Albion Mill in London.

Source: John Robison, *A System of Mechanical Philosophy. With Notes by David Brewster.* 4 vols. Edinburgh: John Murray, 1822, p. ii, plate v. (Courtesy of Aberdeen University Library.)

onlookers and powerful witnesses like Sir Joseph Banks, autocratic president of the Royal Society (ch. 1). The young Banks was a pivotal figure in the relations of science and government; a natural historian by vocation, he studied and supported the mechanisation of cotton manufacture, spinning and weaving; he had already installed a Watt pumping engine at his Gregory Mine and was fascinated by any improvements in power that might develop the economic potential of his own estates (seen as the empire in microcosm).[79] A 'great advertisement for the firm' it may have been, at home and abroad: Banks promoted a national industry of steam-driven mills into the 1790s; Boulton and Watt offered advice to foreign governments wanting similar mills. But the cost of innovation was high. Privately Boulton and Rennie reported broken parts, bad workmanship and dangerous actions which continued for months.[80]

Notoriously, this showcase faltered when the mill was burnt down in March 1791, amidst rumours of arson and to the great 'rejoicings of the mob'.[81] (In the same year Boulton and Watt armed their employees at Soho against rioters, as Priestley was driven from his house and laboratory.)[82] Undoubtedly there was opposition from those who saw the beginnings of milling monopolies, centred on seaports, and distant from the local wind- and water-powered mills. Boulton's ambitions to advertise the new engines and to generate support for a charter for the Mill had encouraged him, and Rennie, to make it a public spectacle but here he was at odds with Watt who preferred internally to 'make it a mystery' whilst it had the 'external appearance of business'. Watt told Boulton in April 1786: 'It has given me the *utmost* pain to hear of the many persons who have been admitted into the Albion Mill merely as an Object of Curiosity.'[83] Apart from the serious waste of Boulton's time, at the time of writing, Watt felt that the poor condition of the mill was 'more likely to do us hurt than good as engineers'; bad management, if visible to visitors, 'must hurt the credit of the Company'. Sitting in Birmingham, Watt had heard that they were looked upon 'by the serious common sense man as vain and rash *adventurers*' and 'that our talking of what we can do is construed into either a want of ability to perform it or the foolish cry of Roast beef' (announcing one's good fortune); puritanical, he was shocked at the thought of a 'Masquerade at the A M'. Whereas Boulton, and members of the Lunar Society, had been willing, in special circumstances to open their doors to travellers, Watt looked to another form of confirmation of engineering reputation: 'everything which contributes to render us conspicuous should be avoided, let us be content with *doing*'. Forgetting, perhaps, the vital role that aristocratic connections had played in Boulton's extension of the 1769 patent, Watt chided Rennie for his enthusiasm for advertisement: 'Dukes & Lords & noble peers will not be his best customers.'[84]

Between 1775 and 1795 Boulton and Watt did not manufacture entire engines but instead directed customers to approved contractors, in the

Midlands and like John Wilkinson further afield, with the skills and resources to cast and forge most of the parts from their drawings.[85] Soho's contribution was the crucial regulator valves. Skilled engine erectors trained with Watt and latterly working with a manual of printed instructions then constructed the engines. Key roles, then, for the partners were first to design and then to co-ordinate practical expertise distributed throughout the country. Watt had acted like this from necessity when, as one of many builders of Newcomen engines, he had sent drawings for parts to be cast at Carron.[86]

Boulton had hoped that an entirely different way of working would ensure the reputation of the engine. Writing to Watt in February 1769, he had speculated that success would require 'mon[e]y, very accurate workmanship, and extensive correspondence'; moreover, the key to 'keeping up the reputation' of the engine was not to surrender its construction to dispersed and inaccurate empirical engineers but instead to create a central 'manufactory'. In that one place Boulton would 'erect all the conveniences necessary for the completion of the Engines', train 'excellent workmen', furnish them with 'excellent tools' affordable only in the context of mass production, and thus 'serve all the World with Engines of all sizes'. These engines would be both cheap and made 'with as great a difference of accuracy as there is between the Blacksmith and the mathematical instrument maker'.[87] Reputation, accurate workmanship, concentration and regulation of specialist skill, investment in tools abridging and enhancing human labour, and networks of correspondence expressing influence and trust underwrote Boulton's vision of the engine, from first demonstration to domination, in quantity, of a world market.

In the event, Watt fought shy of these remarkable system-building plans, likely because of his absolute preference for consulting work rather than direct management. It was a question of 'credit'. In the 1770s, as a canal surveyor, Watt was keen to ensure accruing 'credit' to himself and he knew that in doing this work his employers would, indeed, have 'credit of me'. Direct supervision of men and construction could more easily lead to problems and even personal disgrace. Work on the Monkland Canal, for example, had reinforced Watt's distaste for 'bustling and bargaining with mankind' and encountering 'workmen' he could not discipline; it left him worried about being caught out for lack of 'experience' and fearful about being 'disgraced' whilst at work he was 'unfit for'. Thus, as he confided to his friend Small:

> I would rather face a loaded cannon than settle an account or make a bargain. In short I find myself out of my sphere when I have anything to do with mankind; it is enough for an engineer to force Nature, and to bear the vexation of her getting the better of him.[88]

Nervous of his reputation, and preferring to remain remote from civil engineering works actually carried out, Watt was not the man to take direct charge of the day-to-day running of a large-scale factory and its unruly staff, even one producing his own steam engines.

In fact, it was only in 1795–96, as Boulton and Watt approached retirement, that their sons Matthew Robinson Boulton (1770–1842) and James Watt junior (1769–1848) established a Soho Foundry equipped to make complete engines. Echoing Watt's own comments, the historian Eric Robinson observes that 'the fathers built not only machines but also men'.[89] Following the educational agendas of Darwin, Edgeworth and Priestley, these sons and successors would be crafted as mathematically disciplined, organisationally adept 'captains of industry' (to use the phrase later employed by Carlyle). Watt designed his son to double as manufacturer and engineer. To that end he must write well, without waste of paper or needless ceremony (in readiness for business); he must draw; he must have practical experience (in carpentry and at Wilkinson's foundry); to the skills of accountancy he should add those of mathematics, natural philosophy and chemistry. Languages ancient and modern would be consolidated in Geneva and Freiberg (home of the famous School of Mines); in those towns, thanks to Watt's growing reputation, he should fraternise with men of influence and report on useful innovations. He must benefit from Watt's advice on early rising, 'swearing, economy, talkativeness, drink, tips and paying "your share of the reckoning" '.[90] Novels engendered absent-mindedness and the theatre eroded morality; fencing taught the younger Watt bodily control and the ability to present himself well in company.

Boulton junior, too, was to be bred up liberal and genteel, to lose his provincial accent, to declaim with clarity, and, like Watt junior, to study abroad that he might act as agent for his father. Well coutured but not showy, free from gambling debts, punctual, neat, honest, industrious, able to dance, ride and converse politely: Boulton junior, if he would but listen to his father, might express those qualities of distinctness, orderliness and exactness vital to laboratory, manufactory and counting house alike.[91] These were the men that would take over the business as the new century dawned, manufacture engines complete and, trained to industry, generate more business than ever their parents had done.[92]

Throughout the 1780s and 1790s Boulton and Watt battled to protect their profits from the actions of the 'pirates' who used, or manufactured, engines to their design without paying a premium or who created engines infringing the broad specifications of the patent of 1769 (renewed in 1775).[93] The developing steam culture both trained innovators and created new opportunities for pirates. An engineer like William Symington (ch. 3), once trained to erect Watt engines, adeptly created an alternative in 1784 that avoided Watt's patent.[94] Manufacturers with inside knowledge occasionally ventured into piracy. One result of having parts of the engines made outside the Soho foundry was that other mechanical engineers had the most intimate knowledge of the construction of the engines. Astonishingly, it transpired that John Wilkinson, who had been so important in making the Watt engine viable by providing accurate cylinders, was himself manufacturing and selling complete engines.[95]

The case of the innovators classified as infringers is more complex. It is difficult to resist the conclusion that these engineers had produced engines which were, if anything, more efficient than the Watt engine with separate condenser; and that, more worryingly, the patent Boulton and Watt had taken out was so vague as probably not to stand in a court if contested. This put Boulton and Watt in a difficult position: prosecuting pirates of their own engines was a relatively easy option; prosecuting those who had 'improved' forms of engines could only lead to a questioning of the validity of the patent itself. Nevertheless, Boulton and Watt went ahead – and in winning, eventually, the various cases, they netted themselves vast back-payments in royalties.

The irony here is that by blocking innovation in engine design, rather as Savery had done with his patent from 1698 to 1733, the men popularly held responsible for the engine that powered the industrial revolution may have promoted atrophy in mechanical engineering.[96] Watt himself wrote in 1785 of being ready 'to cease attempting to invent new things, or to attempt anything which is attended with any risk of not succeeding, or of trouble in the execution. Let us go on executing the things we understand and leave the rest to younger men, who have neither money nor character to lose.'[97] Clearly then, invention was a matter of high risk, both in financial terms but also in terms of the reputation that came through established character; having once been successful, there was no need to risk a fall in reputation by embarking on unproven adventures. This stance carried over to innovation within the firm itself, since at least one of Watt's assistants, William Murdock, was actively discouraged from developing a carriage powered by steam between 1784 and 1786, even as Watt, in a bid to silence his most loyal and talented servant, 'resolved to try if God will work a miracle in favour of these carriages'.[98] Forty years later, railway engineers found another way of harnessing steam to locomotion (ch. 4).

As Boulton and Watt fought these battles, they also developed markets for steam, both nationally and, eventually, internationally. They erected over a hundred engines in the ten years after 1775 and more than 180 in the following decade.[99] An effective marketing strategy was to satisfy existing demands at home: twenty-one pumping engines went to Cornwall by 1785. Another was to encourage aristocratic patrons (like the King of Naples, unconcerned at high mark-ups) or high-profile 'improving' public works abroad, although always with great caution about preserving rights to the invention. For example, members of the Batavian Society of Experimental Philosophy, having failed in 1776 to regulate Rotterdam's canals with a Newcomen engine, obtained in 1786 the permission to use Watt's inventions in the Low Countries for fifteen years. Boulton and Watt obtained similar 'privileges' in France and Spain and entered into discussions with

America, Prussia, Belgium and Austria/Bohemia: whilst Watt capitalised on his connections in science, Boulton redeployed his parliamentary skills, approaching influential individuals (including the Spanish King), and established engines in key public works as advertisements. They attempted to work by proxy, offering engines to individuals in other countries who would then be responsible for preventing Watt engines to be made by others there, but tentative agreements depended on international trust which was in short supply: as Boulton remarked, 'when a man takes a partner he puts it into his power to ruin him'; international piracy was as great as or greater than piracy at home and there was less chance of legal come-back. Agents in the ports of London, Liverpool and Bristol acted as intermediaries for a growing number of international orders, especially after 1810.[100] Steam power, too, became a vital part of colonial business – and power – in the plantations of the West Indies.

This dissemination and export of steam power, then, required expert marketing and highly skilled assistants. Watt had little experience in the first, but Boulton knew – and cared – about advertisement, could draw upon a network of agents in Europe, and was used to seeing fashions, sold to nobility uninterested in economy, trickling down to larger publics. For skilled assistants, Boulton and Watt could draw upon men like William Murdock and John Southern, who carried with them 'tacit knowledge' of engine building, erection and maintenance. The practical knowledge of steam engineering was developed hands-on and transmitted by direct experience. Thus, before the mining engineer George Symington (a relative of William Symington) began to construct a Watt pumping engine in Dumfriesshire he chose to work beside one of Watt's 'erectors' on another engine; even then, William Murdock supervised the final stages of the installation in 1779.[101]

For engines sold abroad, in deals expertly greased by Boulton, the partners had a vital, complementary, resource in skilled mechanics. For prestige engines, they sent men of similar status: James Watt junior to supervise the first Spanish engine, a key London agent for a corn mill at Nantes, a trusted Soho mechanic to erect the King of Naples' engine, Boulton's nephew (attired in velvet on Boulton's instruction, and apparently with a sales budget of £2,000) to promote the St Petersburg Mint engine. The high status – and high salaries – offered to Soho-trained mechanics abroad led to much absconding, as expatriates chose to develop skills and business empires abroad.[102]

The same went for the transfer of technology to culturally distinct environments, especially those lacking the systematic infrastructures of material supply, transport and expertise that nurtured a British steam culture.[103] Making steam a robust, international, imperial possession meant an export of cultures of skill as well of nuts and bolts – or valves and cylinders. Watt knew this only too well – since he had received offers from foreign governments (France, Russia) to start over in an environment that would fully resource his innovative exploits, for the benefit of the empire-builders and emissaries of Enlightenment of other nations. John Robison had settled in St Petersburg (1770–74) to instruct Catherine the Great's Navy in shipbuilding and navigation. Thus even Watt's friends succumbed to an eighteenth-century brain drain which threatened to sap Britain of its empire-making capacity.

Superseding steam: 'gaz', 'caloric' and electro-magnetic rivalry

The potency of steam, as mediated by its voracious advocates, in transforming so many aspects of manufacture and transportation made it hard to counter the gloss of 'progress' given to the practical arts. That 'progress', typified by steam's displacement of atmospheric engines, by steam's supplanting – true to Carnot's prediction – of animal, water and wind power, and by steam's successive development – according to Carnot, by a chain of British innovators – raised the spectre of steam itself being superseded. Carnot had been concerned with heat engines in general, and not just those powered by steam. He knew full well about the engineers at work on rivals to steam; and indeed, engineers interpreted his commentary as a licence to investigate these and other challengers.[104] Steam culture engendered talk of progress; that talk encompassed, tolerated and even encouraged claims that the reign of steam was finite.

Watt's Lunar Society friend Erasmus Darwin, for example, had claimed that he could provide power for the potter Wedgwood using a new form of vertical windmill.[105] Fantasies of machines developing power without end had not dwindled with the advent of Newtonian science.[106] From the early nineteenth century new engineering periodicals were full of news of perpetual motion machines, guaranteeing motion or even power for free, and thus threatening to disrupt the most robust economy. The London-based *Mechanics' Magazine* tolerated, when it did not actively encourage, extended submissions from engineering hopefuls convinced they had found a method of harnessing gravity or some other natural force of power for nothing.[107]

Other machines occupied a liminal zone as candidates for power that remained as small-scale experimental shows or that, as large-scale commercial ventures, were short-lived. Marc Isambard Brunel and his son Isambard Kingdom Brunel laboured long on an engine which for the elder Brunel was a plausible extension of previous ventures. He had been working since the beginning of the century with heat engines exploiting bubbling water (Cagniard Latour's 'buoyancy' engine),[108] hot air (in the 1810s) and steam (in the 1820s).[109] The working substance of the new so-called 'gaz engine' was 'carbonic acid gas' (carbon dioxide).[110]

Like Watt, the Brunels were in regular discussion with prominent scientific figures. They communicated with leading savants at London's Royal Institution:

Michael Faraday, who had liquefied the gas (and others) in March 1823, whilst engaged in experiments under the direction of his mentor Humphry Davy; and Davy himself, who noted that little heat was required to transform the liquid into gaseous form and speculated on how that might be the basis of a heat engine.[111] Royal Society connections gave Marc Brunel ready access to both figures. In May 1823 Brunel met Davy and they discussed the use of carbonic acid gas 'as a power'. Two receivers held the condensed gas and could be alternately heated and cooled by water; an expansion vessel leading out of each receiver fed into the top (or bottom) of a power cylinder containing the piston; as the gas in one receiver was cooled and the gas in the other was heated there was a difference in pressure on the piston which would drive this 'difference engine'.

At around thirty-five atmospheres, the pressure differential was far greater than that used in stationary steam engines – and appeared to promise an engine that would supersede steam. But containing high working pressures of up to sixty-five atmospheres was a serious technical problem, equivalent to that faced by Watt in making cylinders steam tight; Brunel called on Faraday to discuss the possibility of a tough 'alloy of iron with nickel' for the receivers before opting, instead, for parts made of high quality gun metal.[112] The Brunels began two years of intensive experiments. Development continued sporadically until 1833, with the younger Brunel, then building a reputation for masculine endeavour, primarily exposed to the manifest dangers of the work. Although little noticed by subsequent commentators, Isambard Kingdom Brunel clearly saw this engine, one of his 'châteaux d'Espagne', as a possible route to the fortune and fame he aspired to from an early age – not least because he anticipated it would 'supersede steam altogether as a motive power'.[113]

Connection with the Royal Institution, and with Faraday, bolstered the reputation of the engine, allowing it to be read as the trustworthy practical counterpart to elite London science. Faraday's involvement continued long after Marc Brunel's 1823 visit.[114] The engine's projectors offered Faraday a demonstration of progress with the 'Forcing pumps' (in September 1823),[115] and a staged display to 'prove' a set of tubes prone to leakage under pressure (February 1824).[116] By April 1824 the 'Proving Apparatus' had been charged to forty-five atmospheres – but not enough, apparently, to condense the gas.[117] With a revised schedule, this confidence-generating demonstration would be ready for Faraday to see on Christmas Day of 1824 – or perhaps a fortnight later.[118] At last, in mid-January, having relocated household and apparatus to Blackfriars, Brunel announced he had 'succeeded in making my vessels light and sound, after making continued and almost daily alterations and trials for the last four weeks'.[119]

Although the image presented to Faraday was one of confidence, steady progress and single-mindedness, in private, Isambard was distracted by his father's notorious Thames Tunnel (from Rotherhithe to Wapping) and by plans for the Bermondsey docks and even a 'canal across the Panama'; the 'Gaz Engine' was a source of 'great anxiety', might only be 'tolerably good' even with considerable investment, and was just one of many grand projects which 'may take place!' As the family struggled on with no carriage, horse or footman and 'only two maid-servants', Brunel lurched from more 'castles in the air about steam boats that go fifteen miles per hour' and plans for 'a large fortune, building a house for myself, etc. etc.' through ideas of contentment with 'a small house' and a 'gig' to worries that, if his father died, or the Tunnel failed, he might, as he put it, 'cut my throat or hang myself'.[120]

By June 1825 the experimenters, having mastered the failure of packings and the bursting of parts charged to ninety-six atmospheres, condensed the gas.[121] This practical fact, costing £1250 to demonstrate, was a vital point in the transition from philosophical 'proving apparatus' to large-scale 'differential engine'.[122] Like Watt, Marc Brunel marked the transition from proving apparatus to large-scale trial by securing a patent (in July 1825 for Britain, with Irish and French patents to follow).[123] He traced the principle of the engine directly back to Faraday, and indeed to the Royal Society, as the venue for the announcement of his 1823 results; but the patent described a practical engine in detail. Power would come through the action of a double-acting piston moving up and down within an oil-filled cylinder; each end of the cylinder was connected, by a tube, to another 'pressure vessel' half filled with oil; another pipe would connect the top of each pressure vessel to a cylindrical container for the liquid gas; through that container closely packed tubes would pass. With hot water passing through the tubes in the first liquid-gas container, and cold water through the tubes of the other, a pressure differential would arise and this would be transmitted hydraulically, by the oil, to the working cylinder, thus raising the piston; reversing the water flow would force the piston down. With the aid of a crank, rotary power could be generated – of the kind needed for factory machinery.

With their commercial rights protected, an advertising campaign began. Marc Brunel's friend and biographer, Richard Beamish, claimed that the 'benefits anticipated from the gas machine strongly excited the hopes of the scientific as well as the commercial world'.[124] If this was true, it was in part because Faraday agreed to present it in February 1826 at one of the 'evening discourses' delivered to the genteel members of the Royal Institution. Marc Brunel offered Faraday specially made drawings, a copy of the patent specification, and a face-to-face explanation of its functioning.[125] Once on stage, Faraday emphasised Brunel senior's 'three years of exertion', the role played by the Royal Institution's laboratory as the site of the original liquefaction of the gas, and Brunel's object: 'an engine which should rival in power and utility the steam-engine'. Reports in the Institution's *Quarterly Journal of Science* detailed the vast pressures achieved (120 atmospheres) whilst maintaining 'perfect security' without 'a single accident', the large quantities of condensed gas now available ('a pint and a-half') and plans to put together a large-scale engine with a 4 ft stroke.[126]

Such talk, especially Faraday's talk, promoted confidence in a new mechanical power with a more than usually intimate connection to elite experimental science. Work on the engine continued whilst other projects drained time: as water flooded into the Thames Tunnel from the river above, progress on the 'Great Bore' ceased, not to be recommenced until 1835. Until then Isambard continued work on other projects for his father and, increasingly, looked for his own commissions – but with the aid of an assistant (Withers) he continued to struggle with the gaz engine, the apparatus having been transported by boat and set up at the abandoned tunnel works in Rotherhithe: 'Here I am at Rotherhithe renewing experiments on Gaz – been getting the apparatus up for the last *six months*!! Is it possible? A 1/40 of the remainder of my life – what a life, the life of a dreamer – am always building castles in the air, what time I waste!'[127]

Even in 1824 the Brunels had pondered the suitability of their nascent power for marine engines – effectively competing with the V-frame steam engine Marc had patented in 1822. In 1832 the Admiralty, with whom Brunel had had a fruitful if uneasy relationship for many years, showed flickers of interest in supporting the gaz engine. By September 1832 the Lord Commissioners had been persuaded to peruse it.[128] A committee to investigate the viability of the application of condensed gases as mechanical agents was to include two chemists and the elite civil engineer John Rennie (a younger member of the dynasty founded by John Rennie of the Albion Mill) but Faraday himself refused to take part. He suggested to John Barrow, Second Secretary to the Admiralty, that his unique status 'as discoverer' of the principle might compromise his impartiality; it was better that the committee reported first, after which he would submit an independent evaluation – aware that the minute investigations of 'the apparatus' might reveal 'some philosophical point which [he] should feel inclined to pursue alone'.[129] In the event, Brunel's track record, his diverse connections with aristocratic patrons, and Faraday's interest elicited all of £200 'to aid Brunel in obtaining a practical result'. In total, the Brunels invested £15,000 of their own money.[130]

At the end of 1832 the gaz engine competed with Isambard's Woolwich Docks and Clifton Suspension Bridge: the common factor here was investment without immediate profit. It was becoming clear that the success of the engine depended not merely on scaling up from philosophical apparatus to practical machine but on questions of economy. Measurements of fuel consumption had convinced mine owners that Watt's engines surpassed the economies of Newcomen machines and Watt continued to make experiments on the properties of steam into the 1780s; during the winter of 1832–33, Marc Brunel used new funds to make protracted experiments to measure 'carefully & accurately' the pressures and temperatures of the (repeatedly purified) carbonic acid liquid/gas in order to determine the likely economy of the gaz engine. A chief aim was to deliver to Faraday figures on the latent heat of the gas.[131] The conclusion, related by Isambard Brunel, was crushing: 'no sufficient advantage on the score of economy of fuel can be obtained. All the time and expense, both *enormous*, devoted to this thing for nearly 10 years are therefore *wasted*...It must therefore die and with it all my fine hopes – crash – gone – well, well, it can't be helped.'[132] Beamish drew an appropriate moral from this adventurous but ultimately wasteful tale:

> the beautiful theory, which had given so much promise, and which had been hailed as the harbinger of a new era in practical mechanics, was found incapable of realising those economic conditions by which alone it could be rendered commercially valuable.[133]

Of course, Watt could not evaluate the 'gaz' engine, since Faraday's liquefaction experiments took place in the years just after his death. But Watt did know about attempts going on in his native Scotland to harness not the rare, and physically extreme, commodity of liquid carbonic acid gas but rather the ubiquitous substance of atmospheric air.[134] These trials originated with Robert Stirling who laboured in a workshop attached to the manse in Galston where he was minister: Stirling patented the engine in 1816 but his church connections brought little direct credibility to such practical engineering endeavours. Later, his brother James Stirling continued the experiment on a commercial scale at the Dundee Foundry which, as a professional engineer, he managed. The idea was to use a cylinder in which air was heated and cooled whilst a special apparatus, the 'economiser', moved back and forth through it. The name of this apparatus left no doubt as to the purpose of an engine which, like Watt's, purported to lessen radically the consumption of fuel.

Always an avid and attentive reader, in 1817 Watt complained that this habit had finally damaged his sight – and he had given up reviewing the 'Scientific Journals'. This left him ill-informed about experiments in heat engineering; but when presented with a pencil sketch and a verbal account of the new 'Air Engine', he could offer scant encouragement. Curious to know the advantages of air over water he was told that air would be better 'for the purposes of locomotion' (there being no water to carry or find in remote areas). Also, Stirling aimed at the 'economizing of fuel by arresting the heat and making it available another time': such, allegedly, was the function of the economiser. Unimpressed, Watt would give no opinion until 'Mr. Stirling has got the engine mounted on wheels, and travelling along the road...Mr. S will discover that he has many difficultys to overcome, before he gets it into general use, as I experienced in my own invention.'[135] A younger Watt would surely have approved the idea of economising fuel; a jaded and ageing Watt did not endorse competition from a radical engine to the firm of Boulton and Watt still operated by his and Boulton's sons. But into the 1840s James Stirling would insist on the improvements his engine offered over the steam engines; thanks to the wide publicity given to large-scale and long-term practical demonstrations in Dundee, European mechanical engineers and natural philosophers took a keen interest in hot air as an alternative to steam.[136]

The Stirling brothers, in the sedate and plausible accounts they gave of their engine, could not begin to compete with the practised hyperbole of John Ericsson. His technological showmanship and self-promotion more than equalled in extravagance that of his contemporary, Isambard Kingdom Brunel.[137] Ericsson came to England from Sweden early in the nineteenth century and rushed to explore audacious projects in mechanical and heat engineering – like his 'flame engine', or his 'combined rotary steam-engine and water-wheel'.[138] Much admired by the editors of the *Mechanics' Magazine*, it was never entirely clear whether his projects crossed the border from bold but legitimate innovation to unacceptable fantasy. Ericsson would be a serious contender at the Rainhill locomotive trials (ch. 4). But his controversial 'caloric engine', patented in April 1833, just after the Brunels had abandoned their gaz engine, divided domestic publics and skilled engineers alike. Brunel senior, who early in 1834 examined it in the company of the then Home Secretary, Earl Spencer, judged it unworkable and its principles spurious, a verdict which denied Ericsson the government funding he, like Brunel, had sought.[139] For Ericsson's many supporters, however, it was the caloric engine, and certainly not the gaz engine or any other pretender to the title of first prime mover, that would supersede steam. How could it fail to do that, if there was any truth in Ericsson's claims that it might generate work with the consumption of little fuel, far less than steam – or perhaps no fuel at all. After all, Ericsson's engine was equipped with a device, comparable to the Stirlings' economiser and dubbed the 'regenerator'. This apparatus, according to its extensive press, had the useful function of reusing heat (caloric) again and again whilst the steam engine wastefully threw away its caloric after each stroke – or so Ericsson claimed.[140] This attractive possibility of power for nothing, however, violated Carnot's theoretical claim that there was a limit to the amount of power which might be developed from a quantity of heat and as that claim became wider known, it undermined Ericsson's credibility amongst practical engineers and men of science alike.

Ericsson had insufficiently sustained success in convincing onlookers in England that his engine really did achieve the economies of fuel he claimed for it, despite – perhaps even because of – the fact that on 14 February 1834 Faraday presented it, as he had the gaz engine, at a Royal Institution Friday evening discourse. Accounts of that display varied dramatically and Ericsson had not been at hand to push for a consensus in his favour. As with the Brunels, other projects (the screw propeller for steamships) beckoned; and when it transpired that in August 1835 debts to his Bond Street tailor and others amounted to £12,000, his moral and economic probity was so seriously in question that gauges of his practical and philosophical acumen were in danger of sinking catastrophically.[141] Since man and machine were so deliberately and closely allied, the reputation of the caloric engine fell with the reputation of its flamboyant promoter. Thus, Ericsson left Britain for North American soils – less worked out by tradition and more fertile for mechanical inspirations.

Ericsson persisted with his caloric experiments far longer than Watt and the Brunels. In 1853, twenty years after filing his British patent, Ericsson could claim that, equipped with his engines, the 2000-ton, 250 ft ocean-going ship *Ericsson* – amply embodying his personal ambitions – would cross the Atlantic by caloric. It was a call transmitted rapidly across the ocean, and discussed at length in the *conversaziones* of the engineering Old World (especially at the Institution of Civil Engineers), attended by men just then beginning to build large iron ships with huge ranges – but aware of the danger of finding a shipping empire dominated by other powers launched in the New World. I.K. Brunel, by now re-invented as audacious and controversial engineer of the Great Western Railway (ch. 4), author of ever-larger ocean-going vessels, and projector of the enormous steam-powered *Great Eastern* (ch. 3), strategically placed Ericsson and his engine in the camp of the deluded seekers of perpetual motion. Commentators were sadly denied the vital practical proof Ericsson needed when his ship sank, in mysterious circumstances, after trials whose results were much disputed, and before ever making the extended voyage engineers routinely demanded as practical proof.[142]

Astonishingly, the resilient Ericsson survived the further implosion of confidence that might have resulted from the demise of his ship; and, against the expectations of persistent doubters, caloric engines resurfaced over the next decade as small, omnivorous power units which were versatile, easy to run and maintain, safe to use and cheap to operate. Despite Ericsson's original hopes for them as the premier marine power, they turned out to be well suited to the American print trade, then mechanised and sending out huge runs. There was more than a little irony here, given Ericsson's insatiable appetite for publicity – and his penchant, equal to the mature I.K. Brunel's, for media manipulation. Later in life, Ericsson would be feted in the United States primarily for the military machinery, notably the 'monitor' series of ships he produced in the years of the Civil War, and for his 'visionary' views on solar power. But in the 1850s, he was looking still not to the sun but to the atmosphere for profit, power and reputation.[143]

Trade publicity for the Massachusetts Caloric Engine Company from 1859 was utterly uninhibited in its praise of Ericsson's Caloric Engine, 'one of the greatest boons which the ingenuity of man has ever bestowed upon his race'. Thus the New York *Printer* allowed its view of caloric innovation to be repeated:

These [caloric] innovators upon steam are at length brought to perfection, and will shortly supersede the steam engine in every department of the mechanic arts where a powerful motor is required.[144]

The trade pamphlet insisted that wind and water were 'unreliable and fluctuating powers'; steam was 'more certain, but dangerous' and attended with 'numerous perils'. But the 'common atmosphere', claimed Ericsson's publicists, was now at work in a 'harmless, controllable, certain, and universal Motor'. The progressive world of practical art was littered with bold but short-lived experiments (of which the gaz engine was just one example). Sensitive, perhaps, to that issue, the Caloric Engine Company now asserted emphatically that Ericsson's engine was 'no longer a subject of experiment, but...a perfect, practical machine, daily at work in numerous and diversified uses, with undeviating success'.

Although power-engineering projectors were ultimately concerned with questions of economy, they knew that other factors fuelled the adoption of, or failure to adopt, their machines. Certainly the caloric engine promised economy over the steam engine (using, according to its promoters, only a third of the fuel); but arguments also abounded that it offered a beneficial revision of regimes of skill, maintenance, insurance and supply. Whereas steam needed an established aristocracy of specialist engineers trained in theory and practice as Watt and others demanded, caloric engines (more democratically) were the province of the 'mere child'. In a throwback to Smith's 'fire-engine', this 'simplest machine, as a motor, ever invented by man' was being tended by 'one of the boys in the office', 'an apprentice', and a warehouse porter. From his earliest experiments, Watt had been nervous about accidents stemming from high-pressure steam – even though later it was high pressure that promised improved economy; the Brunels faced head on the question of bursting tubes; but for a low-pressure caloric engine, filled with non-scalding air, these problems evaporated. The caloric engine, being 'no more dangerous than a common stove', was cheap to insure.[145] A constant question for steam engineers, in factories and then, from the late 1820s, on the passenger and freight railways (ch. 4), was to ensure a plentiful supply of water. The caloric engine required none. As one enthusiast succinctly remarked: 'It saves us an engineer, it saves us fuel, and it saves us from all apprehension of having that useful portion of our person, our head, blown off.'[146]

Much of the pamphlet was taken up in attempting to add credibility to these and other assertions of superiority – almost to the extent that Ericsson's publicists did 'protest too much'. There were confidence-inducing illustrations of small single- and double-acting engines (carefully regulated by governors, and ready to transmit rotary power to belt-driven machinery). There was an engraving of a substantial multi-storey factory, in South Groton, adjacent, ironically perhaps, to a passenger and freight steam railway line.

There were lists to indicate that by 1859 a hundred or more caloric engines had been constructed, with cylinders of anything from eight to thirty-two inches, for companies and individuals, from Boston to Baltimore, from Cincinnati to Cuba, and from Jamaica Plain to Providence.

If the testimonials were to be believed, those engines were then at work in industries so various as to astonish the most hardened industrial commentator. They were moving hoisting gear at goods depots. Caloric, not steam and not electro-magnetism, was driving the 'double cylinder' and 'Hoe cylinder' printing presses, capable of thousands of impressions an hour, snapped up by the *Yankee Notions* newspaper, the Brooklyn *Daily Transcript*, the Baltimore *Price Current*, the *Cincinnati Press* and the *Ogdensburg Democrat*. Fashion followers take note: Watt's plan to pump water by steam from the Seine to the Palace of Versailles fell foul of the Revolutionary turmoil;[147] but Ericsson's engines were powering domestic pumps at the country seats of numerous novelty-hungry gentlemen. Railway stations, still saddled with thirsty steam engines, were nevertheless opting for caloric pumps (which delivered the same work in half the time of the steam variety). Just five years after the demise of the *Ericsson*, J.A. Ronalds of Pelham boasted a 50 ft paddle-wheel yacht engined by Ericsson and, should anyone be in doubt, christened *Caloric*; the 36 ft *Marie Louise* was reported to have averaged 7 miles an hour on a 60-mile journey, up the Hudson River; one Bostonian gentleman, moving with the times, had commissioned a 70 ft caloric yacht driven by screw propeller.

Low-grade fuel, not expensive coal or coke, kept these engines at work in 'all the forms of human industry': tanning, cabinet-making, wire-working, lathe- and foundry-work, ship-building, coffee-grinding, tobacco-working, skirt-making and glass-cutting. Ericsson's power was as fit for the homeopathic pharmacy as for the plantation. According to these willing testimonials, caloric provided a power of 'universal application, wherever a limited, economical, safe, independent, and self-managed motive power is desired'.[148] Domestic, and yet also democratic, it seemed to be proving wrong the predictions of those traditionalists willing it to fail. So, at least, was the case if one believed the puffing of Ericsson's allies in South Groton, unconstrained in all but one respect: the idea that the engines used no fuel, thanks to the regenerator, had now been quietly dropped.

That claim, prominent in the propaganda of the 1830s and surfacing again during the trials of the *Ericsson*, had been a claim too far for more sober engineers in Great Britain. There, cautious promoters of a revamped Stirling engine deliberately distanced themselves from such 'enthusiastic' and unmeasured talk. Chief amongst these rationalist engineers were men who eschewed the flashy publicity and show-making that typified the works of Ericsson and, to a lesser extent, the Brunels, but who had no shortage of experience in the steam-powered technologies of rail and the ocean liner. These men were James Robert Napier and Macquorn Rankine. Napier was a key figure in a family with roots in Glasgow and London famous for its success in mechanical engineering and, more recently, marine engineering (ch. 3). He was particularly attracted to the idea that the natural philosophy of Glasgow College might have practical payoffs. This it seemed to be doing in the words and work of Rankine who in 1855 had migrated from the career of civil engineer (in railways, surveying and electrical telegraphy) to the regius chair of civil engineering and mechanics. From that position, Professor Rankine made it his personal mission to effect a harmonious bridge between worlds of 'theory and practice', construed as impractical thought and untutored action, through 'the science of the engineer'.[149] In the practical and technical realm he was an omnivorous consumer and a prolific and increasingly authoritative producer, especially in a monumental – and indigestible – series of textbooks taking in applied mechanics, heat engines, mechanical engineering, civil engineering and shipbuilding (Conclusion). Rankine's somewhat fragile scientific reputation then rested primarily on his explorations in the new science of thermodynamics in the late 1840s and in the early 1850s.[150]

Few, however, doubted Rankine's status when it came to drawing out the specific practical implications of the new theory of thermodynamic engines (or engines deriving work from heat) – a theory which had been developed precisely to explain, in general terms, the engines (having steam, air and other working substances) which so astonished Carnot in the 1820s and William Thomson in the 1840s. In public, Rankine expressed general, if not dispassionate, opinions about the optimal future course of practical engine design, seeing it as a quasi-religious quest for the perfection of actual engines, where that perfection was to be measured not by Watt's standard – he too had talked of a perfect engine – but by the limits now securely established by thermodynamics and the science of energy. In private, Rankine was convinced, by theoretical considerations rather than the public demonstrations of James Stirling in Dundee or the antics of Ericsson in New York, that the future of heat-engineering lay in the hot-air engine of James and Robert Stirling. As he told Napier in February 1853:

> If Captain Ericsson's ship . . . should succeed in crossing the Atlantic, of which there seems no reason to doubt, it is evident that public opinion will turn in favour of air-engines, and that a demand for them may be expected to arise at once; and it is not unlikely that for marine purposes, and in all districts where either fuel or water is scarce, and perhaps also for locomotives, they will ultimately supplant the steam-engine altogether.[151]

Napier and Rankine complemented each other, much as did Boulton and Watt, although the precise distribution of skills, resources and ambitions was somewhat different. Napier had money, mechanical expertise, considerable technical resources and a desire, carefully nurtured by Rankine, to use the engines for marine purposes. Rankine had technical, theoretical and pre-sentational skills, but was short of cash and wanted an ally who would adopt the engine directly. It was Rankine who considered how to 'improve' the Stirling engine; in a strange echo of Watt's experience, he may, through his connections with Glasgow College, have had access to a model of that engine apparently deposited there by the Stirling brothers, used for teaching in the natural philosophy class, and evaluated approvingly by William Thomson early in his professorial career. Thomson indeed directly com-pared Stirling to Watt, and played up the now-famous Glasgow College connection. Like Boulton and Watt, Napier and Rankine worked with experimental engines, dealt with the issues of patents, and sought important patrons and customers for their engines; like the Brunels, they used a growing network of societies and institutions – here the British Association for the Advancement of Science – to try to generate enthusiasm for their work, particularly in the marine market (ch. 3).

The historical irony of these events was that what would eventually become the new scientific orthodoxy, in a 'science of energy' that reformed and restructured the empire of nineteenth-century physics, might well in Rankine's hands, and with the support of William Thomson and James Joule, predict that the air engine, in some form, was the new power destined to 'supplant the steam-engine altogether'; it might, further, tame the enthu-siasm of those engineers, like Ericsson, whose discursive pronouncements and media manipulation threatened to destablilise public trust and to sink the project of the air engine without hope of salvage. But it could not overcome specific practical difficulties encountered in grappling with the new technology of the hot-air engine quickly enough and cheaply enough to make it worthwhile for its British backers to continue its development in the face of innovations in marine steam engines which promised rapid returns (ch. 3).

In parallel with these experiments into 'gaz' and air were investigations into another natural power briefly tipped to challenge steam. In 1820, the Copenhagen-based savant H.C. Oersted had shown that a current-carrying wire had an associated magnetic effect. By 1824–25 William Sturgeon had produced an 'electro-magnet', with a soft-iron core, that manifested spectacular power in public forums, celebrating the workman's artistry rather than the philosopher's eternal principles.[152] Sturgeon's own *Annals of Electricity, Magnetism, & Chemistry; and Guardian of Experimental Science* gave wide publicity to the various machines developed from the middle third of the 1830s in Europe and the United States to answer the question whether the related forces of electricity and magnetism might, like steam, be harnessed by man to make a moving power.[153]

Hermann von Jacobi, professor at the Russian Imperial University of Dorpat (now Tartu in Estonia), was convinced that Oersted's 'discovery... promise[d] a new career to the practical mechanic'.[154] Steam provided these mechanics with an exemplary career trajectory. As with Watt, that career meant breeding up a skilled practitioner, the electrician. It also meant transforming demonstration models (the kind of apparatus Jacobi called 'an amusing plaything' enriching 'physical cabinets') into large-scale engines. Further, to compete with steam, it required the creation of an engine with the rotary power demanded in the factory. Princeton professor Joseph Henry's earliest electro-magnetic engine had 'a beam suspended in the centre, which performed regular vibrations in the manner of a beam of a steam engine';[155] Jacobi began with devices for 'rectilineal motion' but by 1834 he had his first 'magnetic machine for continued primitive circular motion' which like Watt's steam engine, avoided 'injurious and destructive jerks' to connected machinery. Jacobi's solution to the problem of rotary motion was the 'commutator'.[156] Although sober in his claims, and careful to root them in elite electrical science, Jacobi pointed to the safety, simplicity and cheapness of this 'new mover', qualities which placed its superiority over steam 'beyond a doubt'.[157]

In October 1836 Sturgeon himself described an 'electro-magnetic engine for turning machinery' which he claimed to have constructed in the autumn of 1832 and exhibited at the Western Literary and Scientific Institution, off London's Leicester Square, in March 1833. In subsequent demonstration lectures the engine 'excited a great deal of curiosity', perhaps because Sturgeon drew out comparisons with the steam engine by using it to draw wagons and 'carriages on a railway', to saw wood, and to pump water 'on about the same scale as we see pieces of machinery put into motion by the large models of steam engines'.[158] By April 1837 he had achieved, albeit in a 'rude state' not yet ready for public unveiling, two of the key goals of steam promoters: 'propelling a boat, and also a loco-motive carriage, by the power of Electromagnetism'.[159]

In America, the untutored Vermont blacksmith Thomas Davenport produced rotary motion from electro-magnetism in July 1834; he showed his machine to Professor Henry in July 1835 and walked away with a certificate of originality – vouching for his credibility as inventor. To move the invention out of the philosophical cabinet, Davenport turned to Ransom Cook who offered mechanical expertise, advertising skills ('bringing the subject before the public in the most effectual way') and the impetus to patent the machine in the United States (in February 1837) and to try to secure his rights to it in Europe.[160] The *Journal of the Franklin Institute*, which reprinted the patent, hinted that Davenport had been greedy (the patent covered the application of 'magnetic and electromagnetic power, as a moving principle, for machinery, in the manner above described [in a specific machine], *or in any other substantially the same principle*')[161] and diluted his claims to originality with reports of experiments already made by 'distinguished philosophers' in Europe or of machines sketched in London's *Mechanics' Magazine* in 1833. But Yale professor Benjamin Silliman assigned Davenport priority of invention for a 'galvanic machine of great simplicity and efficiency'; Sturgeon – as editor of *Annals* and electro-magnetic rival aware of the thirst for power in the factories – merely qualified this to make Davenport the 'inventor of a method of applying galvanism to produce *rotary motion*'.[162]

Davenport's work had been extensively reviewed especially in the American newspapers and in a pamphlet of 94 pages, published in New York, that outlined the machine's history, echoed its positive press and placed it in a context of science (culled from Mary Somerville). But Silliman felt that journalists had been unguarded and Sturgeon worried about the 'erroneous notions' formed about the 'celebrated machine' in England (where models of Davenport's engines were also displayed).[163] At issue for Silliman was whether the bragging of the journalists would stop men from coming forward to fund 'this interesting research' and allow the necessary 'persevering experiments' to continue.[164] Projectors, and their allies, trod a fine line between legitimate reportage and confidence-destroying enthusiasm.

In order to moderate advocates' audacious claims, and to lay down a cautious and trustworthy evaluation, and to inspire financial support, Silliman offered renewed sanction from the sphere of science (already represented by Henry). He examined two of Davenport's engines in person; the direct observations of a reliable disinterested observer punctured the wonder-mongering of circulation-conscious journalists. Davenport's skilled associate Israel Slade transported the first machine (with permanent magnets) to New Haven for Silliman's perusal: Silliman described it as a 'philosophical instrument', operating with 'beautiful and surprising effect', but remained concerned about the limits which existed to extending its power on a large scale. Thomas Davenport visited Silliman personally in March 1837 to show him a second machine (transcending limitations on scale by using electro-magnets): Silliman was yet more impressed. Within days he rushed into print his verdict that electro-magnetism could generate rotary motion 'cheaply and certainly', 'indefinitely' and to a limit as yet unknown. His conclusion was that its 'investigation should be presented with zeal, *aided by correct scientific knowledge*, by *mechanical skill*, and by *ample funds*'. Science and art, the 'handmaids of discovery', would then receive 'a liberal reward'.[165] Silliman was clearly concerned to generate trust and then, quite literally, to see Davenport capitalise upon it.

Where Sturgeon set electro-magnetic motors to do the work of steam in popular lectures, Davenport and Cook worked to achieve effects on a larger scale. By December 1837 they had a turning lathe with 'astonishing strength' and could drill through iron, steel and wood of dimensions similar to those found in practice.[166] They undertook experiments in New York with a machine designed to propel a Napier printing press which, according to the Franklin Institute's editor would be the true test of its value. The proprietors of a joint stock Electro-Magnetic Association sought funds from 'public spirited individuals' to 'carry on experiments' for the 'benefit of mankind'.[167]

Davenport promoted his machines with an exhibition in New York, approached government agents and succeeded in having an electro-magnetic motor witnessed in Washington by the President. Cautious 'not to be considered an enthusiast', Davenport nevertheless predicted 'that soon engines capable of propelling the largest machinery will be produced by the simple action of *two galvanic magnets*, and worked with much less expense than steam'.[168] The key issue, in evaluating whether such a power could 'be substituted for that of steam', would be the 'economy' of a large-scale engine; but already it seemed likely that 'the cost of operating the electro-magnetic apparatus, will be much below that of the steam engine'.[169]

Others speculated that even on a small scale, the electro-magnetic engine might find numerous roles rivalling steam: an engine of 'the power of a man only...would be equally valuable with the steam engine', assuming the cost of fuel was less than that given to a man for his labour, and would thus 'produce as great, if not a greater change, in the economy of the useful arts, as has been produced by that instrument'. The new engine would consume fuel only when it was at work; it could do the jobs for which steam engines were too cumbersome; it would not need the skilled attention of an engineer; it would be ready at any moment for 'our lathes, our grindstones, our washing machines, our churns, our circular saws, and a catalogue of other things', supplying the wants of the farmer, mechanic and house-keeper. Comparing it to the new railways, he asserted: 'We could no more submit to live without [this new power] than we can now submit to travel at the slow rate of ten miles an hour, an event which we have learnt to think one of the miseries of human life.'[170]

The railway mania of the mid-1830s only recently having subsided (ch. 4), commentators identified an international euphoria for electro-magnetic engines, reflected – and fuelled – in Sturgeon's *Annals*, and participated in by the young James Joule. From January 1838 he advertised an engine which 'may be used to turn any kind of machinery'[171] and by May 1839 admitted he had been convinced, like many other electricians, that electro-magnetism would 'ultimately be substituted for steam to propel machinery'.[172] Most emphatic amongst the apologists for the new power was Dr Taylor who in 1841 informed British audiences of electro-magnetic events in Germany, Russia and America threatening to depose steam not simply from various roles as a motive power, but as a symbol of empire:

> Mighty and vast is the influence which Steam ... has exerted on all the relations, as well of nations, as of individuals. It has served to facilitate the intercourse between lands the most distant, and has confessedly exerted an influence on the face of empires, more incalculable in its effects than any previous invention to which the mind of man has given birth. For England, especially, it has served to consolidate the varied elements of its

power, and by annihilating space and time, brought the geographical parts of the gigantic empire of Britain into such immediate proximity, that it may be said to have diffused new life and vigour throughout all the veins of its giant members, and given a nervous strength to its arm, which must long perpetuate its sway.... That Steam is about to be superseded [however] can hardly admit of question.[173]

It was by 1841 a cliché for Taylor to portray steam as a force which had transformed the relations of nations and individuals, and which, through its application to ships and railways, had annihilated the space and time otherwise dispersing the 'gigantic empire of Britain'. But Taylor built on the familiar notion that electricity was the stimulus to animation by making steam the vital force of the great organism of empire, which 'diffused new life and vigour throughout all the veins of its giant members' and gave 'nervous strength to its arm'.[174] Invigorated with steam, the British Empire's life could only be lengthened. Yet this was a power which was dangerous, costly and, since utterly dependent upon a fuel supply frequently inaccessible, fatally flawed.

A newer agency, the electro-magnetic engine, was waiting to power up the sprawling body of the empire – not only with lathes, printing presses, factories and sawmills (there was already one powered by electro-magnetism in Bavaria) but with new electro-magnetic ships and galvanic railways. In 1837 Jacobi had worked with the Russian government to 'render electro-magnetism applicable to the working of machinery, and particularly to the propulsion of ships'. By September 1838 a 28 ft eight-oared galleon equipped with paddles, like a steam packet, and powered by electro-magnetic engines whose 320 pairs of battery plates left little space on deck for the fourteen passengers, achieved 'three English miles per hour' and made way against the current of the Neva.[175] Jacobi wanted to build on this and put together an engine rated at no fewer than 50 horse-power for a much larger vessel.

By June 1837 the Rev. N.J. Callan, professor of natural philosophy at the controversial Roman Catholic College at Maynooth, had constructed his own electro-magnetic engine which he had become convinced, by 'experiments', 'might be successfully applied to the working of machinery of every kind'. Callan planned an engine of up to two horse-power to propel a carriage at 7 or 8 miles an hour. With a railway system consolidating in England, and with projectors speculating on a skeleton of Irish routes (ch. 4), Callan's experiments seemed already to indicate that 'an electro-magnetic engine as powerful as any of the steam engines on the Kingstown Railway' might be made for £250, would weigh no more than two tons, would cost under £300 per annum to work and maintain and thus: 'the expense of propelling the railway's carriages by electro-magnetism, will be scarcely one fourth of the cost of steam'.[176] In 1841, the engineer Stoehrer arranged with one of the new German railway companies for his electro-magnetic engines to 'propel a train of waggons with the usual number of passengers, from Leipsic to Dresden'.

He insisted the total cost was six shillings instead of the usual £5 for steam. Given these financial projections he was astonished that England 'possessing, as she does, the most enormous capital for the purposes of experiment... remain so indifferent on a subject of such vast importance'.[177]

If not in England, then in Scotland, a young electrician posed a similar question to railway capitalists. After 'years of labour, both mentally and bodily', Robert Davidson secured the support of staff at King's College, Aberdeen and especially Patrick Forbes. Knowing of Jacobi's experiments, Forbes wrote to Faraday in October 1839 to sing the praises of a 'countryman of our own' who by then had a small lathe and an electrically driven wagon big enough for two, both devices directly witnessed by himself and Professor John Fleming. Forbes had been watching Davidson's progress for two years; by asserting he was bound to create 'a highly useful, efficient, and exceedingly simple moving power' the professor gave the artisan a testimonial of value and authenticity equivalent to that Henry had given Davenport in the United States.[178]

A reporter in *Chambers's Edinburgh Journal* (probably Robert Chambers himself) was so impressed by Davidson's demonstrations at Edinburgh's St Andrew's Square, and by his 'machine for printing handbills', that he had asked him to 'fit up an apparatus for turning our printing machinery': here, said the author, 'has been manifested that great desideratum – something to supersede steam power', in a safe, small, easy to manage, home-grown rival to Jacobi's engine and having about 'a man's power'.[179] In 1839, Davidson's sponsors appealed not to printers but to 'railway proprietors in particular'; to allow him to 'languish in obscurity' would be a 'reproach to our country and our countrymen'; the railway companies, then, should finance a company to transform his experiments, bringing the engine into operation on a 'great scale'. The *Aberdeen Constitutional* sniped at the foreigner Jacobi's achievements and declared Davidson's machine, now seen by thousands, 'the desideratum that has been wanting to perfect the power of locomotive agency':

> Were a company formed to carry out the principle, we have not the slightest doubt, but, in a very few years, there would not be a locomotive used without a magnetic propeller; and, considering the number of accidents that are daily occurring on railways, it is especially to be desired that it should supersede the steam engine there.[180]

Davidson reappeared at the Egyptian Hall, Piccadilly where, under the patronage of the Royal Scottish Society of Arts, he staged an exhibition of 'electro-magnetism, as a moving power'. There, for the price of a shilling (or sixpence for children under twelve), one could see a locomotive engine carrying passengers on a circular railway, the printing machine (which printed up the handbills for the show), a turning lathe and a sawmill, not to mention a vast electro-magnet and the new galvanic telegraph (ch. 5). In 1842

Davidson's small locomotive, the *Galvani*, ran along the new Edinburgh and Glasgow Railway with carriage at about 4 miles an hour.[181]

Such accounts augured the displacement of steam from established industries and from the new steamships and steam railways. But, recalling Silliman's caution against confidence-eroding bragging, other projectors reined in such rhetoric by subjecting the electro-magnetic engine to the theoretical and practical discipline of elite science, as Carnot had recently done for heat engines. Speaking at the British Association meeting at Glasgow in August 1840, Jacobi had emphasised the continued importance of the question of electro-magnetic engines, the progress made with his electro-magnetically powered boat – and the need to replace the many 'mere trials' and 'irregular attempts' with 'legitimate' investigation on 'scientific foundation'. Like Silliman, Jacobi wanted to tame the rhetoric of wild projectors but not to call a halt to electro-magnetic progress: 'The future use and application of electro-magnetic machines' appeared to Jacobi 'quite certain, especially because the trials and vague ideas which have hitherto prevailed in the construction of these machines, have now at length yielded to the precise and definite laws' typical of those observed by nature in matters of cause and effect.[182]

Joule, once full of 'expectations that electro-magnetism would ultimately supersede steam', now accepted the sobering limits placed upon the velocity of rotation in such engines by Jacobi's practical science. Joule made no apology to his Manchester audience in February 1841: he had 'neither propelled vessels, carriages, nor printing presses. My object has been', he belatedly insisted, 'first to discover correct principles and then to suggest their practical development'. Reasoning from principles and practical measurement to compare the likely duty (in pounds raised to a height per pound of zinc consumed) of an electro-magnetic engine with that datum of steam efficiency, the Cornish engine, Joule saw little chance of the new engines becoming 'an economical source of power'. Indeed, now he asserted that such engines would only ever be used in 'very peculiar purposes'.[183]

If steam stood for progress, it stood for the possibility of some contender – if not gaz, caloric, or electro-magnetism then for some other power – to usurp steam. Yet steam survived and indeed thrived. Steam's hardiness came from constant innovation within steam culture, often in conflict with the principles of caution, necessary to maintain business confidence, adhered to by Watt after his frenzy of invention in the early 1780s. Thus, *pace* Carnot, Watt was lukewarm about the high-pressure steam engines introduced, especially by Trevithick, from 1801 after Watt's first patent lapsed.[184] Preferring low pressures, Watt did little to exploit the expansive action of steam (in which a fixed quantity of steam expanded against the piston whilst its pressure fell) or the possibilities of 'compounding' (where high-pressure steam is pushed against a piston in a small cylinder before being transferred to a larger cylinder in which it developed more work).

It was another of Carnot's heroes, Arthur Woolf, who again waited for the lapse of Watt's legal monopoly before he patented the compound engine in 1804. By 1811 Woolf was claiming, like many projectors in power engineering, that his engines did the work of a Watt engine with a faction (a third, to be precise) of the fuel; his was the engine to develop with optimal economy the power of steam, unlocking nature's storehouse of fuel. Reports at Cornish mines in 1815 confirmed the astonishing efficiency in engines manufactured, as Boulton had wanted Watt's to be, to the highest accuracy and finished like showroom ornaments. Nevertheless, during the 1820s controversies raged over the value of Woolf's engines when they turned out to be harder to maintain than their single-cylinder predecessors.[185] If it was unclear whether or not compound engines would supersede Watt's, marine engineers of the 1850s and especially the 1860s found a role for compounding in ocean steamships (ch. 3).

One aspect, then, of the failure of the challengers to steam was that particular steam engines were themselves superseded – by others more suited, through economy or durability or a host of other factors, to developing environments. This endurance gave time for the literary reconstruction of Watt as a hero of eighteenth-century technology, crucial to the mechanisation of industry, and above all to the genesis of steam power on which British economic power rested.

The worship of power: constructing icons of steam

Throughout the nineteenth century, Watt and the steam engine were simultaneously distilled in visual images, material objects and words. Print makers disseminated portraits taken from the half-dozen originals made during his lifetime (dating from 1793 to 1815, and including amongst the artists Sir Henry Raeburn). Images of artefacts associated with Watt's invention were widely reproduced, just as were images of Newton and his apple:[186] R.W. Buss, for example, imagined the young Watt idling away his time gazing at the tea kettle under the glare of a disapproving aunt;[187] other Victorian images – like James E. Lander's 'Watt and the Steam Engine' – showed Watt in similar guise, or in his twenties, in his workshop, struggling to understand the Newcomen model in Glasgow College.[188] A popular school print of Watt 'discovering' the principle of the separate condenser appeared in Meiji Japan during the last third of the nineteenth century, designed to stimulate young Japanese entrepreneurs (apparently successfully).[189]

Busts of Watt, modelled by Sir Francis Chantrey, gravitated to places of memorial and of scientific and engineering power. Thus James Watt junior ensured a bust of his father was displayed at the Royal Society of London and another went to the Paris Academy of Sciences. He erected a statue in memory of Watt at the family home in Handsworth, and there was a further image at Glasgow University. The inhabitants of Greenock ordered their own statue to be situated in the library Watt had liberally endowed. A public subscription financed a monumental seated statue, again augmented from Chantrey's bust. Copies of the statue found their way, controversially, into Westminster Abbey as part of the Watt memorial (with Brougham's epigraph), into Glasgow's George Square, adjacent to the merchant district, and into affluent centres of steam-powered textile production, like Leeds.[190]

In a pattern followed with in-house narratives of the early railways (ch. 4), or pamphlets issued by key protagonists in electric telegraphy (ch. 5), friends and sponsors authored – and policed – early accounts of the steam engine and its purported inventor, thus ensuring the dominance of the 'Watt version'.[191] The increasing 'worship of power' through this period, and the ambivalence of 'mechanism' or 'machinery' – as source of national wealth, but as indicator of spiritual bankruptcy – meant that 'steam' and 'Watt' were given many and various valuations.

The historiography of steam had been elaborated in extraordinarily antagonistic circumstances. Watt's own 'A Plain Story' of 1796, outlining the genesis of the steam engine, began life as a submission to patent lawyers.[192] The Edinburgh-based *Encyclopaedia Britannica* published articles on 'steam' and 'steam engine' around 1797 by one of Watt's key allies, John Robison – and thus confirmed Watt's status as progenitor of steam's revolutionary improvement.[193] When Richard Phillips included Watt in his *Public Characters* (1802–03), he asked for Watt's own approval and changes.[194] It was Watt that later revised Robison's accounts of 'steam' and the 'steam engine' and saw that rival contributions were downplayed. With Rennie's urging, it was James Watt junior who assembled materials for a biography and in 1823–24 published an obituary, securing his father's reputation, in Macvey Napier's updated *Encyclopaedia Britannica*. Even John Farey's canonical treatise on the steam engine, published in 1827 after Watt's death, betrayed its author's debt to Watt's living heirs.[195]

Watt's death in 1819 had, if anything, increased the volume of plaudits. In June 1824 men of science, commerce, politics and war gathered to demand subscriptions for a public monument to Watt. It was an extraordinary consensus extending from advocates for artisan education to economic liberals in an age of self-conscious 'industrial revolution'.[196] Sir Humphry Davy, President of the Royal Society, extolled Watt's 'profound science', elevated him above Archimedes, and saw in him the exemplification of 'the practical utility of knowledge'. Watt had 'enlarged the power of man over the external world, and both multiplied and diffused the conveniences and enjoyments of human life'.[197] No surprise, then, that 'Loud applause' greeted Robert Peel's admission that Watt's discoveries had led to his 'direct personal benefit': it was Watt that brought a 'new life and spirit' to the 'Cotton Manufacture of this Country' (the 'source of our National wealth'); Peel wanted a monument to bear general testimony to the 'dignity of human nature . . .

ennobled by discoveries like those that give subsistence & dispense comfort to thousands while they widen the limits & add to the strength of the Empire'.[198] Henry Brougham would later distil these comments in the epigraph which graced the monument eventually erected in Westminster Abbey, making Watt an individual who 'directing the force of an original genius/Early exercised in philosophic research/To the improvement of/The steam engine/Enlarged the resources of his country/Increased the power of man/And rose to an eminent place/Among the most illustrious followers of science/And the real benefactors of the world'.[199]

To these transformations in science, manufactures, civilisation and empire were added thrilling victories securing national identity against past – and future – foreign threats. The Prime Minister, Lord Liverpool, asserted that the steam engine meant the 'success of a [military] campaign, or even a war' would no longer depend on 'contrary winds'. Whilst classifying the steam engine as 'a moral and irresistible lever in pushing forward the grand cause of civilization', President of the Board of Trade Huskisson insisted it would have been impossible to sustain the immense expense of the last war with France without Watt's inventions. Likewise Sir James Mackintosh, M.P., said that it was Watt's discoveries which had 'enabled England to sustain the most arduous and dangerous conflict in which she has ever been engaged'.[200] Sadi Carnot was thus not alone in seeing steam as the key to British power – commercial *and* military.

A more ambivalent assessment of Watt from one more dubious of validations, like these, stemming from practical utility, came from the social critic Thomas Carlyle. Shortly after the public meeting praising Watt, Thomas Carlyle compared the engineer to Marcus Brutus: certainly Brutus was more virtuous than Watt of Soho; but 'for its *utility* the steam-engine was worth five hundred deaths of Caesar'.[201] In his essay on 'Signs of the times' published in the *Edinburgh Review* in 1829, Carlyle wrote: 'In all senses, we worship and follow after Power; which may be called a physical pursuit.' Carlyle identified the age in which he lived as 'not an Heroical, Devotional, Philosophical, or Moral Age, but, above all others, the Mechanical Age'. This was the 'Age of Machinery, in every outward and inward sense of that word.' As a consequence, 'Nothing is now done directly, or by hand; all is by rule and calculated contrivance.' Craft skills had been displaced by mechanism: 'on every hand, the living artisan is driven from his workshop, to make room for a speedier inanimate one'. Thus the human fingers of the weaver had given way to iron fingers. Even sails and oars had yielded to mechanical power: 'Men have crossed oceans by steam; the Birmingham Fire-king has visited the fabulous East.'[202]

Carlyle's implicit indictment of 'the Birmingham Fire-king', and all other manifestations of 'the Age of Machinery', for their destruction of the mysterious, poetic and spiritual sides of man did not extend to James Watt. In the same essay, he referred to the Watts, father and son, as advancing science (alongside Newton and Kepler) in 'obscure closets' and 'workshops' through 'the free gift of Nature' and not through institutions and other forms of social mechanism; writing on 'Chartism' he again depicted Watt 'with blackened fingers, with grim brow... searching out, in his workshop, the Fire-secret'. Such a distinction between the transcendental and heroic 'creative genius' of James Watt and the more ambivalent uses to which his creation had been put, however, rarely troubled nineteenth-century power worshippers who freely invoked the name of Watt as earlier generations invoked their saints.[203]

The chief celebration of Watt's life came not, in fact, from Britain but from France, in the words delivered in December 1834 by François Arago, Perpetual Secretary to the Paris Academy of Sciences. In the preparation of this 'the most important Eloge ever written by Arago', the savant had called at the Soho works; he had visited also Port Glasgow with the statistician Dr James Cleland – curious to find there a monument to Watt not in a house (sadly demolished) where he had carried out experiments but in the steam engines then at work.[204] As well as conversing with acquaintances and friends, Arago had corresponded frequently with James Watt junior and they discussed the project face-to-face in Paris.[205]

In the autumn of 1834 Arago probed Watt's relatives and friends for 'familiar anecdotes' to 'illustrate his early genius', which raw materials they readily provided. James Watt junior relayed an account sent to him by James Gibson of stories told to him by Jane Campbell which had, in turn, been told to her by her aunt Marion Campbell: Watt's childhood friend Marion, it later transpired, had written down this narrative of events fifty years after their childhood encounters. There were several candidates: Watt the hilarious childhood raconteur, Watt the budding problem-solver covering hearth and walls with mathematical workings – and, of course, a story 'not exactly' as Marion Campbell told it, 'but the facts are the same', tracing Watt's 'first idea of the power and elasticity of steam' to a bored child watching steam issuing from a boiling tea kettle and condensing on a filched plate. Later, James Watt junior attempted to force Arago to discard Gibson's later embellishments and adopt Marion's original dictation. Thus supplied – and constrained – Arago wove the kettle into his account.[206]

In fact, Arago's Watt was not without warts: his weakness in succumbing to the 'stupid neglect of capitalists' and giving up on the engine for canal work in the 1760s might have denied humanity the benefits of steam;[207] only his wife, apparently, had from 'lassitude, discouragement, and misanthropy' rescued a man with only two admitted pleasures: 'idleness and sleep'.[208] Otherwise Watt's life had been so dull as to need apology to an audience used to richer fare; it was an existence manifesting, rather, assiduous labour, 'very delicate experiments', study and meditation, seasoned with 'candour', 'simplicity', 'love of justice' and 'inexhaustible benevolence'.[209]

Where Watt had been roused to sterner actions it was in fighting against those 'interests', 'obstinate partisans' of ancient tradition and 'jealous' persons who pretended to deny credit – both national and personal – where it was properly due.[210] For Arago, Watt was the rightful successor, not to Savery – who, ignored by miners, had dallied in country houses and gardens – but to a woefully neglected French national hero, Papin; Watt was evidence that those from humble spheres – from 'all classes of society' and not merely from the 'titled house' – had played their part in the genesis of the steam engine and in the elevation of 'the British nation to an unheard of height of power' even if Watt's inferior 'caste' had seen him (shamefully) denied a peerage in his own land.[211]

Arago cultivated the image of Watt as a 'philosopher', just as Watt himself had done – seeing that reputation as vital to his commercial success.[212] A Fellow of the Royal Society of Edinburgh (1784) and of the Royal Society of London (1785),[213] Watt adopted a letter-seal engraved with a human eye, and the word 'OBSERVARE', emphasising his readiness, as befit one of Adam Smith's 'men of speculation', to improve all that he saw.[214] Arago placed Watt at the summit of a ladder of philosophical steam engine improvers with its feet in the classical antiquity of Hero of Alexandria. The budding philosopher's childhood observation of the kettle showed him, like Newton, achieving great things by 'always thinking of it'; his clear statement of the composite nature of water paralleled the similarly incisive and incontrovertible elucidation from experimental facts shown in Newton's optics. Like Bacon, Watt was always armed with facts, a reader, as much as an observer, whose workshop, *pace* Carlyle, was no obscure garret but rather a 'kind of academy' open to all for the discussion of all issues in art, science and literature.[215]

Arago claimed to have travelled throughout Britain asking more than a hundred individuals, of all classes and political persuasions: 'What is your opinion of the influence which Watt exercised upon the wealth, the power, and the prosperity of England?'[216] His answer, echoing that of the speech-makers of June 1824, emphasised Watt's paramount and patriotic role in preserving the 'independence and national liberty' of Britain against foreign foe.[217] Watt's benevolence had been manifested not through direct military action, but rather by providing the wealth behind military success and, furthermore, the ideas so respected, Arago claimed, by military commanders from Alexander the Great onwards. The precise degree of that prosperity which had preserved the nation was made concrete in a calculation facilitated by James Watt junior's data on the number of engines sold from Soho – and elsewhere. That data yielded evidence of Watt's super-human creative – and regulatory – power. He was:

> [The] creator of six or eight millions of labourers, – of assiduous and indefatigable labourers, among whom authority is never required to repress either coalition or commotion, and who labour for a halfpenny a-day; – this man, who, by his brilliant discoveries, afforded England the

means of supporting a most furious struggle, during which her very nationality was at stake...[218]

So brilliant – and so stark – was the image Arago painted that he found it advisable to respond to those who, more vehement than Carlyle, insisted that Watt's inventions were 'instruments of evil', leading to 'social calamity'.[219] Had machinery simply concentrated 'prosperity' in the hands of the capitalists, and were its benefits outweighed by the denial of work, leisure and autonomy to the 'working classes'? Arago's geometrically inspired *reductio ad absurdum* arguments, and his generalities drawn from the frequently repeated 'experiment of the substitution of machinery for manual labour', responded to the rhetoric of those who, in England in 1830, shouted '*Death to machinery!*' For Arago, machinery, powered by steam, was a great levelling force, bringing to the poor the blessings hitherto known only to the rich.[220]

Arago saw Watt, assisted by his steam engine, 'penetrating...into the bowels of the earth', clearing 'spacious galleries', procuring 'inexhaustible mineral riches'; he would twist alike the 'colossal cable' securing vessels at sea and the 'laces and airy gauzes'; he would drain marshes and irrigate scorched soils; power, previously found at the foot of the cascade, would now be located in the 'centre of towns'; industry would be concentrated, productions would be cheaper, population will increase, 'well fed, well clad, and comfortably lodged':

Transferred to our ships, the steam-engine will replace an hundredfold, the efforts of the triple and quadruple banks of rowers, from who our fathers required an extent and kind of labour, ranked among the punishments of the greatest criminals. With the help of a few bushels of coals Man will overcome the elements, and will make light of calms, contrary winds, and even storms... [The] steam-engine, conveying in its train thousands of travellers, will run, upon railroads, more swiftly than the best race-horse, loaded only with its diminutive jockey.[221]

With a still more fervent manifesto, Arago concluded his eloge: 'I do not hesitate to declare my conviction, that, when the immense services already rendered by the steam-engine shall be added to all the marvels it holds out to promise, a grateful population will then familiarly talk of the ages of Papin and of Watt!'[222]

3
Belief in Steamers: Making Trustworthy the Iron Steamship

> Quick and reliable navigation in steamships may be seen as an entirely new advance that we owe to the heat engine. Already, as a result, we have been able to establish prompt, regular communications across the seas and on the great rivers of the Old and New Worlds. We have been able to traverse wild regions that were previously almost impenetrable, and to bear the fruits of civilization to parts of the globe where they would otherwise have remained unknown for many years to come. There is a sense in which steam navigation brings the most widely separated nations close together. It tends to unite the people of the world and make them dwellers, as it were, in one country. In diminishing the time taken and in lessening the fatigue, uncertainties and dangers associated with voyages, are we not, in effect, greatly reducing distances?
>
> – *Sadi Carnot writes from post-Napoleonic France of the potential consequences of steam navigation for the nineteenth-century world (1824)*[1]

Carnot's optimistic appraisal of the civilising and enlightening power of steamships serves as an appropriate epigraph for a chapter on steam at sea in an age of Victorian empire. Thirty years on, however, his faith that steam-powered vessels 'lessened the dangers of voyages' would have rung hollow. The year 1854 was a disastrous one both for transatlantic steamships and for iron sailing ships out of Liverpool. On 1 March the iron screw steamer *City of Glasgow*, constructed on the River Clyde by Tod & Macgregor four years earlier to their own account and now owned by the ambitious Inman Line of Liverpool, left her home port with some 480 passengers and crew bound for Philadelphia. She never arrived. No trace was ever found of a vessel much admired as the exemplar of progress and economy compared to the generously subsidised wooden paddle steamers of the Collins and Cunard lines. Six months later, her larger consort, *City of Philadelphia*, ran ashore near Cape Race, Newfoundland, on her maiden voyage from Liverpool, leaving Inman with just one ship. Then, on 27 September as the BAAS was in full session in the fresh neo-classical splendour of St George's Hall, Liverpool, the Collins Line steamship *Arctic* foundered with the loss of perhaps as many as 350 passengers and crew after colliding with a small French steamer near the same cape.[2]

Even twenty years later in 1874 the grim statistics from the principal marine insurers, Lloyds of London, offered little consolation to offset the harrowing stories of individual losses. Out of 11,569 British ships of over 100 tons registered in that year, 593 (5 per cent) were lost, of which 281 (2.4 per cent) foundered or disappeared at sea, that is, by 'Act of God' through stress of weather, and the rest (2.6 per cent) were lost by stranding or collision, that is, largely by navigational error.[3]

In 1850 the UK possessed 3.3 million tons of sailing ships and 167,398 tons of steamships. By 1860 the total of sailing ships had risen to 4.1 million tons and steamers to 452,352 tons. In 1870 the figures were 4.5 million tons for sail and 1.1 million tons for steam, the former reaching a peak in that year and the latter rising to 1.9 million tons by 1874.[4] Although the figures show that investment in steamers between 1850 and 1874 had increased markedly, the continuing rise in total tonnage of sailing ships demonstrates the persistence of this technology. In this period, projectors of steamships, especially over long-distance ocean routes, could not assume an inevitable triumph of steam over sail. Indeed, during the closing decades of the nineteenth century the construction of much larger, steel-hulled sailing ships – known as 'windjammers' and well-adapted to the carriage of bulky, low-value commodities such as coal, grain or nitrates – ensured that sail would continue as a serious rival in these trades until the Great War.[5]

Both scholarly and popular accounts of the history of ocean steam navigation typically work in terms of progressive 'developments' (economic and technological) which are assumed to be inevitable. On this view, 'sound' technologies and managements will succeed; unsound ones will fail because that, in essence, is the way things are. In particular, iron and steam are usually represented as bringing about a spectacular revolution in shipping, both merchant and naval, in the middle years of the nineteenth century with major consequences for European empires.[6] Following new standards of scholarship set within the cultural history of science and technology in recent years, this chapter challenges these traditional assumptions with a historiographical narrative that integrates in local contexts the often-ignored 'failures' that counteract a story of heroic and progressive 'success'.

William Schaw Lindsay (1816–77) published his monumental four-volume *History of Merchant Shipping and Ancient Commerce* shortly before his death. Born in Ayr, Scotland, Lindsay received much of his education from an uncle, Rev. William Schaw, after whom he had taken his name and who was a minister of the local United Presbyterian Church.[7] At the age of twenty, after five years at sea, Lindsay became master of a Sunderland-owned sailing vessel. In due course he turned ship-owner himself and for a time in the 1850s became a well-known advocate of iron sailing ships fitted with auxiliary steam power to assist in entering and leaving ports as well as in calms at sea. While Member of Parliament for Tynemouth and North Shields (1854–59) he campaigned for the removal of all state restrictions on maritime trade and, following retirement from ship-owning in 1864,[8] worked to express his agenda in the form of a definitive history of shipping:

> To trace the origin of navigation, and to detail the numerous steps by which the merchant vessels of the great trading nations of the world have reached their present state of perfection; to record those discoveries in science and art connected with navigation, which enable the mariner to cross the ocean without fear and with unerring certainty; to dilate upon those triumphs of man's genius and skill whereby he can bid defiance to the elements; and to enter in these pages the names of the men who have benefited mankind by their maritime discoveries, or by affording greatly increased facilities for intercourse between nations, is to me a task of the most gratifying description.[9]

This confession of faith in a Whig history of merchant shipping meshed with a profound commitment to British maritime commerce in the context of the Victorian Empire. Lindsay therefore hoped that 'from the vast stores of knowledge bequeathed to us, we may leave more lasting records of our maritime commerce than either Tyre or Carthage, and that the improved civilization and extensive colonial possessions of Great Britain may render her pre-eminence at sea and her commercial greatness much more enduring than the once celebrated maritime city of the Phoenicians'.[10]

As well as providing such a record of progress – especially the progress of British merchant shipping – Lindsay hoped to 'inculcate lessons of use for the future'. Although he wrote of merchant vessels in terms of 'their present state of perfection', he wanted his readers – especially his readers in the maritime communities of ship-owners, shipbuilders and merchants – to recognise that 'failure' still marred the historical record. The 'perfect ship' remained an ideal rather than a reality:

> To construct a perfect ship is itself a problem of the highest order, to which the attention of mathematicians and the knowledge, skill, and tact of naval architects have of late years been constantly directed, with as yet no examples of complete success, however much the ships of our own time surpass those of our forefathers. Nor can the construction of safe, effective, powerful, profitable, and durable engines and boilers for marine purposes be a matter of easy determination, as shown from the fact, that there are still continual failures, revealing many difficulties yet to be overcome. Again, the means of propelling the vessel through the water suggests

questions as to the resistance of fluids, which hydro-dynamic science has hitherto failed fully to resolve. Finally, the combination of all these, so as to bring about to the greatest advantage the effect desired, is a still more arduous task which the skill of the naval architect, the mechanician [engineer], and the sailor, even when combined, has not yet overcome. To the perfecting of our steam-ships we must still continue to apply ourselves, if we would maintain the high maritime position we now hold...[11]

An early candidate for such maritime perfection was the *Great Britain*. As Lindsay observed with hindsight in 1876, 'there were [in the early 1840s] still many unbelievers in the suitability of iron, for the construction of sea-going vessels, and still more who had no faith whatever in the value of the screw, [and thus] this second step in advance on the part of the directors of the Great Western Steamship Company led to much discussion among scientific men, and created many evil forebodings as to the ultimate fate of the *Great Britain*, all of which, she, however, falsified'.[12] Later maritime historians have frequently represented the ship as marking an epoch in ocean steamship technology. The projectors themselves, however, successfully engaged the contemporary press in a campaign of superlatives. *The Illustrated London News*, for example, extolled the ship 'of unparalleled vastness... which circumstance, conjoined with her peculiarities of material and construction, must render her completion an important event in the records of engineering and mechanical skill'.[13]

With lines furnished by William Patterson of Bristol, builder of Brunel's wooden paddle steamer *Great Western*, the *Mammoth* (as the *Great Britain* was originally known) took more than five years to construct. In 1843 the press reported her naming by the Prince Consort with masts and funnels (but no engines) placed in position for the sake of appearances. But as one contemporary observer noted, 'the papers say she was christened by the Prince hurling a bottle of Champagne at her but in reality Mrs Miles tried to perform the operation, and being clumsy or nervous instead of throwing the bottle let it drop out of her hands into the water; so the vessel was not christened at all, perhaps it is a bad omen'. The ship finally emerged in 1845 as a six-masted iron steamer driven by a screw and measuring over 3200 tons.[14]

The *Great Britain*, however, created a saga which fell far short of profit, still less of perfection. With a mere 60 out of a capacity of 360 passengers aboard, the maiden voyage from Liverpool to New York took an unmemorable 15 days, with subsequent passages scarcely any faster. The fifth outward voyage in September 1846 ended on the sands of Dundrum Bay, Co. Down, where the *Great Britain*, holed amidships, languished until the following August, protected by extensive defences instituted by Brunel. The causes of the stranding remained highly controversial. In particular, if the navigational errors had been due to the effect of so much iron on the compasses, then the viability of iron ships must be in doubt (ch. 1). On the other hand, proponents of iron ships emphasised the amazing durability of her iron hull to withstand eleven months of wave action.[15] But her interior – and the finances of her owners – had been ruined by the winter gales on the Irish Sea coast. Purchased by her Liverpool agents at one-fifth her building costs, the ship received new machinery and fittings in Liverpool in the period 1850–52. In contrast to the spectacle of her early years, the *Great Britain* completed some 32 round voyages to Australia (about one per annum) before being humiliatingly reduced to sail-only in the 1880s. Serving as a Falklands storage hulk, the ship finally reached Bristol in 1970 for restoration.[16]

Networks of steam

Cross-channel travel in the opening decades of the nineteenth century could be a slow and tortuous ordeal. An undergraduate at the University of Glasgow in the period 1812–14, for example, spent four days aboard a small sailing vessel en route from his Co. Down farmhouse to his studies on an uncertain passage of a 100 miles, at the mercy of winds and tides, carrying limestone from the North of Ireland to the Clyde's ironworks.[17]

Since the late eighteenth century, however, projectors had been conducting 'experiments' – a term that conferred 'scientific' status on these ventures – with steam-engined boats on the rivers and canals of Britain, France and America. Already a seasoned experimenter with 'steam boats', a Scottish mining engineer, William Symington (1763–1831) gained a new patron, Lord Dundas, in 1801. As a major shareholder in the recent Forth and Clyde Canal across the Scottish lowlands, Dundas funded the installation in the canal-boat *Charlotte Dundas* of a single-cylindered steam engine (combining Watt's double-acting arrangement with Pickard's connecting rod and crank) driving a stern-wheel (with the crank attached to the axis of the paddle-wheel). Capable of towing two 70-ton barges on the canal against a strong gale, the vessel provided a widely publicised practical demonstration of the power of steam. Indeed, the Duke of Bridgewater was so impressed that he ordered eight similar craft for his own canal. Before the details of the agreement had been completed, however, the Duke died and the order lapsed. Lindsay hints that Symington's subsequent poverty was in part connected with a dispute over the steamboat patent which he had taken out in 1801.[18]

In the young United States of America, the existence of large river systems tempted projectors such as John Fitch, John Cox Stevens and Oliver Evans to come up with vessels capable of moving against the current.[19] Seen by the American civil engineer Robert Fulton (1765–1815), the *Charlotte Dundas* prompted him to order a steam engine from Boulton and Watt (ch. 2) and a boiler from London. Ordering a wooden hull from the yard of Charles Brown in New York, Fulton fitted the engine and boiler to the *Clermont* which in August 1807 steamed up the mighty Hudson River from New York to Albany, a distance of about 150 miles in some 32 hours at an average speed of just under 5 miles per hour. He wrote to the *American Citizen* that 'the success of my experiment gives me great hope that such boats may be rendered of much importance to my country'.[20] Although rivalries and jealousies left Fulton to die (in Lindsay's account) 'very poor and almost broken-hearted', he had initiated a spectacular fashion for travel by often-luxurious riverboat on North America's natural arteries.[21]

Fifty years later, Fulton's inaugural voyage for passengers had assumed almost mythic significance in the United States. 'The seventh day of the coming August [1867] will be the sixtieth anniversary of Robert Fulton's steamboat voyage from New York to Albany', Charles Francis Adams (great-grandson of John Adams, second President of the USA) wrote in the *North American Review*. Everyone, he claimed, had read the story and shared the excitement of the voyage of the little steamer, 'unsuspected pioneer of future commercial marines and navies'. It was a story, Adams insisted, to rival that of 'the famous night which preceded the discovery of America'.[22]

Back in Scotland, however, Henry Bell (1767–1830) projected a steamboat service which, though located principally on the river Clyde, would soon be represented as marking the beginnings of steam at sea. Still largely a shallow Scottish salmon river up to the last quarter of the eighteenth century, the Clyde passed through Glasgow, the west of Scotland's ancient commercial and ecclesiastical centre with its medieval cathedral and university. Until then, Glasgow could be reached only by barges. These craft had to be sailed, rowed or poled upstream under favourable tidal conditions from Port Glasgow, established in 1668 some 15 miles downstream as the town's gateway to deep-sea trade with Britain's North American and Caribbean empire. From the 1770s, the artificial confinement of the river by the con-struction of stone jetties at right angles to the current, the eventual joining of the ends to make a new river bank, and much artificial dredging, turned the old shallows into a ship canal without locks. But deep-sea ships dependent on winds and tides still struggled to reach the town itself.[23]

Bell named his steamboat after the spectacular comet of 1812. Featuring much later in Leo Tolstoy's *War and Peace* (1865–68), it was 'said to portend all manner of horrors and the end of the world' and coincided with Napoleon's ill-fated advance on Moscow.[24] The advent of the *Comet* too would be read as prophetic, the auspicious herald of the beginning of Britain's maritime empire of steam. But her projector possessed no guarantees of future success. Bell had served an apprenticeship in a shipyard at Bo'ness on Scotland's east coast, settled in Helensburgh on the north side of the Clyde, tried unsuc-cessfully in 1803 to interest British, continental and American governments in steamship projects, and finally set up Helensburgh Baths in 1808 for the benefit of Glasgow's health-conscious mercantile classes. Fulton, meanwhile, had opened a correspondence with Bell, perhaps as a result of the latter's com-munication with the United States Government. Fulton urged Bell to call on Messrs Miller and Symington in order to obtain a drawing and description of their last steamboat for forwarding to Fulton. Bell later argued that this correspondence prompted him to 'set on foot a steam-boat, for which I made a number of models before I was satisfied'.[25] Designed in part to facilitate the passage to his Helensburgh establishment, the *Comet* had a wooden hull 40 feet in length built by John Wood of Greenock, a boiler constructed by David Napier's Camlachie Foundry in Glasgow, and a four horse-power engine designed by Bell with a vertical cylinder driving the paddle-wheels (Figure 3.1).[26]

Beset by problems of the Clyde shallows, the *Comet* lacked motive power. One former passenger recalled much later that on account of a 'ripple of head wind' the 10-mile passage from Greenock to Bowling – at the western entrance to the Forth and Clyde Canal – took four hours. As a consequence, the ebb tide rendered further progress impossible and the passengers had to complete their journey to Glasgow on foot. On other occasions it was common for passengers, 'when the little steamer was getting exhausted, to take to turning the fly-wheel to assist her'.[27]

Lindsay also noted the 'prejudice raised against steam by rival interests' on the Clyde – most likely by those watermen who made their living by rowing, poling or sailing the shallow-draft vessels on the river. Lacking adequate returns, therefore, Bell took to employing her 'for some months as an excursion boat on the coasts of Scotland and Ireland, extending his cruises to the shores of England when the weather permitted, to show the superior advantages of steamboat navigation over other modes of transport to the public, many of whom viewed her with feelings of mingled awe and super-stition'.[28] In these ways the *Comet* provided practical proof in home waters of the potential of steam at sea.

David Napier's skills as blacksmith and iron-founder served him well as he increasingly turned to marine engines in the wake of the *Comet*'s modest beginnings as a passenger-carrying steamboat. By 1815 he had started steam navigation on the Thames, where concentrated opposition from watermen and bargemen was more intense than on the Clyde.[29] Having constructed his first marine engine in 1816, Napier made many careful observations of ship performance both at sea and with a model before producing in 1818 the 90-ton, 32 horse-power *Rob Roy* (wooden hull by Denny of Dumbarton and engine by himself) for the first steamer link between the Clyde and Belfast and the first cross-channel service anywhere in Britain or indeed in the world. Links between the Clyde and Liverpool and between Holyhead and Dublin also commenced within a year. On such routes, steam rapidly displaced sail for the carriage of passengers and mail: there were no watermen to oppose the new technology.[30]

When David Napier established a Glasgow shipbuilding yard in 1821, his cousin Robert took over the foundry and soon acquired a reputation for reliable marine engines, beginning with a side-lever engine for the local paddle steamer *Leven* in 1823. Side-lever engines became the standard for cross-channel packets, such as those of the Isle of Man Steam Packet Company. Robert Napier too often held an interest in those companies

THE FIRST STEAM BOAT, THE COMET, BUILT BY
HENRY BELL, 1811
WHO BROUGHT STEAM NAVIGATION INTO PRACTICE IN EUROPE.

Figure 3.1 Henry Bell's *Comet* was one of the icons of the history of steam power.

Source: George Williamson, *Memorials of the Lineage, Early Life, Education, and Development of the Genius of James Watt.* Edinburgh: Constable, 1856, facing p. 234. (Courtesy of Aberdeen University Library.)

which he supplied with marine engines. As part of a tightly knit Glasgow business community with strong Presbyterian values of hard, useful work and disciplined Sabbath observance, Robert worked closely with the entrepreneurial George Burns (the son of a famous Glasgow minister) who quickly began to build related networks of cross-channel steamship routes extending to Ireland, Liverpool and the Western Isles.[31]

Around the year 1824 the brothers George and James Burns entered the coasting trade between Glasgow and Liverpool by becoming the Glasgow agents for the Liverpool-based firm of Matthie and Theakstone who operated six sailing smacks between the two ports. The initial face-to-face encounter between fellow Scots George Burns and Hugh Matthie is revealing. Although encountering strong competition for the agency, Burns refused to give up a commitment to attending local civic events when Matthie unexpectedly arrived to meet him. During the deferred meeting, held on the following day, Matthie stated that the rival firm was supported 'by a round-robin of recommendations from the most influential people'. But his parting shot to Burns was that Matthie looked 'to personal fitness as of the first importance.... it will be given to the best and most capable man I can get'. Having established face-to-face trust and demonstrated an unwillingness to ingratiate himself, Burns won the agency and soon after the firm of G. & J. Burns acquired Theakstone's 50 per cent share in the smacks.[32]

George Burns had witnessed for himself the *Comet* at Glasgow's Broomielaw departure point for her journeys downriver. While staying at Bell's Helensburgh Baths he had also learnt of her wreck on the West Coast in 1820, and had a stake in the steamer *Ayr* which collided with, and sank along with 70 lives, the second *Comet* five years later. Yet, despite these inauspicious signs, Burns inaugurated his own Glasgow–Liverpool steamship service with three ships in 1829–30. Characteristically, he at first defied popular superstition by announcing sailings from Liverpool and Glasgow on Fridays in order to avoid Sabbath sailings. When commercial requirements dictated Saturday departures, he took seriously Matthie's jest that he might provide chaplains to accompany the ships. As a result, Broomielaw wits accused the master of the second such voyage as 'Sailing in a steam chapel'. Above all, the evangelical Burns seems to have regarded his steamship ventures as a Divine 'calling', paralleling the 'call' of young men to the Scottish ministry or of a minister to a different parish.[33]

Early vexation with his Irish Sea venture, however, came with the establishment of a rival Liverpool–Glasgow service at the hands of another Scot, David MacIver, based in Liverpool. Determined to break Burns's monopoly on steam, MacIver set up the City of Glasgow Steam Packet Company with capital from a wealthy Glasgow cotton broker, James Donaldson, and with engineering support from Robert Napier. MacIver himself 'vowed that he would, if possible, drive the Burns's off the seas'. MacIver later told George's wife Jane that he 'had travelled in the *City of Glasgow* backwards and forwards between Liverpool and Glasgow, going down himself into the engine-room to superintend the firing of the furnaces, in order that he might leave nothing undone' in order to conquer Burns. Unable to match the Burns's reputation and profits, however, MacIver eventually agreed to George Burns's offer to combine the fleets on a division of revenue ratio of two-fifths (MacIver) to three-fifths (Burns). The arrangement was honoured and a powerful new bond of trust built between the Burns, MacIver and his brother Charles, James Donaldson and Robert Napier. This would be the community into which a stranger from Nova Scotia, Samuel Cunard, would step in 1839.[34]

Leviathans of the deep

For most of the first half of the nineteenth century, ocean trade was overwhelmingly dependent upon sail. To contemporaries, the finest and fastest sailing ships to be found on any ocean belonged not to Britain but to the United States of America and in particular to the north-eastern coastal states from Maine to New York. The fast and reliable sailing packets of the Black Ball Line, for example, ruled the North Atlantic passenger and mail routes. The skills of their builders and of their crews found a new challenge in the 1840s when the first of the American-built and American-owned clippers began carrying China tea for the London market. Literally launching for the first time a global economy, these proud and profitable ships would carry passengers and freight from America's north-east coast to California by way of Cape Horn, sail from San Francisco to the Chinese river ports for the arrival of the new season's tea, race to London by way of the Cape of Good Hope, and complete their epic earth-circling voyages with a westbound transatlantic passage.[35]

British builders, located mainly in Greenock and Aberdeen, eventually replied with clippers constructed of durable hardwoods such as teak shipped from India and the colonies of South-east Asia. Crucially, the hardwood hulls were less prone to leaks than the softwood hulls of the American clippers and so offered greater protection to the delicate and valuable chests of tea. Performance too impressed the public when in 1856 the *Lord of the Isles*, built by Scotts of Greenock, beat two of the fastest American clippers in a race from China to London. In 1866, three British clippers, all built by Robert Steele of Greenock (*Taeping*, *Ariel*, and *Serica*), arrived in London on the very same day, *Taeping* winning by a hair's breadth what became known as 'the Great Tea-Race'. Three years later, the Dumbarton-built newcomer *Cutty Sark*, her frames built of iron and her skin of teak, joined the ranks of these much-celebrated clippers.[36]

In contrast to such confident – and widely publicised – displays of skill on the part of designers, builders, masters and crews of these new generations of long-distance clipper ships, the prospects for ocean steamships looked extraordinarily uncertain. In the case of the ocean-going iron steamer, recent studies have begun to challenge assumptions about the inexorable 'progress' of the new technology between 1845 and 1914.[37] The advocates of iron steamers on long-distance ocean trade routes – John Scott Russell, Isambard Kingdom Brunel, William Fairbairn, and others – often aligned themselves with the fashionable sciences of the time, especially at the annual meetings of the BAAS. There, advocates promoted the iron steamer as the very symbol of human progress through scientific knowledge and technical skills.[38] But, at least in the eyes of their detractors, iron steamers possessed peculiar propensities to run ashore, go missing with all passengers and crew, or simply absorb vast amounts of capital without benefit to gullible shareholders. Through its magnetic idiosyncrasies, iron threatened to make navigation by magnetic compass wholly untrustworthy (ch. 1). Wooden steamers fared little better. In 1852 the Royal Mail Steam Packet Company's *Amazon* – at over 3000 tons, the largest wooden merchant steamship built up to that time and costing over £100,000 – was on her maiden voyage when the paddle-wheel bearings apparently generated enough heat to cause a fire that, fanned by gale-force Atlantic winds, consumed the ship together with both mail and passengers.[39]

Projectors, builders and owners of early steamships dreamed of fame and fortune but performance fell short of promise. Using sail-aided-by-steam in 1819, the American *Savannah* crossed the North Atlantic. But no passengers trusted the new technology enough to make the potentially epoch-making passage. Indeed, President James Monroe could not be persuaded even to travel from Charleston to Savannah in the cause of national pride and progress. Only by the late 1830s did investors have the confidence to contemplate subscription to regular passenger and mail services. The first purpose-built Atlantic steamer to be completed, Brunel's *Great Western*, fulfilled her designer's ambition to extend his Great Western Railway westwards across the ocean (ch. 4) (Figure 3.2).[40]

In fact, the rival British & North American Steam Navigation Company, initiated by an American (Dr Junius Smith), had been formed in 1835, several months before the Great Western Steamship Company and with twice the proposed capital. Thanks to Isaac Solly, former chairman of the London and Birmingham Railway, lending his name – and credibility – to the project, investors confidently purchased shares. London builders Curling & Young won the contract for the *British Queen* – some 500 tons larger than Brunel's ship and subsequently named in honour of Victoria's accession. The Company, however, lost their short lead when the Glasgow engine builders went bankrupt. Robert Napier agreed to take over the contract at an increased price, but the Company had to charter the *Sirius*, a cross-channel steamer, to fill the gap and secure the title to the first transatlantic steamship service – arriving in New York only a few hours before the *Great Western* which had left several days later. The much-delayed *British Queen* finally began work in

Figure 3.2 Isambard Kingdom Brunel's wooden paddle steamer *Great Western* struggles against Atlantic storms in 1846.

Source: Illustrated London News 9 (1846), p. 273. (Courtesy of Aberdeen University Library.)

July 1839 but proved consistently slower than the *Great Western*. A larger consort, the 2400-ton *President*, joined the service in 1840. In March 1841 the six-month-old ship, largest steamer in the world, sailed with 136 passengers and crew from New York for Liverpool but never arrived, believed to have been the victim of hurricane-force winds and seas. An anxious public watched and waited for several weeks, their trust in the new technology waning by the day. As a direct result of this first North Atlantic steamship disaster, the Company immediately ceased trading.[41]

Meanwhile, *The Times* had reported at the end of 1838 that the British & American Steam Navigation Company had contracted with John Laird of Birkenhead (ch. 1) for an iron steamer. Although never built, the report may have influenced the Great Western's decision to opt for a project of enormous novelty and risk: the construction of the *Great Britain*.[42]

Critics of Brunel's projects in general and of the *Great Britain* in particular abounded. Secretary – and a key promoter since 1822 – of the Liverpool & Manchester Railway (ch. 4), Henry Booth had a commitment to all things mechanical that found expression in the new Liverpool Polytechnic Society (established in 1838 'for the encouragement of useful Arts and Inventions'). The Society's principal aims embraced 'Mechanical Engineering of all kinds' and asserted its significance to Liverpool which 'has taken a prominent part in the establishment of those wonderful exhibitions of mechanical power which are to be found in the working of modern Railways...there is every probability that, in a few years, this port will be no less distinguished as the great centre of steam navigation'. Members of the Polytechnic Society, including Booth himself and the iron shipbuilder John Grantham (President), had a shared enthusiasm for iron hulls, screw propulsion, and the economy of high-pressure steam. In March 1844 Booth presented to the Society a paper, typical of the style and content of such local enthusiasm, 'On the prospects of steam navigation'.[43]

He began by summarising the results of a series of experiments on floating bodies (representing boats) moved various distances along a 40 ft trough of water by a cord passing over a pulley and attached to falling weights. Conducted with fellow-member Alfred King (engineer of the Liverpool Gas Works and possessing, according to Booth, 'well-known accuracy in mechanical arrangements and details'), the experiments were directed to an investigation of the increase of power required to increase the speed of vessels.[44]

Emphasising the importance for ship construction of having in view a 'precise object', he argued that one such object, 'the grand desideratum' for 'a great commercial country like England', was to 'convey the mercantile mail bags [from Liverpool to New York] in a week'. Yet 'we have done almost nothing to effect this all-important object'. His solution was based on his (and others') conclusions from 'accurate' experiments with models, namely, that greater length in proportion to beam would make possible greater speed for steamships whose 'power is below [and thus, unlike masts and sails above, *increasing* the stability of the hull], steady and uniform'. Iron, moreover, 'affords the means of giving our vessels that strength and stiffness which it would be very difficult to accomplish in a timber-built ship'.[45]

Booth concluded, however, in contrast to the design of Brunel's *Great Britain*, that a vessel successfully combining high speed (for the purpose of carrying mails) *and* large cargo was impossible:

> it must be borne in mind that the vessel is an *implement* with which to accomplish an object, and the efficiency of the *power* employed depends on the *form* of the implement more than on the *size*. By treating the question mechanically, we treat it *economically*, which, in steam navigation, is of paramount importance. Let the implement be of the true form – that is, let it be mechanically constructed – and let the power be proportioned to its capacity, and I do not see why ... [a] steamer of one thousand tons burthen should not make its passage through the water as rapidly as the Leviathan [*Great Britain*] of three thousand tons. The larger vessel may, indeed, accomplish *cargo* as well as speed, but it will be accomplished at enormous cost, and, in the present discussion, speed, and not cargo, is our object.[46]

For Booth, then, the 'enormous *expense* of [running] the huge steam-ships' which year after year 'have been increasing in size and costliness [of construction]' was a matter of the 'most grave and anxious consideration'. Indeed, he observed, 'without Government co-operation and aid, how few of our steam-packet [mail] companies have derived a fair and moderate return for capital invested'. To counter this trend towards ever-increasing wastefulness of the nation's wealth and coal, Booth urged 'that some new element of economy should be discovered'.[47]

Booth had thus entreated his Liverpool audience to approach steamship questions mechanically and to handle them economically, to regard steamships as implements with specific objects in view, and to search for some new element of economy. As we shall see later, a young protégé, Alfred Holt, took these entreaties to heart after a five-year apprenticeship with Edward Woods (1814–1903), a resident engineer on the Liverpool & Manchester Railway and also a Polytechnic Society member.[48]

When, at the behest of local scientific societies – most notably the Liverpool Literary & Philosophical Society – the BAAS returned to Liverpool in 1854, iron steamship questions appeared as matters of both local and national importance (ch. 1). Doubts over the trustworthiness of Airy's compass methods were only one symptom of a much wider concern with the trustworthiness of the large ocean steamer. Leaders of the BAAS attempted to provide scientific authorisation of the new iron steamers, viewed as a product of the new experimental and mathematical sciences of hydrodynamics and strengths of materials. William Fairbairn (whose heavy engineering activities extended for a time to iron shipbuilding on the Thames) and John Scott Russell (celebrated naval architect and now Thames shipbuilder) in particular played major roles in the BAAS, whether in terms of Committees focusing on data collection and reduction or on the public stage.[49]

In this latter role, Scott Russell addressed the assembled Merseyside public, including civic leaders, merchants, shippers and ship-owners, in the newly opened St George's Hall, symbol of Liverpool's status as a modern Athens. With the north entrance hall allotted to the Mechanical Science Section, 'an excellent theatre has been maintained by placing the president's table on the ground floor, near the entrance doors, and arranging forms upon the steps in front'. Often the poor relation of Section A (Astronomy and Physics) at the top of the hierarchy of the BAAS sciences, Section G (Mechanical Science) in fact received almost verbatim coverage in the *Liverpool Mercury*, the port's prominent liberal and commercial paper. With Scott Russell as Section President, and Fairbairn and Booth as Vice-Presidents, 'Mechanical Science' was indeed likely on this occasion to command immense authority and attention not least because the Section Committee included prominent iron shipbuilder John Laird of Birkenhead, hydraulic machinery and armaments manufacturer W.G. Armstrong of Newcastle and William Scoresby (ch. 1).[50]

The President began by informing the audience that the Section 'had received an almost unprecedented number of valuable and important publications' which had been divided into different subjects to be taken on different days. Pride of place went on the first full day to 'the reading of all important papers upon the subject generally of ships, of navigation, and of steam navigation in particular – a subject in which the port of Liverpool was naturally most deeply interested'.[51] Above all, however, the President himself quickly became the centre of attention on account of his role as builder of the largest iron steamship ever constructed.

A large audience, 'amongst whom were many of the fair sex', assembled to hear Scott Russell's address 'On the progress of naval architecture and steam navigation, including a notice of a large ship of the Eastern Steam Navigation Company.' The President had been specifically requested by the Association and Section officers upon arrival to allude 'to one of the mechanical subjects with which the history of that association was in some important degree connected' and to illustrate this progress by explaining the construction of the large ship now under construction on the Thames. In his address he therefore confidently linked the BAAS to the *Leviathan* (later named *Great Eastern*) under construction in his Millwall yard on the Thames.[52]

Russell first reviewed the change – which he as naval architect had long advocated – away from slow and unwieldy hulls with their maximum beam (width) about one-third of their length from the convex bow to fast hulls with maximum beam at a point much further aft and 'a fine, hollow line [concave] at the bows'. These were the lines of the fast American and English clippers, achieved by those who, like himself, had 'consulted' and 'cross-question[ed] nature' through experiment.[53]

The speaker also singled out for mockery critics of Brunel's earlier ships: those like Dionysius Lardner who had doubted the credibility of Brunel's *Great Western*, exemplar for Cunard's early wooden paddle steamers, and later critics of the *Great Britain*, whose design they had blamed for her stranding rather than her captain. Such remarks were of course not likely to win the speaker support from local shipmasters. Now the unnamed giant steamer, 675 feet in length and 83 feet in breadth, 'would supersede smaller ships, in being able to carry sufficient coal to serve for the voyage to and from Australia, an immense sum being saved in the item of coal expense... Besides her coals, the great leviathan would carry 600[0] tons of goods, 600 first-class passengers, and 1000 second class.'[54]

Builder's plans, showing the ship in elevation, reveal not a conventional ship but an enormous machine for the production of motive power, resembling nothing so much as a factory with five chimneys, each carrying away the waste products of five sets of furnaces surrounded on almost all sides by immense stocks of coal. Indeed, Brunel himself viewed the ship as 'a new Machine' requiring a new kind of commander whose 'attention must be devoted exclusively to the general management of the *whole* system under his control'.[55] In his conclusion to the BAAS, however, Scott Russell invoked not the imagery of the factory but the name of God and science to confer additional authority on the project. He thus alluded 'to the amazing discoveries which had taken place, not only in shipbuilding but in every branch of science, and concluded by a warm appeal to his audience to follow out the truth, and to do their best to carry out the works of God'.[56]

In the ensuing discussion, other 'gentlemen of science' rallied to the cause of the leviathan. Fairbairn in particular expressed himself satisfied that his earlier doubts as to the longitudinal strength of the vessel had been overcome when shown the drawings by Mr Brunel himself: 'from those drawings and the principles on which the vessel was constructed, he would [now] not have the slightest hesitation in saying that she would be perfectly stable and strong, and fully able to carry out her objects'. It had indeed been Fairbairn's Millwall yard (now part of Scott Russell's) that had provided the facilities for experimental testing of a scaled-down version of Robert Stephenson's proposed Britannia Bridge over the Menai Straits in North Wales, designed as a box-section girder with cellular structure top and bottom to give maximum strength during the transit of heavy railway traffic (ch. 4). The *Great Eastern* followed much the same 'girder' design, incorporating a cellular double bottom below the waterline, in contrast to the traditional backbone (keel), ribs (transverse frames) and skin (timbers or plates) of ordinary ships, whether of wood or iron.[57]

Others in the audience generally 'coincided' with Scott Russell's views. A Mr Webster highlighted the positive role of the BAAS in lending scientific authority to the large ocean steamer: 'the principles...laid down had been fully discussed in the physical section of the association 15 or 20 years ago, and it must be interesting to the members to know that the knowledge which was promulgated through their books, and which was discussed in their sections, had been realised by shipbuilders and commercial men, and which had led to those astounding results that the crossing of the Atlantic was now performed as regularly and certainly as a train going to Edinburgh'.[58]

The popularity of the strong maritime theme in Section G at Liverpool, however, only increased the chance of controversy with disparate voices given centre stage in succession. These disparities both resonated with the Scoresby–Airy dispute over compass correction (ch. 1) and dramatised, in the theatre of Section G, major debates about the role of experience and mathematical science in naval architecture. Thus a powerful note of dissent emerged on the final day of Section G's gathering, by which time illness had prevented the President from continuing in the chair. The *Mercury* reported a Captain Andrew Henderson's presentation 'On ocean steamers and clipper ships and their descriptive measurement' as a paper 'the more valuable, inasmuch as the views it enunciated were based on long experience and practical acquaintance with those elements whose violence and fury so often overwhelm theories and wreck mechanical science'. Tracing both 'the progress of shipbuilding in Liverpool, and the character of the vessels which visit its port, from the bluff bowed and short sailing ships to the famed "liners" of the present day', Henderson pointed out 'their defects and the improvements he would introduce into them'. His arguments were illustrated by comparisons of the relative speed, tonnage, length and breadth of the 'most celebrated clipper mail steam-ships which now sail the waters'. But it was his opinion that the 'modern Ark' of Mr Scott Russell would be effectually barred by her huge size from 'combining all the requisite qualities of an ocean steamer'.[59]

In a longer summary of his paper published in the *BAAS Report* (1854), Henderson opened with the declaration that 'actual experience at sea [must be considered] the only guide in the application of science to the improvement of our shipping'. For comparison, he noted that the 'largest and longest vessel yet constructed [in the present age] is the "Himalaya", built on the Thames by Mr Mare, for the Peninsular and Oriental Company, measuring 3528 tons, and having a ratio of length to breadth of 7.41; her length is 341 feet, and her breadth 46 feet'. On the other hand he explained that 'Holy Writ records that Noah's Ark, launched on the waters of the flood 2340 years before Christ' would equate to 11,905 tons, with a ratio of length to breadth of 6. The 'modern Ark' had a tonnage almost double that, at 22,942 tons, a ratio of length to breadth of 8.19, and a length twice that of the *Himalaya*. To drive her required engines 'of a nominal horse-power of 2600 horses'.[60]

Captain Henderson's scepticism regarding the leviathan derived from his view that, while the 'speed, economy and comfort promised in this vessel' had attracted a large share of public attention, 'little consideration has been given to the subject of *its security in heavy* SEAS, its capabilities for navigation, and the effect of an ocean wave on a vessel of such extraordinary length'. The speaker claimed to have himself seen waves of a height of 50 feet from crest to hollow which would make it impossible for a vessel of this size 'to be kept under command in such a sea should any accident happen to her machinery'. Here he also invoked the authority of Dr Scoresby (ch. 1) whose investigations had shown 'the elevation of the Atlantic waves in a hard gale was as much as 43 feet' and had estimated 'the usual gales in the Atlantic as producing a wave 25 feet in height and 200 feet long'. And he cited examples of three American steamers as victims of such ocean conditions:

The former [*Frome*] was lost; the Vanderbilt fell into a trough of the sea, and was struck by a wave which carried away all her bulwarks, made a clean sweep of her deck-houses, and put out all her fires. The San Francisco, a large steamer, bound to California, caught in a north-west gale in the Gulf-stream, being disabled got into the trough of the sea; she became perfectly unmanageable, and part of the crew and 150 passengers were washed off her, the decks actually being stove in, they having no command over her when under sail alone.[61]

Such criticisms reveal the highly contested nature of engineering projects in the Victorian period. But with regard to maritime technologies, the *Great Eastern* was indeed the most controversial project of the century. Brunel had conceived his 'Great Ship' on paper towards the end of 1851 and the following year persuaded the Eastern Steam Navigation Company, prospective rivals to P&O for services to the Far East and Australia, to undertake such a project. Brunel and Scott Russell had known each other since the Bristol meeting of the BAAS in 1836 and it was at Scott Russell's yard that the 'Great Ship' had begun to take shape in February 1854.[62] In fact, Scott Russell – his reputation well established as a naval architect – had been a relative newcomer to shipbuilding, having taken over the iron shipbuilding yard of William Fairbairn in the mid-1840s after it had proved highly unprofitable for the famous Manchester engineer. Prior to construction of the 'Great Ship', Scott Russell had launched about fifteen ships, none exceeding 1500 tons and including two schooner yachts for Robert Stephenson (ch. 4). His Royal Yacht Squadron waistcoat gilt buttons, combined with frock coat and top hat, ensured a gentlemanly appearance and reputation.[63]

Historians of technology have spilt much ink over the fraught relations between Brunel and Scott Russell. From the perspective of W.S. Lindsay – both a contemporary of the two men and an eye-witness – the share of the credit divided as follows:

Mr Brunel, having originated this conception, communicated it, at the outset, to Mr Scott Russell, and suggested the construction of a steam-ship large enough to carry all the fuel she might require for the longest voyage; and Mr Scott Russell shared with Mr Brunel in the merit of carrying out these views. The idea of using two sets of engines and two propellers (paddle and screw) is solely due to Mr Brunel, as was, also, the adoption of a cellular construction, like that at the top and bottom of the Britannia Bridge . . . These main characteristics distinguished the *Great Eastern* from all other vessels then afloat. Her model and general structure were in other respects identical with those of the ships built by Mr Scott Russell, on the principal of the 'wave line', which he had systematically carried out during the previous twenty years.[64]

Allocation of credit for specific features of the 'Great Ship' – such as cellular construction or hull-form – was one thing. Turning the project from promises on paper to the fulfilment of those promises in the form of practical action was quite another. Both men commenced the project with high public credibility. Scott Russell brought to bear his reputation as Britain's most distinguished naval architect of the time. Brunel's Great Western Railway (ch. 4) and his earlier two ships had, despite controversy, achieved much public acclaim. But the 'Great Ship' project, beset both by increasing mistrust between Brunel and Scott Russell and by an ever-declining level of public confidence, failed at almost every stage to fulfil its promised actions. During construction in the yard in 1856, for example, Brunel raised with Scott Russell 'the state of things as regards "stock". There are 2400 tons [of iron] to be accounted for.' The mutual mistrust indeed extended to estimates, differing by 1000 tons, over the weight of the ship, crucial to calculations over the sideways launching. When, however, Brunel reported the alleged shortfalls to a meeting of his directors, he had given the lie to his shipbuilder and effectively called into question Scott Russell's trustworthiness as a gentleman.[65]

During the summer of 1857, as the 'Great Ship' neared launching, Brunel accompanied engineer Robert Stephenson and ship-owner Lindsay on a thorough inspection of the ship. In response to Brunel's question as to what he thought of her, Lindsay replied that 'she is the strongest and best built ship I ever saw and she is really a marvellous piece of mechanism'. Dismissing Lindsay's opinion about her build as telling him nothing new, Brunel doggedly sought instead the ship-owner's answer to the question: 'If she belonged to you in what trade would you place her?' Lindsay's reply was not what Brunel wanted to hear:

Turn her into a show . . . something attractive to the masses. . . . She will never pay as a ship. Send her to Brighton, dig out a hole in the beach and bed her stern in it, and if well set she would make a substantial *pier* and her deck a splendid promenade; her hold would make magnificent salt-water baths and her 'tween decks a grand hotel, with restaurant, smoking and dancing saloons. . . . She would be a marvellous attraction for the cockneys, who would flock to her in thousands.[66]

Recalling the encounter some years later, Lindsay noted that 'Stephenson laughed, but Brunel never forgave me.' Much of the expert engineering and commercial scepticism issued to the proprietors before and during construction derived, indeed, from a prediction of 'the mercantile disappointment to which extravagant expectations, as to the combination of high speed and great length of voyage without recoaling, would . . . lead'.[67]

Mistrust soon contaminated the popular perception of a project that had at first seemed to promise a new age in global transport. As thousands of spectators, including the press, waited for the launch, the *Great Eastern* remained stubbornly on dry land. The wider public were witness to some serious miscalculations, only remedied after nearly three months of additional work and the bankruptcy of her original owners. When eventually the ship went to sea in September 1859 without a conventional series of trials, a feedwater tank, designed by Brunel to preheat the boiler water, surrounding the lower part of the first of her five funnels exploded. Scalding water killed five stokers in the boiler room below. As the press reported the subsequent enquiry, Brunel himself, already seriously ill, died.[68] More delays ensued, with a postponement of the maiden voyage – to New York – until June 1860. There were only 35 fare-paying passengers – and 418 crew members – aboard. By the following year, passenger numbers had slowly improved to over 400 but a gale two days out of Liverpool put the steering gear and both paddle-wheels out of action, leaving the 'Great Ship' at the mercy of the seas for 48 hours.[69] These, and other, perceived failures to fulfil promises as a passenger ship ensured that public confidence in the vessel never recovered. Only with the Atlantic telegraph cable project did the *Great Eastern* prove herself in a role very different from that for which she had been designed (ch. 5).

Clyde-built

David Napier left the Clyde in 1836 to set up a yard on the Thames close to where Fairbairn had established an iron shipbuilding 'factory' on the Isle of Dogs the previous year. John Scott Russell, as we have seen, joined his fellow expatriate Scots in the mid-1840s and, especially for the building of the *Great Eastern*, acquired Fairbairn's yard and part of Napier's. With engineers such as Maudslay Sons & Field and John Penn & Sons already established as builders of marine steam engines, the Thames seemed for a time more likely than the Clyde to become the country's leading centre for state-of-the-art iron shipbuilding.[70] In the 1830s, then, there was no inevitability about the Clyde's rise to pre-eminence in iron shipbuilding.

Robert Napier, however, remained firmly in Glasgow. Beginning his working life as blacksmith and millwright, he quickly acquired a reputation for the trustworthiness of his marine engines. Embedded within a cultural network which included Scotland's most famous preacher and theologian of the time, Thomas Chalmers, Napier constructed marine engines which embodied Presbyterian values of hard work, simplicity, reliability and economy. Building on his practical experience with coastal steamships, Napier engined two steam yachts which won the first two places in a race staged by the Northern Yacht Club. This kind of practical demonstration, made possible through a sporting culture of wealthy gentrified patrons, could greatly enhance the trustworthiness of the designer or engineer if promises were fulfilled, and, equally, carried few of the potentially damaging risks attached to similar demonstrations with commercial craft should accidents occur.[71]

By the mid-1830s, Napier had gained a prestigious contract from the venerable East India Company to engine the *Berenice*, one of the first two steamships built for the Company to carry the precious mails from Bombay to Suez from 1837. Highly privileged, the East India Company, effectively ruling India on behalf of the British Crown, held an official monopoly on trade with India until 1813 and with China until 1833. Long associated with 'Old Corruption' in the minds of critics, the Company itself came under a new generation of utilitarian reformers represented by Lord William Bentinck, the Governor-General of India (1828–35). The aims of these reformers included attacks on corrupt practices and the raising of 'a middle class of native gentlemen' through 'rational', utilitarian education. The change from traditional 'East Indiamen', large sailing ships carrying the valuable freight to and from India via the Cape of Good Hope, to mail steamships, formed part of the new order of 'improvement'.[72]

In fulfilling this contract, Napier extended his social network to include the Secretary of the East India Company in London, James Melvill. Meanwhile, at much the same time Napier took over the contract for engining the world's largest steamer, *British Queen* (above), and received his first Admiralty contract for marine engines. His reputation rose rapidly. By 1841 he established a site at Govan for iron shipbuilding. His capacity too for encouraging young engineering and shipbuilding talent meant that a whole generation of Clyde shipbuilders – William Denny, John Elder, James and George Thomson, and A.C. Kirk for example – owed their craft to the Napier works.[73]

Melvill was already acquainted with Nova Scotian merchant and ship-owner Samuel Cunard, who, from his Halifax base, had been an agent for the East India Company since the early 1820s with distribution rights for China tea throughout British North America. When Cunard arrived in England to promote a transatlantic steamship company and secure the contract for the carriage of the Admiralty mails, Melvill was in a strong position to introduce him to Napier. Ignoring dark warnings from associates in England that by building on the Clyde he would have 'neither substantial work nor completed on time', Cunard soon entered into an unassuming Glasgow-centred partnership that provided the capital and skills for the British & North American Steam Packet Company (later known as the Cunard Steamship Company). Most importantly, the partnership won the mail contract, much to the chagrin of Brunel's Great Western Company.[74]

With Quaker origins and a strong loyalty to Britain, an earlier generation of the Cunard family had migrated from the United States to Nova Scotia in the wake of the American War of Independence. Samuel's parents, however, attended the Episcopal (Anglican) Church and his wife had strong Scottish Presbyterian roots.[75] Samuel himself learnt the business of shipbroking in Boston and around 1815 established a packet service linking Halifax with Newfoundland, Boston and Bermuda. His commercial interests ranged widely over whaling, land, ironworks, canal projects, lumber, fishing, and coal mines giving Cunard power and influence in diverse fields. Having already associated himself with a project for transatlantic steamers, he arrived in England in February 1839 with the goal of tendering for a recently advertised Government contract for the carriage of the Royal Mails by monthly steamer from an English port to Halifax with a link onwards to New York.[76]

Cunard's strategy was to secure a deal with a shipbuilder first and then, contract in hand, to approach the Government for the mail contract in the knowledge that two recent contenders, including Brunel's Great Western Company, had failed to satisfy its stringent conditions with regard to frequency. He also had the ear of influential members of Prime Minister Lord Melbourne's Cabinet, Lords Lansdowne and Normanby. But his initial contacts with Napier were probably brokered by a Glasgow merchant, William Kidston, who had spent a large part of his life in Halifax before returning to the Clyde to develop his shipping and ship-owning interests.[77] In February 1839 Cunard stated his requirements to Kidston & Company:

I shall require one or two steamboats of 300 H.P. and about 800 tons. I am told that Messrs Wood [Port Glasgow shipbuilder] & [Robert] Napier are highly respectable Builders and likely to be enabled to fulfil any engagement they may enter into. Will you be so good as to ask them the probable sum for which they would engage to furnish me with these boats in all respects ready for sea in twelve months from this time? I am told that the 'London' [engined by Napier] is a fine vessel, but I have not seen her. I shall want these vessels to be of the very best description and to pass a thorough inspection and examination of the Admiralty. I want a plain and comfortable boat, but not the least unnecessary expense for show. I prefer plain work in the cabin, and it saves a large amount in the cost. If I find these gentlemen are likely to meet my wishes, I will immediately proceed to Glasgow and make the necessary [arrangements] with them.[78]

When Cunard and Napier met soon afterwards, they quickly agreed on a plan for three wooden paddle steamers, each with a capacity for 60–70 passengers and each costing around £32,000. 'I have given him the vessels cheap and I am certain that they will be good and very strong ships', Napier told Melvill. By early May 1839, this agreement, together with his Government contacts, enabled Cunard to secure the Admiralty mail contract worth £55,000 per annum over seven years. The strategy proceeded wholly upon trust between the different interests: Cunard had raised no capital, Napier had built no ships, and the projectors had inaugurated no service. Moreover, the Admiralty demanded four rather than three ships, regular sailing dates with penalties for delays, and carriage of military forces at special rates.[79]

Faced with little enthusiasm for the project in England, Cunard travelled north to meet Napier again, this time to raise the capital for the construction of four larger ships. In order to speak frankly and confidentially in a face-to-face setting with the aim of establishing trust among prospective gentlemanly partners, Napier and Cunard dined with George Burns and David MacIver.[80] Initial difficulties centred on the Admiralty's penalties which would be imposed for longer-than-specified voyage times. But these problems quickly gave way, at a breakfast meeting at Napier's house, conveniently located close to his engine works and the Clyde, to the formation of a partnership founded on trust in Napier's marine engines and on their capacity to meet the Admiralty's tough schedule. In turn, Burns and MacIver attracted the support of some nineteen other Glasgow merchants to form 'the Glasgow Proprietory in the British and North American Royal Mail Steam Packets, established for the purpose of carrying mails, passengers, specie [coins] and merchandise between Britain and certain North American ports'.[81]

In the division of labour among the partners, Napier took charge of the construction of the engines while sub-contracting the hulls to Port Glasgow shipbuilder John Wood. Burns acted as Glasgow agent and kept a watchful eye on construction. The MacIver brothers acted as Liverpool agents and played the lead role in managing the ships on a voyage-to-voyage basis, laying down strict guidelines for masters and officers. Samuel Cunard's Nova Scotian pedigree, his puritanical dislike of extravagance and ostentation, and above all his lack of dogmatic sectarianism undoubtedly facilitated his social networking, especially in bringing together in the transatlantic project investors from early Victorian Glasgow. Cunard himself established the Halifax and Boston branches of the Company while also sailing with the first ship, aptly named (especially with empire loyalism in mind) *Britannia*, from Liverpool on 4 July 1840. Eight years later the Line introduced its first direct service to New York.[82]

Although the Cunard partnership introduced a pair of iron screw steamers in the early 1850s, its preference (and that of the Admiralty) was for wooden paddle steamers until Napier constructed the iron paddle steamer *Persia* in 1856. This 3300-ton ship marked the beginning of construction of Cunard steamers at Napier's own iron shipbuilding yard in Glasgow. The arrangements prompted the prestigious New England quarterly magazine *North American Review* to affirm in 1864 that 'In the machinery especially, the best material, the most skilful mechanism, and the most approved designs are made use of; the engines are always put together and thoroughly tested in motion before going into the vessel.' In service too, everything 'is kept in the best repair.... A visitor at one of the company's works near Glasgow saw several boilers lying about which had been condemned simply because of their age...The company do not wait until a boiler explodes before deciding whether it is defective.'[83]

The context for the review was a post-mortem on the dramatic demise of the American Collins Line, launched with four new ships around 1850 and with the support of large United States Federal Government subsidies. Superior speed, better sea-keeping qualities, and vastly more luxurious interiors attracted admiration on both sides of the Atlantic. But within half-a-dozen years of its foundation, the Line had lost half its fleet, one by collision and the other without trace, but both with loss of life totalling over 500 persons. The *North American Review* attributed the fate of the Line to its extravagance, both financially and in terms of speed. In contrast, the reviewer hailed the Cunard Line as an exemplar of safety and reliability: 'Never in advance of the times, but never far behind them; never experimenting, but always ready to adopt any improvement thoroughly tested by others; avoiding equally extravagance and parsimony...the success of this Company, taking all things into the account, has never been equalled.... the strength of the vessels, the discipline of the crews, and the seamanship of the commanders were made available promptly at the moment when everything was depending on them.'[84]

The Line's special dislike of experimenting with its ships set it apart. On the one hand, as we have seen, Napier could draw advantage from demonstrations in the form of steam yacht races – sponsored by the sporting elites – if they succeeded, but would incur few risks, financial or moral, if they failed. On the other hand, Section G of the BAAS did much to promote a culture of 'experiment' as integral to science-based engineering practice. With Brunel and Scott Russell in the maritime vanguard of such a culture, commercial rivals were quick to seize on any perceived 'failures' as evidence of extravagance, lack of appropriate sobriety and discipline, and even commercial incompetence.[85] Robert Napier's reputation – and indeed that of his Clyde shipbuilding and engineering proteges – grew on the basis of delivering tried and tested products whose promises could be relied upon. By 1874, indeed, Lindsay could write of a very different Clyde from that of Henry Bell half a century or so earlier: 'For miles on both sides of the river stupendous shipbuilding yards line its banks, employing tens of thousands of hardy and skilled mechanics...Along those banks there is now annually constructed a much larger amount of steam tonnage than in all the other ports of Europe combined, those of England alone excepted.'[86]

As a consequence of its 'long-proved [record of] reliability and safety', the Cunard Company had grown from its original four vessels in 1840 to nearly 40 vessels in the mid-1860s and had continued to attract the support of the British government in the form of mail contracts. The record was one of consistency. In Lindsay's judgement in 1876, over a period of more then 35 years *'neither life nor letter entrusted to their care has been lost through shipwreck, collision, fire, or any of the too frequent causes of disaster, during the numerous voyages made by the Cunard steamers across the Atlantic'*.[87] In the second half of the nineteenth century, indeed, the Line's reputation for safety had become legendary. Mark Twain, for example, extolled its virtues in the *New York Tribune* (1873):

> Cunard people would not take Noah himself as first mate till they had worked him up through all the lower grades and tried him ten years... They make every officer serve an apprenticeship under their eyes in their own ships before they advance or trust him.[88]

On realising the advantages of steam power

The meeting of the BAAS in Liverpool in 1854 captured contemporary concerns with the ships upon which the nation and empire increasingly depended for wealth and food. There appeared to be no prospect of agreement among scientific men, engineers, projectors, owners, builders, shippers, insurers or diverse publics on the best design of ocean vessels, whether with respect to size, speed, material or engines. Just over two years earlier news had reached Liverpool of the destruction by fire at sea of the brand new wooden steamer *Amazon*.[89] Now the BAAS met under the shadow of recent losses of two state-of-the-art Inman Line iron screw steamers as well as the iron sailing ship *Tayleur*. Intense controversy, as we have seen, already surrounded Brunel and Scott Russell's projected *Leviathan* where size and science as yet only *promised* to overcome all obstacles.

Such disagreements extended to arguments over the forms of motive power where science, theoretical and experimental, might be thought to provide some definitive answer. But here lay even more discordant communications of promised actions. In late April, John Ericsson's 'caloric engined' ship, the *Ericsson*, had sunk in shallow water while on trials out of New York, thus leaving the promise of marine air engines as yet unrealised.[90] Among steam engineers too, disagreements were rife. Ignorant of the fate that had overtaken the Collins liner *Arctic* only two days before, for example, the New York engineer Thomas Prosser addressed Section G 'On the use of surcharged steam, particularly in reference to some experiments which were made previously to its introduction into the American steamer *Arctic*, with some remarks on the great advantages possessed by steam over every other medium, and suggestions for an improved application of it to ocean steamers.'[91]

Drawing a sharp distinction between high temperatures and high pressures in heat engines, Prosser's paper is a striking illustration of the extent to which engineers had not taken up the Carnot–Clapeyron theory of the motive power of heat (reformulated by Macquorn Rankine, Rudolf Clausius and William Thomson in 1850–51).[92] Prosser argued that 'the great source of difficulty and danger in the use of steam is to be found in its high temperature rather than in its high pressure, and that, consequently,...that medium which requires the highest temperature to produce a given pressure is the most difficult and dangerous to deal with'. We could, he went on, talk nonchalantly of air compressed to ten times the highest pressure 'ever thought of in steam...merely because it is cold'. Indeed, 'Only yesterday our worthy president [Scott Russell] spoke even of 80,000 lbs. to the square inch, which is 800 times more than I shall have to ask for at present.'[93]

The question Prosser addressed was the 'economy and propriety of using surcharged or heated steam...as a means of reducing the enormous amount of coal consumed by ocean steamers (particularly by the owners of the Collins' line)'. The experiments consisted of passing a portion of the steam through wrought iron pipes within the furnace, thereafter to be reunited with the remaining steam prior to entering the cylinder. Such an arrangement had since been fitted to the *Arctic* but, after failure of the connections, awaited her return to New York 'so that this interesting question may be decided by an experiment of no ordinary magnitude and at no ordinary expense'.[94]

Prosser's view, however, was that success was unlikely. He interpreted 'surcharged' steam in terms of dangerously high temperature for pressures of about 6 lb. above atmospheric pressure, temperatures which had led to a burning of the felt covering the steam pipe ('the same as getting the boiler red hot for want of water') and which did not appear to be 'particularly efficient methods of developing the energies' of the steam. In contrast, Prosser recommended 'the use of high-pressure steam, worked expansively, and condensed without a vacuum for ocean steamers'. Such high-pressure steam working expansively, he claimed, always meant safe and dry, rather than scalding and wet steam. With a steam jacket around the working cylinder, then, it was 'more than probable that far greater economy will be attained, without increase of danger, than by any system of "surcharged steam" it is possible to devise'.[95]

This BAAS paper was a promotion of the marine *steam* engine, working expansively at high pressure and exhausting into the atmosphere as in a railway locomotive. It also provided a revealing view of air engines, implying that such engines were safe if, when working at very high pressures, the temperatures were low. Such was not the view, however, of promoters of the new thermodynamics. Combining their respective practical and theoretical skills, Napier and Rankine materialised in their projected air engine the principles of the new thermodynamics which, in contrast to Prosser, made high temperature (strictly, greater temperature difference) the key to greater economy from a given quantity of fuel (ch. 2). Rankine took upon himself the task of preparing a paper 'On the means of realizing the advantages of the air-engine' for Section G. While unable to attend in person on account of illness, Rankine ensured that the paper was well publicised in the national press, including *The Times*, and that it was furnished to the Section.[96]

By showing that the use of air as the working substance of an engine carried none of the potentially lethal dangers – such as scalding – inherent in the use of steam at high temperatures, Rankine's paper exposed the fallacy of Prosser's assumption that high temperatures were inherently dangerous while high pressures were not. In Rankine's analysis high temperatures and high pressures were related variables in any heat engine: the efficiency of any heat engine depended on the temperature (or pressure) *difference* between 'boiler' and 'condenser' regardless of the medium employed. But dangers could be increased or decreased according to the nature of that medium: wet steam, for example, at ordinary temperatures possessed scalding properties from which dry steam at higher temperatures was largely immune.

In the later *BAAS Report* it was Rankine's abstract that appeared in full while Prosser's was relegated to three lines. Despite his physical absence, therefore, Rankine had successfully infiltrated the BAAS and manipulated the press with regard to the new science and its promised embodiment in an air engine. *The Times*, indeed, had already published most of the abstract on 30 September:

> The lecturer [first] explained the two fundamental laws of the mechanical action of heat and their application to determine the efficiency of theoretically perfect engines, working between given limits of temperature. It was shown that as the efficiency increased with the distance between those limits, and, as it was easy to employ air with safety at temperatures far exceeding that at which the pressure of steam would cease to be safe and manageable, the *maximum* theoretical efficiency of air engines, consistent with safety, was much higher than that of steam engines; for example, at the temperature of 650° Fahrenheit, at which the air-engine had been successfully worked, the pressure of steam was 2,100 lb. on the square inch, while that of air was optional, being regulated by the density at which the air was employed.[97]

Second, Rankine classified the 'various causes of waste of heat and power in steam-engines'. *The Times* account did not detail these causes which had to await the full version of Rankine's paper. Published on both sides of the Atlantic within a year of the Liverpool BAAS meeting, one printing appeared in the *Edinburgh New Philosophical Journal* (1855) and the other in the *Journal of the Franklin Institute* (1855) which had also recently carried a paper on the use of 'surcharged steam' in the *Arctic*. Rankine now detailed those causes of waste which 'affect the relation between the expenditure of heat and the action of the working elastic substance upon the piston, – in other words, the *indicated* power of the engine'. Thus restricted, five classes of such waste could be identified: first, 'Imperfect communication of heat from the burning fuel to the working substance'; second, 'Imperfect abstraction of the heat'; third, 'Communication of heat to or from the working substance at improper periods of the stroke'; fourth, 'Any expenditure of the heat in elevating the temperature of the working substance'; and fifth, 'Imperfect arrangement of the series of changes of volume and pressure undergone by the working substance during the stroke.'[98]

Rankine noted that in actual steam engines the various means of diminishing the first three causes were already being practised to a great extent. Tubular boilers (of which Henry Booth had long been a leading advocate) presented a large surface to the products of combustion and thus helped reduce dramatically the first kind of waste. Improved condensation assisted in the reduction of the second. And elimination of conduction and radiation from cylinder, steam-passages and boiler reduced the third. The closely connected fourth and fifth classes of waste required an understanding of the action of steam in a perfect engine. For Rankine, a perfect engine was an ideal engine in which the amount of work – obtained from the passage of a given amount of heat through a given temperature difference – could be used to restore precisely that temperature difference. This ideal heat engine operated very much like a 'perfect' waterwheel whereby the work produced by the fall of a given quantity of water through a certain height might be used to restore exactly that quantity of water to its original level. As with Watt (ch. 2), the perfect engine then served as a standard against which *actual* engines could be measured.[99]

In the light of this classification of 'waste', *The Times* reported that Rankine compared the 'actual efficiency of steam-engines…with their *maximum* theoretical efficiency, and also with the *maximum* actual efficiency which might reasonably be expected to be attainable in the steam-engine by means of any probable improvements'.[100] In the fuller *BAAS Report* he gave the consumption of bituminous coal of a specified quality per horse-power per hour as 1.86 lb. for a 'theoretically perfect engine' working between temperature limits 'as are usual in steam engines', 2.50 lb. for a double-acting steam-engine 'improved to the utmost probable extent', and 4.00 lb. for 'a well-constructed and properly worked double-acting steam-engine on an average'.[101]

The Times further explained that Rankine had similarly classified the causes of waste of heat and power in air engines and compared the actual efficiencies of Stirling's and Ericsson's respective air engines (ch. 2) with the efficiency of theoretically perfect engines working between the same limits of temperature.

The former consumed 2.20 lb. compared to 0.73 lb. for a perfect engine and the latter 2.80 lb. compared to 0.82 lb. 'The results', Rankine announced, 'proved that an air-engine had actually been made to work successfully, and to realize an economy of fuel considerably superior to that of ordinary steam-engines, and, in fact, to surpass the utmost limit to which it was probable that the economy of double-acting steam-engines could ever be brought.'[102]

In particular, Stirling's air engine 'as finally improved, was compact in its dimensions, easily worked, not liable to get out of order, consumed less oil, and required fewer repairs than any steam-engine'. Resistance to the air-engine, on the other hand, derived from two sources. First, its advantages had not been so great 'as to induce practical men to overcome their natural repugnance to exchange a long-tried method for a new one'. And second, both types of air engine had been represented 'by some persons' as examples of power created out of nothing, viz., 'perpetual motion'. Yet it was evident that in practice, at least two-thirds of the fuel was wasted by both engines. The goal, therefore, should be to approximate to the 'theoretical extent of the economy of the air-engine – an extent far exceeding that to which the economy of the steam-engine is restricted', that is, to 'REALIZE THE ADVANTAGES OF THE AIR-ENGINE'. To achieve this goal, the most obvious and powerful causes of that waste of fuel needed to be identified and removed. Rankine and Napier thus announced their own version of an 'improved' air engine, while admitting to delays in its construction. Its great promise, as yet unfulfilled, lay in its compactness, in 'the saving of fuel' and in 'the important and incontestable advantage – that, even should an air-receiver burst (which is very unlikely), the explosion would be harmless, for its force would not be felt beyond the limits of the engine itself, and hot air does not scald'.[103]

Projectors, however, did not in the end take the air engine path as a solution to providing motive power for ocean ships. Not persuasively demonstrated in practice by either Ericsson or Napier and Rankine, these air-engine projects looked too much like experiment, too quick to latch on to scientific theories unproven in practice. Not even the credibility of Napier and Rankine, stronger by far than the easily dismissed showmanship of Ericsson, could sustain air-engine projects in the face of demands from cautious ship-owners who were well aware of the damage that extravagant and impractical scientific experimentation could do to their reputations among their publics. Shipbuilders and marine engineers thus turned to another type of engine, far less radical and firmly within steam-engine practices, but capable of novel forms.

In Rankine's estimates, double-acting steam engines, with an average coal consumption of 4 lb. per horse-power per hour, might be improved by attention to the above causes of waste to about three-quarters of their theoretical maximum or 2.5 lb. per horse-power per hour. The performance of a new kind of Clyde-built marine steam engine, the compound engine, had just begun to move the consumption downwards from a previous low of 4 or 4.5 to 3.25 lb. in 1854. Its designer, John Elder, was a close friend of Rankine.

One-time pupil of Lewis Gordon, the professor of engineering at Glasgow College, John Elder was part of a network of marine engineers and academics which included William Thomson, Rankine and James Robert Napier. Under the direction of his father who had been manager of Robert Napier's engineering works since 1821, Elder served a five-year apprenticeship. After gaining further practical experience in England, he returned to Napier's works as chief draughtsman in 1849. Relinquishing his position with Napier, Elder entered into partnership with a firm of millwrights from 1852.[104] As Randolph, Elder and Co., the firm began marine engine construction and two years later fitted a compound steam engine, designed and patented by Elder, to the 764-ton iron screw steamer *Brandon*, intended for a new service, announced in Liverpool's *The Albion* newspaper in March 1854, between Limerick and New York. The *Brandon* made only one round voyage from Europe to New York before being taken up by the British Government as a Crimean war transport. But at the end of 1854 the *North British Daily Mail* reported that it was 'gratifying to learn that the result, both as to speed and economy, has equalled the expectations of her engineers'. Rankine later recorded her fuel consumption to be 3.25 lb. per indicated horse-power (ihp) per hour.[105]

The *Brandon*'s demonstrable economy quickly prompted orders for engines to suit two ships of the Pacific Steam Navigation Company whose services, along the western seaboard of South America, typified British enthusiasm for so-called 'informal empire'. In the absence of local supplies of fuel, coal had to be shipped out from Britain by sailing vessels. When, therefore, the Crimean War put a premium on such tonnage, Pacific Steam consulted Elder on the prospects for greater engine economy. The outcome was the coastal paddle steamer *Inca* (290 gross tons) with a recorded fuel consumption of only 2.5 lb. of coal per ihp per hour and the larger *Valparaiso* (1060 gross tons) burning 3 lb. The consequent rise in Elder's engineering credibility was dramatic. Up to his death in 1869, Elder supplied some 30 sets of engines and built 22 ships for Pacific Steam (Figure 3.3).[106]

Elder was not the only Clydeside builder engaged in the new technology of high-pressure marine compound engines. Although Scotts of Greenock always prided themselves on their long, continuous shipbuilding history back to the early eighteenth century, family divisions had led to the foundation of a new yard on a green-field site around 1851, operated by Charles Scott and his son John, known as John the youngest ('yst') to avoid confusion with three other John Scotts in the family. John (yst) (1830–1903) had received a gentleman's education at the Edinburgh Academy (1841–46): in fact, he was in the same class as James Clerk Maxwell and Lewis Campbell and one year ahead of the class of Peter Guthrie Tait and Edward Harland.

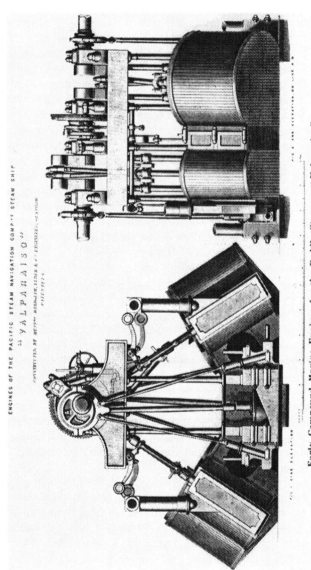

ENGINES OF THE PACIFIC STEAM NAVIGATION COMP'S STEAM SHIP

"VALPARAISO"

Early Compound Marine Engines for the Paddle Steamer "Valparaiso."

Figure 3.3   Diagonal compound engine, designed by John Elder for the Pacific Steam Navigation Company's *Valparaiso* (1865), showing small high-pressure and large low-pressure cylinders.

Source: *The Fairfield Shipbuilding and Engineering Works. The History of the Company; Review of its Productions; and Description of the Works.* London: Offices of 'Engineering', 1909, facing p. 4. (Courtesy of the Mitchell Library, Glasgow.)

Scott matriculated at Glasgow College in 1846, the year of William Thomson's appointment as natural philosophy professor.[107]

Such contexts shaped Scott's commitment to 'scientific engineering' at the new yard which he seems to have managed almost from the beginning. Indeed, in January 1857, Scott launched to his own account a 'revolutionary' new screw steamer, the *Thetis*, with a compound engine built by Rowan of Glasgow and operating at a very high pressure of 115 pounds per square inch (psi). This 500-ton vessel went on trials in May 1857, with delivery on 13 May, a date when, as the surviving correspondence shows, Alfred Holt was at the yard to supervise completion of the first steamer for his brother's firm of Lamport & Holt. Three years later, Rankine reported to J.R. Napier that the engine of the *Thetis*, whose performance he had been measuring, burnt little over 1 lb. of coal per ihp per hour. These impressive results certainly demonstrated the economy of high-pressure steam.[108]

The *Thetis* seems to have run successfully on cross-channel and coastal services before being sold to Mediterranean owners in 1862, and finally scrapped in 1899. But even by the early 1860s there were few compound-engined steamships on ocean routes. Four P&O steamers, innovatively engined by Humphrys, Tennant & Dykes of Deptford with compound engines (operating at a comparatively low pressure of about 26 psi), began work over the period 1861–64 on the Suez to Bombay link in the line's 'Overland' route to India.[109]

In the early to middle 1860s the Admiralty staged a series of experimental trials beginning with a 100-mile run at different speeds to test the performance of the Thames-built HMS *Octavia* (simple engine by Maudslay) and Clyde-built HMS *Constance* (compound engines by Elder). For a given indicated horsepower, the economy appeared nearly equal for the two warships. In late 1865 a much longer trial over 1000 miles to Madeira in heavy weather appeared to demonstrate the superior economy of the *Constance*. Indeed, the *Octavia* and *Arethusa* (simple engines by Penn) exhausted their coal supply and completed the passage under sail. As we shall now see, however, independent attempts were already underway to introduce high-pressure compound engines to long-distance ocean trades.[110]

The coal question

Historians have typically taken as self-evident that the more fuel-efficient compound (high-pressure) steam engine was 'needed' before steamers could displace sailing ships on long-distance ocean trades and before steamships could compete without expensive subsidies. Such self-evidence is wholly retrospective. Designed in accordance with 'scientific principles', for example, Brunel had projected the simple-engined *Great Eastern* as large enough to weather all oceans, accommodate as many passengers, carry as much cargo, and take aboard as great a quantity of coal as would make her eastern voyages both practicable and profitable with or without mail subsidies. Cheapness of

British coal would be the key to her 'success'. Not everyone, however, equated cheapness with 'success' or profit with 'economy'. Writing to Professor William Thomson of Glasgow University in January 1858, the Manchester natural philosopher James Joule identified what he saw as a fundamental flaw in the *Great Eastern*'s philosophy of carrying with her a veritable storehouse of cheap coal to Australia and back: 'I made a calculation what her speed would be when fully laden and find it somewhat less than that of [Cunard's Clyde-built] *Persia*. I think some limit should be placed on the enormous consumption of coal by steamships which is rapidly exhausting our coal mines.'[111]

By the mid-1860s Rankine's work on heat-engine economy was well enough known to be cited and discussed in William Stanley Jevons's *The Coal Question* (1865), a widely read analysis of the future prospects for Britain's wealth based on the insight that coal supplies (the energy sources upon which her wealth was ultimately founded)[112] would inevitably become more and more expensive as accessible sources became exhausted, mines became deeper, or marginal sources produced diminishing supplies. In his chapter 'Of the economy of fuel', for instance, he quoted verbatim from Rankine's 'Air-engine' paper in the *BAAS Report* (1854) comparing the fuel consumption of actual steam engines, those 'improved to the utmost probable extent', and a theoretically perfect engine. In Jevons's reading of the new science of heat engines:

> theory further points out, what practice has partially confirmed, that the work done by an engine for a certain expenditure of fuel is proportional to the difference of temperatures at which steam enters and leaves the engine. From this principle arises the economy of using high-pressure and super-heated steam...The economy already effected in this manner is wonderful....But it is in steam navigation that the improvement of the engine will have most marked effects. Any extensive saving of fuel, saving its stowage-room as well as its cost, will still more completely turn the balance in favour of steam, and sailing-vessels will soon sink into a subordinate rank.[113]

Henry Booth, as we saw, had urged in 1844 'that some new element of economy should be discovered' in place of ever-larger steamers consuming ever-greater tonnages of the nation's precious coal. Furthermore, Booth had identified another characteristic of promised steamship economy, that of targeting specific goals rather than designing vessels that combined every goal in one ship, such as the *Great Britain*'s extravagant combination of speed and large freight capacity. A moral economy of minimising true waste lay at the heart of Liverpool Unitarian ship-owners and merchants worshipping in Renshaw Street Chapel near the city centre. Members of the 'family' included Jevonses, Rathbones, Booths, Lamports, and Holts.[114]

By early 1864 one discourse in the Renshaw Street community centred on the China trade, made famous by the Tea-races among fast sailing clippers, first American and now British, including vessels built by Scotts and other Clyde yards. Far from abandoning the unstable world of steamships in which they had been immersed since the early 1850s, however, the brothers Alfred and Philip Henry talked steamers morning, noon, and night until they had identified a trade 'which we should, as we thought, have to ourselves'. By the autumn of 1864 they had settled on the China trade 'mainly because tea was a very nice thing to carry, and partly, as far as I, at any rate, was concerned, by a remark Sam. Rathbone, when discussing with W.J. Lamport the prospects of sailing ships, had made' that China at least was 'safe for sailing vessels'. In Alfred's recollection, 'the fiend made me say "Is it?"' But it was above all Alfred's confident conviction that 'cargo steamers could be engined so as to go, and pay, on much longer voyages than anyone at that time thought possible' that drove the new agenda. It is also likely that the Holts had in mind a voyage in 1863 of Lindsay's auxiliary steamer *Robert Lowe* up the Yangtse to load tea at Hankow for London, a voyage which earned £10,315 in freight alone.[115]

From the mid-1830s railway projectors and engineers such as Brunel and Fairbairn crossed the boundary between railways and steamships. Holt's mentor Henry Booth too had already moved seamlessly to steamship questions. Alfred Holt's apprenticeship on the Liverpool & Manchester Railway provided him with civil and mechanical engineering skills transferable from land to sea. And in 1854, as he put into coastal service his first new ship, Liverpool's commercial press (notably the *Liverpool Mercury*) was buzzing with reports of problems and prospects for ocean steam navigation under discussion in Section G at the BAAS meeting.

Holt rapidly gained a reputation as the kind of person who could be trusted to solve marine engineering problems. Central to the communication of promised actions were meticulous engineering drawings, some constructed for clients such as Lindsay. Holt greatly enhanced his trustworthiness by means of his own engineering workshop; direct experience of shipbuilding – again, beginning with the role of consulting engineer for Lindsay's auxiliary steamers – on the Clyde; a voyage to the Mediterranean to witness first-hand both a steamship at sea and the nature of the trade; new technical, social and business relations with shipbuilders and marine engineers on the Clyde and Tyne; and perhaps above all by practical demonstrations. This last strategy was one commonly used across our engineering spectrum (chs 2, 4, 5). In the earliest of his demonstrations, Holt reconstructed the machinery of an unreliable coastal steamer, the *Alpha*, which then left astern an entire fleet of sailing vessels bound north under fair winds out of the Mersey. Between 1854 and 1863 he built up a small fleet of steamers of his own, taking care to ensure that each new vessel conformed to his own exacting standards of engineering as well as management at sea and in port.[116]

In 1864 Holt circulated to his small band of shareholders his intention (drafted in the third person) 'as soon as the present steam ship mania has subsided, and good vessels can be built at moderate prices, to construct more steam-ships and enter upon a new trade... [he] hopes to invite those who joined him before to do so again'. This trade was 'a distant one and success or failure is simply a question of consumption of fuel'. Echoing the principal theme raised by Rankine for the 1854 BAAS meeting and invoked by Jevons in *The Coal Question*, Holt explained that the 'steam boat engine as at present used is an exceedingly wasteful machine'. This had 'long been evident to engineers, and many attempts have been made to improve it, some with very considerable success, but the margin of saving still left by the best engines is probably 50% of their actual consumption'. He then asserted that as a general rule he 'dislikes experiments, and has abstained from trying any in his late fleet, not wishing to risk the accident to which a novelty is necessarily liable'. Moreover, he stated his wish that the new ships would not be fitted with 'engines which experience had not warranted him in believing that men of ordinary ability could work successfully on a long voyage'. Either way, 'Commercial failure would be the inevitable result of accident especially at the beginning of a new enterprize.'[117] Avoiding 'manias' – which echoed the railway manias so destabilising of business trust and confidence (ch. 4) – and wary of 'experiment' with his commercial fleet, Holt set out to steer a course between the ceaseless experimenting of a Brunel and the extreme caution of a Cunard. He therefore began to define carefully the limits of experiment.

Holt had just sold his small fleet of steamers that he had built up for a West Indies service. But he had taken care to retain one steamer, the *Cleator*, which he had designed to a very high specification, especially with regard to strength, for the local iron-ore and coal trades. He now proposed 'to occupy the interval from the present time to that in which it will be prudent to commence his proposed undertaking, in making an experiment of a nature to yield trustworthy results... He takes this vessel because she has an excellent hull, and having been under his management for many years he knows her capabilities exactly with her present engines which are a good specimen of the usual [simple] type.' As with his previous venture, he won the confidence of former ship masters who, though their holding might be small in financial terms, did much to enhance the credibility of the enterprise.[118]

The *Cleator*'s compound engine, fitted in December 1864 and designed by Holt himself, operated at the relatively high pressure of 60 psi characteristic of railway locomotives on the Liverpool & Manchester Railway. That pressure contrasted with the reported 26 psi of the Humphrys, Tennant & Dykes 'compound' engines for recent P&O steamers constructed on the Thames. The arrangement for the *Cleator* was almost certainly the same as that designed by Holt and constructed by R. & W. Hawthorn for the Lamport & Holt steamers *Arago* and *Cassini* in 1865–66.[119]

Communication of promised actions eventually yielded to reports of actual performance. Crucially, the Holt brothers were *the* witnesses to that performance, Alfred having the inestimable advantage of knowing the precise performance of the ship's original simple engine which he himself had designed. Alfred recorded the event in his *Autobiography*: 'Having completed the change of machinery in the "Cleator" we (i.e., P[hilip]. H[enry]. H[olt]. and I) went off for a cruise into the channel in her ... [for] about two days ... The experiment left nothing to be desired, the engine worked perfectly, the vessel's speed was improved, and the coals burnt reduced, so that 5 tons did the work for which 8 tons were previously required.' Further passages coastwise to France and deep-sea to Brazil, Archangel and Australia all 'confirmed the results of the experimental cruise'. The results were made more credible by Holt's claim that the *Cleator's* 'old-type engines I [Holt] had designed, and actually drawn, myself, whose speed and coal consumption and performances generally, I knew with absolute accuracy'. It was, he wrote, 'On the strength of this experiment that we built the "Agamemnon", "Ajax" & "Achilles".'[120] These three compound-engined steamers, ordered from Scotts, represented an investment totalling a formidable £156,000.[121]

On the evening of Friday 16 February 1866, and in the hectic few days thereafter, Alfred and his brother George Holt paid a weekend visit to the Tyneside shipbuilder Andrew Leslie who was building the Lamport & Holt steamers. On Sunday morning Alfred recorded that they 'took a walk about the Wallsend [-on-Tyne] neighbourhood'. He confessed to 'economy of fuel' as his 'constant crotchet'. A critical look at local blast furnaces revealed an 'immense waste of heat owing to a constant flow of melted slag, the heat in which I am sure might have been turned to some useful purpose; but I suppose it is cheaper and easier to use fresh fuel, tho' the result cannot be other than the unnecessary diminution of our stock of coal'. Significantly, cheapness equalled waste rather than economy. Or as Jevons expressed the point succinctly in his 'Preface': 'where fuel is cheap it is wasted, and where it is dear it is economised. The finest engines are those in Cornwall, or in steam-vessels plying in distant parts of the ocean.'[122]

The Jevonses and the Holts moved in the same Liverpool mercantile and Unitarian culture. Jevons himself, moreover, was based at Queens College, Liverpool, during the period of the publication of *The Coal Question* and the construction of Alfred's high-pressure marine compound engines. Indeed, just as Jevons was drawing on evidence from actual and theoretical heat engines, so Alfred seemed to be acutely sensitive to the whole issue of 'the coal question'. Alfred's response to waste at Wallsend, located on Britain's largest coalfield, suggests that his 'constant crotchet' had relevance not simply to steam vessels plying distant parts of the ocean. In other words, striving to eliminate waste was not merely driven by commercial pressures to reduce costs or by physical concerns to avoid exhausting bunkers on long ocean voyages. It was indeed for him a universal moral concern, coinciding with the values of the Renshaw Street network in particular and with those of wider nonconformist (especially Presbyterian) communities in northern Britain.[123]

While at Leslie's yard on the Tyne in February 1866, Alfred and George 'sent the [Lamport & Holt] "Humboldt" to sea' under the command of Captain Kidd. Alfred had apparently persuaded Mr Lamport, whose first steamer had been delivered by Scotts as recently as 1857, to adopt a larger version of the *Cleator*'s compound engine. 'I took her from the Builders, Leslie's, on the Tyne with a cargo of coals to Gibraltar & returned to Liverpool via Lisbon', wrote Kidd in his Diary with characteristic frankness. 'This was a test & trial trip with Alfred Holt's new engine. Mr Lamport thought the engine was a failure. Alfred Holt thought otherwise and sent me to prove it was all right in the Humboldt.' In the battle of different readings of the *Humboldt*'s engine performance, Alfred's engineering authority won out over Lamport's scepticism concerning steam at sea.[124]

With the first of the trio, *Agamemnon*, launched in November 1865, Holt circulated a printed letter in mid-January 1866 announcing to shippers that he was 'about to establish a line of Screw Steamers from Liverpool to China'. The communication, however, went far beyond generalities. It promised that the *Agamemnon* would be commanded by Captain Middleton, that she was 'now nearly ready for sea, and intended to sail about the 20th March', though an exact day of sailing 'will be advertised when I obtain delivery from the builders'. He also promised that the 'service I propose these vessels to perform' would involve a non-stop passage to Mauritius via the Cape with an expected voyage time from Liverpool to Mauritius of 39 days. The outward passage would continue with calls at Penang, Singapore and Hong Kong before ending at Shanghai with a total voyage time of 76 days. Additional ports would be included but care would be taken 'to avoid any material lengthening of the voyage'.[125]

In order to make this promised schedule credible, Holt stressed that 'all the steamers have been built on the Clyde; they are of full power; and will steam the whole passage, both out and home'. In addition to their Clyde-built pedigree and their independence of sail, these new steamers had been designed to embody 'Every precaution which experience suggests...to fit them for the safe conveyance of valuable cargo.' Moreover, each captain had 'been many years in my employ, and...well accustomed to the care and navigation of steamers'. In a manner as readily applicable to Henry Booth's goals for the Liverpool & Manchester Railway, Holt therefore summed up his expectations: 'to establish a *reliable line* of steamers, which will carry Cargo [and about 40 cabin passengers], at *moderate rates* of freight, both *safely* and *at tolerable speed*'.[126] Such were the promises: reports of performance had yet to be communicated.

Three weeks later the Ocean Steam Ship Company's first annual meeting took place at India Buildings. Two ships had been launched (the *Ajax* the day before) but neither were yet ready for trials. The 'Managers' (Alfred and Philip Henry Holt) reported:

A more lengthened trial of the principle on which the engines for these vessels are constructed in a vessel [the *Humboldt*] of 1347 tons leads the Managers to be satisfied with the success of the plan. Improvements in detail will no doubt be suggested by further experience but the leading principle that the same speed can be obtained on about half the usual consumption of fuel is established beyond a doubt.[127]

This passage effectively opened the Line's official diary. Unlike Alfred's personal diary, or that of Captain Kidd, there was here little of the early struggles and nothing at all of the scepticism of a ship-owning authority like William Lamport.

At the end of March Alfred recorded in his diary his journey to Greenock to 'see the engine of the "Agamemnon" tried on the 24th [when] the trial came off and was satisfactory'. But all did not go according to plan five days later. As the ship headed into the sheltered Gareloch (to the north of Greenock) for sea trials, she ran aground near high water and remained there until the next tide when lightening and the aid of a tug finally freed her. On 31 March the new ship left her builders for an overnight trial which would take her to Liverpool. Holt privately recorded that on departure the engine was 'heating slightly and not working very well for a few hours but after passing the Cumbraes [where the Clyde widens towards the open sea] this passed off, and shortly after sighting the Mull of Galloway [at the south-west tip of Scotland] we were able to allow her to have full speed whereupon she at once went away 10½ knots which she easily maintained during the trial'.[128]

Matters of fuel economy permeated Holt's diary entry recording the event: 'We found her consumption by very accurate experiment to be 6½ tons in 7 hours & 40 min^s = 20 [tons]. 6 [hundredweight]. 3 [quarters]. 23 [lb.] in 24 hours, and this result as far as I know is not approached by any vessel afloat. I congratulate myself exceedingly thereon and think I have got a very fine vessel indeed.' This consumption approximated to just over 2 lb. per ihp. Off the North Wales coast the pilot came aboard with disastrous news to dampen the sense of promises being fulfilled: Lamport & Holt's new *Arago*, compound-engined by Alfred a few months earlier, 'had been lost on the South Stack [on the west side of Anglesey] two days before'.

By 19 April, a month later than originally projected, the *Agamemnon* was ready for her maiden voyage. Mutual trust between owners and masters was strong, reinforced by Captains Middleton and Kidd dining at the Holts' residence in Edge Hill a few days before departure. The owners' party, including Alfred's wife, brothers William and Philip Henry, and father-in-law, accompanied the ship down the Mersey channel and returned by the tug. 'She went and looked well', Alfred recorded. 'May she be successful.' At this highly significant point, owner and ship were parted, leaving the master with the safe-keeping of the vessel over thousands of miles of ocean.[129]

On 24 October, eight days behind her promised schedule, the *Agamemnon* arrived in the Thames at the end of her long voyage to China and back. Alfred travelled to London to meet the ship just as she entered the Victoria Docks where he 'rejoiced to receive an account of her performances which I consider eminently satisfactory'. Sceptics might indeed have pointed to her late arrival as evidence of technological failure and even friends were aware of just how finely balanced was any judgement of 'success' or 'failure', a judgement which would mean the difference between future commercial prosperity or poverty. A caricature sent to Alfred by such anonymous friends, represented a scene aboard the *Agamemnon* during the homeward first voyage, with Alfred very much under sentence should the steamer fail to fulfil promises.[130]

'Though the steamers of this line now proceed to China by the Suez Canal, their performances were remarkable when engaged in the... [Cape] route', wrote Lindsay in 1876: 'Starting from Liverpool they *never stopped till they reached Mauritius, a distance of 8500 miles*, being under steam the whole way, a feat hitherto considered impossible; thence they proceeded to Penang, Singapore, Hong Kong, and Shanghai, and, though unaided by any government grants, performed these distant voyages with extraordinary regularity.' Since the Canal opened, indeed, it was not exceptional for the *Agamemnon* to reach London from Hankow, 600 miles upriver from Shanghai, in under sixty days and ahead of the mail steamer. Consequently, only from the time Alfred Holt 'thus *practically* demonstrated the great value of such engines, ... have [they] been generally adopted'.[131]

Lindsay's account rendered the long-distance ocean steamer a historical fact. His remarks, indeed, neatly captured the manner in which Alfred and Philip Henry Holt's Ocean Steam Ship Company had on the one hand communicated the promise of trustworthiness and on the other been seen by wider maritime communities to deliver on that promise. The owners' stock of trustworthiness, in relation not simply to business matters but to their new technology of high-pressure compound-engined ocean steamships, had risen dramatically because family shareholders, the Renshaw Street merchants and ship-owners, Liverpool shippers and overseas agents and consignees had all been persuaded that promises of performance were being honoured. The integrity of the owners, it seemed, was being matched by the integrity of their ships. In an industry notorious for its financial and technological instabilities, Alfred Holt and his Company were building an unrivalled reputation for the communication of reliable knowledge, especially as it concerned the viability and trustworthiness of the ocean steamship, both commercially and mechanically.

Unlike Brunel's *Great Eastern* whose dismal track record quickly undermined the trust of shippers and travelling publics despite impeccable credentials from the BAAS elite, the Holts were handling their local audiences of investors and shippers with consummate skill. Appealing to a moral economy of minimum waste and maximum reliability, and contrasting to other ship-owners' seeming extravagance and recklessness, they were able to manage technological accident as well as engineering performance. Kidd, for example, recorded in his diary a highly public stranding of the *Saladin* opposite Liverpool and a whole series of mishaps while he commanded the *Ajax* including a costly 600-mile tow following a lost propeller, another expensive salvage deal due to a broken crank-shaft, a severely damaged stern frame, shortages of fuel on some voyages, and, worst of all, a sinking at Shanghai when the river current rotated the propeller shaft out of its tunnel while it was undergoing maintenance.[132] With the stock of trust now more than sufficient, such 'accidents' were not seen as undermining faith in steamers but as lessons to be learned for the benefit of future steamship practice.

The 'Blue Funnel' fleet made money and expanded. Individual voyage earnings on the *Ajax*'s third six-month round trip, for example, reached an astonishing £12,584, a quarter of her cost. By the close of 1880 the Line had taken delivery of some twenty-six ships, all from Scotts (twelve) or Leslie (fourteen), and annual dividends to shareholders averaged 15 per cent. Here indeed was a 'practical demonstration' of the viability of a long-distance, unsubsidised ocean steamer fleet. It was a fleet that not only offered a scheduled service but which increasingly provided a frequency of sailings on what it termed the 'main line' routes that rendered it a railway system of the sea.[133]

Retrospective assessments extolling the entrepreneurial or engineering genius of Alfred Holt take little account of the highly contingent nature of what he himself later called 'the great venture'.[134] There was nothing inevitable about the fortune generated from steamships trading to China. The new technology itself was a highly unstable and contingent one, as likely as not to end in shipwreck like the *Arago*, boiler explosion, engine failure, loss of men, or simply loss of capital. Alfred Holt and his associates had thus worked within local contexts and networks of ship-owners, shipbuilders, marine engineers, shipmasters and others to render the ocean steamship trustworthy in a period when credibility was far from guaranteed.

In 1877 Alfred Holt made a rare public appearance to deliver a historical 'Review of the progress of steam shipping during the last quarter of a century' to a distinguished audience at the Institution of Civil Engineers. Challenged during the ensuing discussion that he had not 'given honour where honour was due, and mentioned the name of Mr Brunel', he responded with characteristic irony to assign Brunel due credit: 'Mr Brunel deserved great praise, if only for the "Great Britain", an admirably built vessel, in which he made an excellent *experiment* with the screw...but with regard to the "Great Eastern", he [Holt] thought size was her chief peculiarity, and that there was not so much to learn from her as from other vessels. Considering Mr Brunel's genius and the flow of capital his designs attracted, his [Holt's] only wonder was that she was so small.'[135]

Conclusion

Conventional histories of the British Empire distinguish between 'formal' empire in which British rule extended to political control and colonisation as in India and Australia and 'informal empire' in which British commercial and scientific interests stopped short of political annexation as in China or large parts of South America. The distinctive features of the widely scattered island empire of Queen Victoria invite a revisionist account in which the Empire ceases to be regarded merely as the sum of the disparate parts but can be better understood by focusing on those physical and symbolic systems which, for a time, gave it unity amid diversity and made it what it was. Nothing exemplified better these bonds, embodied in space, than the engineering artefacts that were the iron steamships of the Victorian age. With their railway-like character, shipping lines literally reified the maritime empire in geographical space. Names such as British India, Pacific Steam, Peninsular & Oriental, Orient Line, China Navigation, New Zealand Shipping, and Federal all encapsulated this sense of Britannia's rule. And so too did the personal names associated with shipping lines that seemed more like national institutions than private commercial companies: Cunard, Lamport & Holt, Alfred Holt, Elder Dempster, Brocklebank, Bibby, Booth and many more.

From the 1860s, the Clyde and other northern rivers provided the sites for the shipyards of Empire. Failing to fulfil its early promises, its prospects blighted by the spectacle of the *Great Eastern* and by a reputation for high costs, the Thames lost the market for high-class imperial liners by the end of the 1860s and never again regained its early pre-eminence for marine engineering. But the shipbuilding sites were only the central focus of increasingly complex engineering systems which would, by the 1900s, include the business of producing veritable floating cities, constructed of steel, powered by the new steam turbines delivering thousands of horsepower to four propellers, lit by electricity and navigated by magnetic compasses precisely adjusted to compensate for a vast iron and steel environment. The great liners of the North Atlantic were the most prestigious products of the Clyde which boasted ships of every speciality designed to serve the maritime economy of Imperial Britain and the world: refrigerated ships to bring home the meat and fruit, tugs to handle the ocean liners as they approached their destination, dredgers to deepen and expand the ports of Empire and warships to guard the trade routes from all intruders.

4
Building Railway Empires: Promises in Space and Time

Not long after my return from Liverpool [to witness the opening of the Liverpool and Manchester Railway] I found myself seated at dinner next to an elderly gentleman, an eminent London banker. The new system of railroads, of course, was the ordinary topic of conversation. Much had been said in its favour, but my neighbour did not appear to concur with the majority. . . . 'Ah,' said the banker, 'I don't approve of this new mode of travelling. It will enable our clerks to plunder us, and then be off to Liverpool on their way to America at the rate of *twenty* miles an hour.' I suggested that science might perhaps remedy this evil, and that possibly we might send lightning to outstrip the culprit's arrival at Liverpool, and thus render the railroad a sure means of arresting the thief.
> *– Charles Babbage recollects how conversations centred on the 'new system of railroads' could bring together subjects as diverse as bank robbery and science (1864)*[1]

For some three weeks in 1898 construction stopped on one of the British Empire's most ambitious railway projects, linking the Indian Ocean port of Mombasa with the eastern shore of Lake Victoria in Uganda. Neither tribal opposition nor engineering problems played any part in this brief pause in the onward march of imperial civilisation. Yet before progress recommenced, 'twenty-eight Indian coolies and an indefinite number of African natives' had lost their lives, not as a consequence of any internal labour dispute or as a result of technological failure, but from a cause which spread fear across the workforce and which would feed the insatiable appetite of readers back home for heroic tales of empire-builders. As that most imperial of prime ministers, Lord Salisbury, informed the House of Lords: 'The whole of the work was put a stop to for three weeks because a party of man-eating lions appeared in the locality and conceived a most unfortunate taste for our porters. . . . Of course it is difficult to work a railway under these conditions, and until we found an enthusiastic sportsman [Lieutenant-Colonel J.H. Patterson, D.S.O., who commanded the Indian navvies] to get rid of these lions our enterprise was seriously hindered.'[2] In such contemporary readings of the episode, placing minimal value on the lives of imperial subjects, the power of the iron horse, aided by the 'noble' actions of the gentlemanly officer, triumphed over the king of beasts.

This episode, juxtaposed with our epigraph from Babbage's *Passages from the Life of a Philosopher*, captures the two broad themes of this chapter. On the one hand, the story of railways may be read as an account which shows a growth from highly localised interests in the early nineteenth century to the eventual establishment of vast networks, often on continental scales, serving goals of nation- and empire-building simultaneously with the transportation of passengers and freight in ever-increasing quantities. In this account, the Liverpool & Manchester Railway (L&MR), opened in 1830, served as a model for subsequent railways. On the other hand, the story of railways may also be read in terms of a series of contested historical processes in which neither the L&MR nor any of its rivals such as the Great Western (GWR) acted unproblematically as fixed templates for later developments. With the standardisation of gauge, for example, railway projectors and engineers worked for homogenisation in contexts shaped by the new scientific cultures calling for uniformity and economy, especially in relation to mapping and measurement (ch. 1).[3]

Building on these themes, we exploit specific case studies to show the cultural shaping of railway projects and to explore the ways in which railway projectors constructed not only networks of iron and steam but also new cultural systems. Those systems included grand 'central' stations, railway hotels, sleeper cars, and all manner of novel adjuncts to appeal to the nineteenth-century travelling publics. Drawing on the spectacle of railway engineering, architecture and scenery, projectors transformed rail travel from a seemingly unnatural and hazardous mode of transport into a taken-for-granted way of life suited to industrious citizens for whom the saving of time meant the creation of greater wealth. Thus as early as 1832, Babbage explained in his *Economy of Machinery and Manufactures* that 'rapid modes of conveyance increase the power of a country':

> On the Manchester rail-road, for example, above half a million persons travel annually; and supposing each person to save only one hour in the time of transit, between Manchester and Liverpool, a saving of five hundred thousand hours, or of fifty thousand working days, of ten hours each, is effected. Now this is equivalent to an addition to the actual power of the country of one hundred and sixty-seven men, without increasing the quantity of food consumed; and it should also be remarked, that the time of the class of men thus supplied, is far more valuable than that of mere labourers.[4]

Power in motion

At the beginning of the nineteenth century, private travellers and the new industrialists alike depended on an assembly of roads, some like the turn-pikes charging tolls, and all designed with horse, cart and pedestrian – but not heavy machinery – in mind. Voicing the public good, interested indi-viduals and government bodies had invested in roads between centres of population. In eighteenth-century Scotland, for example, Thomas Telford joined other aspirant 'civil engineers' to foster economic improvement by means of an enhanced infrastructure of communication. Both newly built and extended roads, traversing standardised but sturdy bridges, imposed national priorities on a capricious but tameable environment.[5] In Ireland too roads existed, but, even in the fertile north, eighteenth-century travellers complained of the broken and narrow causeways that were the roads converging on market towns through rough and untamed countryside.[6]

From the 1780s in England, coaches carried mail and passengers segre-gated and differentially cosseted according to their ability to pay. By the early nineteenth century a strict timetable, and horses bred for the purpose, delivered respectable folk in safety and at an average of 10 miles per hour.[7] The state of the roads, however, combined with the physical limitations of the horse, were such that it was often faster, and usually cheaper, to travel, transport and trade over water. Most of Britain's traditional centres of trade and population – London, Bristol, Liverpool, Glasgow, Newcastle – had flourished beside rivers which provided, directly or indirectly, access to sea ports. Programmes of improvement, typical of the Enlightenment and promoted by city merchants, saw rivers such as the Clyde dredged and made navigable for sea-going ships (ch. 3).[8]

Where natural terrain meant that rivers could not be 'improved' for the transport of industrial materials, projectors offered a solution. If geography limited the possibilities of a natural river network, commerce dictated artificial systems of communication. Canals literally cut across natural boundaries, dividing and reconnecting the country according to commercial concerns in a landscape restructured, and newly controlled according to commercial priorities. Raw materials entered as the cargoes of horse-drawn canal narrow boats; finished goods flooded outwards, towards the ports and to wider markets at home and abroad.[9]

Unlike most traditional British trading centres, the new manufacturing town of Birmingham in the English Midlands had no direct river access, still less access to the sea. By the end of the eighteenth century, however, Birmingham formed the hub of a booming canal system. Standard locks with a width of seven feet both enabled and constrained trade. This flourishing town and home of Boulton and Watt (ch. 2) had made itself into a remarkable inland port. Its commercial and professional citizens were keen to cultivate Enlightenment values of improvement and philosophical inquiry. The Lunar

Society of Birmingham in particular enthusiastically promoted a 'philosophy' of transport. Members like Erasmus Darwin, forever on the road, toyed with enhanced carriages and yearned for efficient turnpikes, recognising the vital role transport played not just in personal comfort but in the political economy of trade and industry. The potter Josiah Wedgwood and the chemist James Keir relied upon – and also sponsored – canal construction as part of a systematic approach to commercial endeavour.[10]

James Watt too had joined this Midlands culture of Enlightenment philosophy. Although he originally intended his steam engine to challenge the Newcomen pumping engines, Boulton encouraged him during the 1770s and 1780s to make steam a factory power (ch. 2). The high cost of fodder, a shortage of horses, and, by the 1790s, a shortage of manpower during the wars between Britain and France, all supported the argument in favour of mechanization.[11] By the second half of the 1780s the double-acting rotative engine, made intelligent with the governor and monitored by the indicator diagram, was able to generate rotary power. Consumed by the increasingly mechanised textile mills and by other factory spaces, the supply of rotary motion ceased to be monopolised by water power. The managers of mills in Britain, continental Europe, South America and India succumbed to the diverse marketing techniques of Boulton and Watt and their competitors.[12] Inventors and philosophers, however, also raised the possibility of harnessing steam not just to replace motive power in the factory but as a substitute for the horse-drawn carriage.

The horse, like the human, was not a source of unlimited work. Better, surely, than an exhaustible horse was an inexhaustible steam-driven carriage. Philosophers across Europe began to experiment. In 1769, just as Watt patented the principle of the separate condenser in Britain, the military engineer Nicolas Cugnot experimented with a steam-powered three-wheeled wagon suitable for French roads so much straighter and firmer than British ones. Watt's assistant Murdock had an experimental carriage ready for demonstrations by 1785, despite his employer's lack of encouragement (ch. 2). In the 1820s David Gordon established a company to try to poach the mail contract for a steam- or gas-powered carriage. But chief amongst those who puffed up schemes for steam road carriages in early nineteenth-century England was Goldsworthy Gurney.[13]

A gentlemanly chemist and science lecturer, Gurney had heard about Faraday's liquefaction of carbonic acid gas in 1823 and, like the Brunels (ch. 2), was inspired to create a moving power that could supersede steam. His specific wish was to propel a carriage. First, he built an experimental locomotive model running on ammonia gas. Quickly returning to steam power, he experimented with and exhibited a steam-driven carriage; when no capitalist or mechanical engineer took the bait, he decided to re-invent himself as an engineer, as Watt had done. Unlike Watt, Gurney had the personal resources to fit up an extensive workshop from scratch; he built a full-sized carriage in London and patented each design improvement (1825, 1827, 1829). It was a competitive environment, crowded with steam-carriage hopefuls and with promoters of a rival, the railway, as yet unproven. Opponents asked whether steam-carriages would ever cope with the common roads, surmount steep inclines, or negotiate tight corners; they complained of noise, smoke and suffocation. Advocates for the steam-carriage proved it, publicly, against hills and obstacles, silenced it (as far as possible) and proofed it against pollution; they represented it as an individualistic means of personal transport (like the horse-driven carriage) or a familiar one (designed to look like the stage-coach). Moreover, it would be unconstrained by the need to provide, or maintain, its own systematic infrastructure (like the canals or the railways).

Such debates were fuelled by Gurney's excursions, in central London, in busy sites of fashionable spectacle like Bath, and at Hounslow barracks; they were fuelled also when Gurney, like Boulton and like Babbage (with his calculating engine), positively encouraged visitors to his workshop.[14] There was extensive reporting in the *Observer, Gentleman's Magazine, Mirror, Morning Herald* and elsewhere. Although not every commentary was positive – workers outside Bath attacked the carriage as a symbol of the mechanisation threatening their jobs – Gurney's showmanship, much practised as a lecturer at the Surrey Institution, paid off with funds and influential supporters. John Herapath, later a railway aficionado, claimed these trials 'had established the merit of the invention on the only incontrovertible grounds, an experimental basis'.[15] The Duke of Wellington and other military men considered using the steam-carriage to transport military equipment and personnel; Dionysius Lardner, also later a railway critic, applauded Gurney's efforts and insisted that the turnpikes should not obstruct his endeavours; James McAdam worried, in 1828, that the carriages would rip up the roads and he predicted heavy tolls would be levied by the turnpike trusts.

But with these thoroughfares firmly in place, investors began to speculate on a system of steam-carriage lines. These men, amongst them a retired East India Company official familiar with the needs of integrated transport systems, offered substantial development capital in return for the privilege of running steam-carriages from London to Liverpool and thence to Scotland, from London to Bristol and Bath, from London to Exeter and Plymouth, and from London to Holyhead. Even before the carriage itself had been proven, the steam-carriage network had been fully imagined. In 1831 a regular service from Gloucester to Cheltenham, designed as the first stage in a Bristol–Birmingham line, apparently forced coach lines to drop their prices in competition. Soon, however, there was an orchestrated opposition from rival steam-carriage builders, railroad magnates (see below), turnpike trusts, stagecoach operators, all those concerned with supplying and feeding horses, and even the coaching inns. Ironically, the transition from experimental trial to routine practice stimulated the concerted opposition which crushed the life out of the steam-carriage as a nascent culture of transport.[16]

The case of the steam-carriage indicates that there were many imagined transport possibilities, none of them clearly destined to win out in the bustling technological marketplace of the early nineteenth century. Just one amongst many was to take a 'stationary steam engine', like those common in the factories, to reduce its size, and to make it move. But rather than render it a cumbersome attachment to a carriage working on the contested common roads, it might instead operate on its own 'permanent way' of wooden or metal rails, indeed its own 'railroad' or, in what became common British but not North American parlance, its 'railway'. The practice of moving trucks on rails for short distances dated from the sixteenth century, when small trucks (called 'dogs') trundled along fixed wooden rails in the bowels of German and British mines. By the second half of the eighteenth century, both the north-east of England and southern Wales were flecked with short but durable cast-iron tramways linking coal mines with the waterways and ports through which an essential fuel made its way to British factories and homes.[17]

It was a different matter to establish extensive railway lines for purposes other than local practices, directed to the movement of heavy or bulk materials, of mines, foundries and factories. Rivers and most roads represented long-established ribbons of communication which, unlike conventional properties, formed elongated lines of conveyance crossing many boundaries. Indeed, canal projectors had already discovered just how many vested interests could be marshalled against a new line. Thus to create a railway between a factory and a port, or between two towns would at the very least require considerable expense in the purchase of land, against the opposition of vested interests among road proprietors, canal managers and other landowners. Such vested interests pointed to the environmental encroachments of noise, smell and smoke, and the threat of danger from the new technology of mobile steam. Proponents claimed safety and prophesied economic benefit. In order to maintain such a strip of railway, furthermore, meant fencing and policing land – again an expensive and labour-intensive process.[18]

On the other hand there were clear advantages to be derived from placing a steam-driven carriage on a fixed railway. Steam road carriages had to cope with every gradient and sharp turn thrown at them by road-makers who did not have steam in mind. But steam rail carriages could be operated on tracks designed with gentle gradients and curves such that the locomotive vehicle need not make radical changes in the quantity of power it delivered. In other words, such an integrated system could eliminate sudden hills requiring complex gearing or braking mechanisms.

Watt had lacked enthusiasm for high-pressure steam in practice, in part because of questions of safety and in part because he understood his own engine to rely on a pressure difference between low-pressure steam and the near vacuum delivered by the action of the separate condenser (ch. 2). Other engineers, with different incentives, developed the high-pressure engines with greater power for a given cylinder size, or the same power with a more compact engine. In Britain, the key figure was the Cornish power engineer Richard Trevithick, a man whose activities were well known to Gurney, and to a wider public, through professional contacts and through extensive publicity campaigns.[19]

When Watt's much-disputed patent expired in 1800, rival engineers were free to experiment without fear of reprisal from Boulton and Watt's lawyers. Trevithick seized the opportunity. Adapting his experience with high-pressure 'column-of-water' engines, he designed and built high-pressure steam engines. By Christmas Eve of 1801 he had fitted out a road carriage with a high-pressure locomotive engine and challenged it with a steep hill. The roads were too treacherous and the power of the engine too slight. Trevithick therefore turned to experiments with a steam-carriage running on cast-iron tram tracks, or 'railroads' – like those already associated with the coal mines of north-east England and Wales. One such steam railroad locomotive in Wales carried 70 men at 5 miles per hour (mph) from the Peny Darren ironworks to the Glamorgan canal 10 miles away. The problem now was that this mechanical ensemble cracked the rails; but this did not prevent Trevithick from sending further railway locomotives to the mining centre of Newcastle in 1805.[20]

Trevithick was clearly thinking of wider markets when he rented land in Euston Square, fenced it off, and, for an admission price of one shilling, allowed London's show-goers to travel around in a carriage hauled by the steam locomotive *Catch Me Who Can*, transported at some cost to the capital. He was neither the first nor the last to call for public sanction of a new technology. Boulton and Watt had invited Cornish engineers to inspect their new engines in Birmingham and established the Albion Mill as an advertisement to factory owners, not just millers, recommending a shift to steam (ch. 2); later, Gurney made one terminus the fashionable centre of Bath when he ferried tourists along the common roads of England in his latest steam-carriage. Making a show of new technologies paid the bills and, in attracting public interest and building wider trust, might deliver more serious sponsorship.[21]

Other engineers now attempted to harness the steam engine for locomotion. John Blenkinsop's locomotive, with toothed wheels gripping a racked rail, worked well from 1812 on the Leeds to Middleton Colliery Railroad. The next year the colliery viewer at Wylam Colliery, William Hedley, risked a smooth wheel and rail (it turned out that friction was sufficient for traction). Aware of Hedley's work by 1813, the Killingworth Colliery engine-wright, George Stephenson was using locomotives to draw loads of 30 tons at walking pace up inclines of 1 in 450.[22] The history of the locomotive, then,

paralleled that of the Watt engine: a technology which had begun for a specific purpose was gradually applied more generally.

In 1821 the experienced George Stephenson was appointed engineer to
a Stockton and Darlington railway designed to carry freight, especially fuel,
and, as an afterthought, passengers. It was opened in September 1825, the
empire-building motto of the corporation being 'At Private Risk for Public
Service'.[23] The passenger-carrying Surrey Iron Railway of 1803, or even the
much later 'innocent railway' linking Edinburgh with the seaport of Leith
in 1838, continued to use horses rather than locomotives, indicating that
rail and steam were not inevitably, exclusively or even rapidly coupled.[24]
The Stockton and Darlington line used trucks pulled by horses primarily,
indicating the adherence to earlier practice, and the commitment to the
new technology of locomotives was relatively weak. Stephenson's chief task,
however, was to produce the locomotive, named *Locomotion*, which he built
at his newly established Newcastle works. Stephenson saw the railway built
of durable wrought iron bedded down at a gauge of 4 ft 8½ in., a choice,
Babbage claimed, that 'was the result of the accident that certain tram-roads
adjacent to mines were of that width'.[25]

Trying steam locomotion: the Liverpool & Manchester Railway

By 1830, after several years of planning and construction, the Liverpool &
Manchester Railway (L&MR) was in operation for the carriage of passengers
and freight. What made it special to contemporaries was not the novelty of
its constituent parts such as tracks and locomotives, but the fact that it
represented the launch of a new railway culture, a mode of travel and transport in its own right and not merely an adjunct to other, hitherto more
fundamental modes such as the canals, the rivers, or the sea. Earlier railways
had served to supply barges and ships with cargoes such as coal from the
mines, but now the railway engineers and promoters seized the opportunity
to market the railway to the public, both as an investment and as a revolutionary means of rapid travel.

The shaping of the L&MR and especially the events surrounding the
'Liverpool & Manchester Railway Trials' at Rainhill (near Liverpool) in
October 1829 superbly illustrate the issues of paramount importance to the
first builders of local and national railway empires.[26] Those empire-builders
looked, first, to generate practical proof of the viability of a new passenger
railway; they speculated on whether horse, steam or some other power
would be the best form of traction; choosing steam, they tamed capricious
experimental railway locomotives; to gain a public and professional warrant
for their chosen locomotive contenders, they ventured into unstable arenas
of public exhibition. Securing the trustworthiness of the railway meant
certifying the character and reputation of the engineer.

By the late 1820s, Liverpool was a thriving port. It was congested with British exports and emigrants heading out to America and its frontiers, and with imports of raw cotton and unskilled Irish labour. Manchester was a rapidly growing centre of textile production some 30 miles distant. But the service provided by roads and canals in supplying it with raw materials from its nearest deep-sea port appeared inadequate, especially in the light of the transport possibilities opened up by the experiment of the Stockton and Darlington line. That railway was now paraded as the prototype and testing ground for a more ambitious project. Promoters of a line between Liverpool and Manchester, eager to extend existing networks of maritime trade inland, approached George Stephenson. They prepared to rebut the challenges of the canal interests, to survey despite the obstructions of farmers and land-owners, and to force through a bill against parliamentary opposition and those prepared to challenge Stephenson's technical credibility – despite his undoubted experience. Opposition from vested interests neutralised, an act of parliament allowed for the construction of a railway line between the two towns.[27]

Debates about the means of haulage, however, were far from settled. In 1829, John Rastrick reported to the Company's Directors on the merits of the chief contenders: horse, cables worked by stationary engines, and the new 'loco-motive' engines.[28] Perhaps wishing to minimise risk, the promoters were at first minded to adopt cable haulage, using the proven technology of stationary steam. Nevertheless, the locomotives' advocates were offered an opportunity to prove (or rather, to garner consent for) the reliability of their favoured means at a series of public trials to take place at Rainhill.[29] As one commentator later claimed, this trial was 'nominally to prove which out of five machines was best; but in reality it was the crucial test of the practical value in the matter of speed and power of the railway system itself'.[30]

The trials were to take place on a level of one and three-quarter miles; the locomotives must work at a pressure of no greater than 50 psi; they must attain an average speed of at least 10 mph (equalling that of long-distance horse transport); and they must maintain that average over ten double jour-neys, or a distance of 35 miles. This was more than the distance between the two towns. To satisfy the Act of Parliament which enabled the line to be built, the engines must also consume their own smoke (and thus, like Gurney, avoid pollution). As an incentive for the capital outlay of constructing the competing machines, the L&MR Directors offered a prize of £500.[31]

Amongst the competitors, there were four locomotive 'novelties' and one horse-powered alternative, usually expunged from subsequent accounts but crucial in illustrating the vestiges of the comparative element of the trials. The contestants were engineers redeploying skills towards the risky new adventure of locomotive manufacture, in the glare of publicity, when, as Watt's experience had shown, cultures of experimental engineering practice demanded the utmost secrecy. Hackworth's *Sanspareil* was, in the event, more than equalled; Burstall's *Perseverence* could not stand the pace. The *Rocket* of George and Robert Stephenson has entered the folklore of railway history as the clear winner of that notorious event.

But included in the list of hopefuls was the Swedish heat engineer and consummate technological showman John Ericsson, then only twenty-two years of age but already notorious as the inventor of the 'caloric engine' (ch. 2).[32] Ericsson here collaborated with his London partner John Braithwaite, parading the locomotive aptly named, like one of Ericsson's earlier fire engines, *Novelty*. He had the support, also, of the London-based civil engineer (and rival of Marc Brunel) Charles Blacker Vignoles. Since the mid-1820s, Vignoles had been thinking about the best means of railway traction; opting for the locomotive, he invested heavily in Ericsson and Braithwaite's competitive engine which, apparently, relied on Ericsson's caloric system.[33] Although Ericsson had made his mark in the metropolis he worked at a disadvantage: his backer Vignoles was poorly equipped as a mechanical engineer and out of touch with the Stephensons' programmes of locomotive improvement. Ericsson had only six weeks to prepare his engine; and London had no railway on which it could be tested.

When the trials began, on 6 October, the 'whole scientific world was... watching intently for the result'; there were 'vast numbers' of people actually present.[34] To satisfy a technical readership, and to amplify the number of competent witnesses unable to see the events directly, Vignoles wrote extended accounts for the *Mechanics' Magazine* (which was not, apparently, worried about a conflict of interest); the *Liverpool Mercury* and other newspapers satisfied local, and national, curiosity in the latest technical wonder.[35] The *Rocket* appeared on the first day, liveried in yellow and black with a striking white chimney; George Stephenson saw the *Novelty*, as the only possible rival in this technological prize-fight and with concern checked the quality of his fuel and purity of feed water. Burnished with copper and dark blue paint, its machinery better finished than the *Rocket*'s, the Liverpudlians learnt that this beautiful, light and compact specimen was 'the very *beau-idéal* of a locomotive engine'.[36]

Contemporary reports indicate that the *Rocket* offered reliability and adequate speed under load in contrast to the *Novelty*'s tremendous – but erratic – speed. The *Novelty* was the 'Galloway pony' when the L&MR wanted the 'road teamster' that was the *Rocket*.[37] Ericsson's machine managed 35 mph unloaded, and 28 mph under a load of ten tons, albeit for a short distance, before coming to a standstill. At one moment a lack of fuel, at another the clogging of the rails with mud, at the next a burst pipe with no forge close by, conspired to take the machine out of the competition. Corrosion due to intense heat and the repeated failure of a fan, both integral to the design and to Ericsson's 'caloric system', were dismissed as the results of defective workmanship, but Stephenson, evidently unconcerned by the loss of gentlemanly credibility from a Tyneside accent, allegedly told his supporter, the engineer Joseph Locke, 'Eh! Mon, we needn't fear yon thing; it's got no goots.'[38]

The *Rocket*, on the other hand, had a more reliable innovation of multi-tubular boiler.[39] On the very first day the *Rocket* had achieved a speed of more than 10 mph under a load of 12 tons, and 18 mph unloaded, but Vignoles talked of its 'very unequal' velocity and its failure of the test when it 'did not at first thoroughly consume its own smoke'.[40] On the third day it pushed up to an average of 12 mph loaded over the full distance; by then the *Novelty*'s promoters were protesting at a change of rules to include an assessment of the time taken to get the engine's pressure up to 50 psi (a rule-change probably designed to benefit Stephenson's bid). Briefly, the rules and conditions of the trials were abandoned completely and the show took over: *Novelty* resorted to carrying 45 passengers 'at upwards of thirty five miles an hour'; stripped down to an absolute minimum, the *Rocket* pushed its average speed to a little over 20 mph. And although it 'was not pretended that it was to be considered a competition trial', this final demonstration of speed, hitherto the *Rocket*'s weakest point, now stood as 'convincing proof of the qualities of the engine' (Figure 4.1).[41]

This trial did not, quite, finish with the closure of the formal tournament. Ericsson's supporters were keen to remind readers that Stephenson met with frequent difficulties in his earlier locomotives and it was 'only through the experience learnt in such failures, coupled with the increasing skill of the workmen employed...that his machines were little by little raised to the requisite standard of power and efficiency'. Two months had not been enough for Ericsson;[42] but by December, he was still, apparently, trying to win over the Liverpool public, with a rival trial not even at full steam reaching 40 miles an hour, despite the withdrawal of support and equipment from his *'friends'* the L&MR Directors. With Ericsson guessing he might make *'one mile in one minute'*, he told Braithwaite:

> The spectactors appeared highly delighted with our mode of conveyance; and later on we rigged out some of the waggons [*sic*] for passengers (our *friends* having sent away the passage waggon), as the ladies were quite distressed at not being able to get a ride...As the engine passed on its velocity the spectators cheered in a glorious manner. I will send the particulars of more experiments to-morrow.[43]

Of course, however, it was the *Rocket* created by George Stephenson's son Robert Stephenson that triumphed, meeting the carefully chosen – then varied – conditions established by those at the L&MR who designed the show.[44] Now, according to the *Scotsman*, the *Rocket* would 'give a greater impulse to civilization than it has ever received from any single cause since the press first opened the gates of knowledge to the human species at large'.[45] But the losers debated the nature of that triumph, seeing that the

MACHINES.   21

30. A LOCOMOTIVE ENGINE is a machine which has a tendency to change its place by the effect of some action among its parts.

But in all such cases, some action upon external bodies is necessary, in order that the engine may travel ; thus a row-boat advances by the action of the rowers upon the oars ; but this effect requires the re-action of the water. The same is the case with a steam-boat. In like manner a carriage on a road, moved by steam, or any other power acting within it, requires the re-action of the road in order to advance.

31. PROP. *In a locomotive engine which is made to travel by turning a wheel on which it rests ; the force required is the same as if the center of the wheel were fixed, and the resistance of the motion acted at the circum-ference of the wheel.*

The figure represents the original " Rocket " engine of Mr. Stephenson.

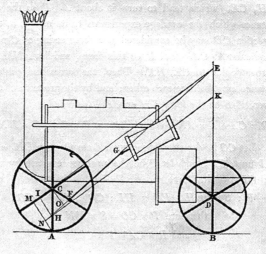

Figure 4.1   William Whewell redefines the locomotive, and especially the *Rocket*, for student engineers in 1841.

Source: William Whewell, *Mechanics of Engineering. Intended for Use in Universities and in Colleges of Engineers.* Cambridge and London: Parker and Deighton, 1841, p. 21. (Courtesy of Aberdeen University Library.)

winner at Rainhill was the locomotive which succeeded in a test administered by interested parties and knowing that the public rhetoric of exhibition, as much as any unequivocal, technical demonstration internal to an engineering profession, had been at play. Although doubts later arose amongst Stephenson senior's contemporaries over his competence in managing large construction projects,[46] theatrical events at Rainhill had done much to launch a railway dynasty – providing that dynasty with an image as iconic as Watt's Newcomen model. This had not been merely a trial of engines; it was a trial of the steam locomotive itself – and in Cardwell's view it was the '*reputations* of Stephenson and his gifted son Robert' that clinched the debate in favour of the steam locomotive.[47]

Amongst their closest associates, the Stephensons had articulate reputation builders with excellent literary and moral qualifications. Henry Booth, secretary to the L&MR has appeared in earlier chapters as a promoter of Greenwich time and of iron steamers (chs 1, 3). But his writings ranged from tragedies to the laws of motion in fluids. Booth shared with the Stephensons and with William Fairbairn a Unitarian Christianity which emphasised self-reliance, dislike of ostentation, and a belief in an orderly, law-governed universe to which humankind had access through science. For Booth, engineering projects, not least the L&MR, 'imitated' divine Providence. Booth's leading role in Renshaw Street Unitarian Chapel, centrally placed within a tightly knit and very wealthy community of Liverpool merchants (ch. 3), gave his every utterance on the new railway both power and probity.[48]

Booth was a master communicator of promised actions. His *Liverpool and Manchester Railway* (1830) was a first-hand account, written from the commercial perspective of a Liverpool patrician, of the project's construction and its promised commercial and moral performance. As befitted a radical reformer appealing to the mercantile classes, Booth portrayed the parliamentary campaigns that preceded construction as a struggle for public good over private interests (echoing the Stockton and Darlington Railway's motto), as a contest between aristocratic Establishment and mercantile reformers, and as a victory of free trade over protection:

> Two noble Lords, the Earls of Derby and Sefton, a part of whose estates the railway crossed, made common cause with the Canals to prevent the establishment of a Railway. On the part of these noblemen, it was contended that the sanctity of their domains would be invaded, and the privacy of their residences destroyed, by thus bringing into their neighbourhood a public highway, with all the varied traffic of coals and merchandise and passengers...[49]

In contrast to this alleged opposition from aristocratic Establishment, the 'Free Traders' (among whom Booth numbered himself) eventually attained an outcome 'satisfactory to all who contemplate with pleasure the commercial prosperity of the country, or who take an interest in marking those great steps in the progress of mechanical science, the successful study and application of which, to the arts and manufactures, have contributed in no small degree to raise this country to its present pre-eminence in wealth, power, and civilization'.[50] By implication, the same mechanical science that had delivered the nation's pre-eminence would underpin the future promise of the L&MR.

Because the concept of rapid transportation of goods and passengers by railway was so fresh in 1830, Booth's book aimed to persuade readers of the trustworthiness of the system. Strikingly, he conducted his readers on a 'virtual' excursion from Liverpool to Manchester. They would soon learn that this project was not the fragile whim of a lone inventor but rather an integrated system, planned with precision by experts in surveying, mechanical science, and commerce. After leaving the Moorish archway at Edge Hill, 'The traveller now finds himself on the open road to Manchester, and has the opportunity of contemplating the peculiar features of a well-constructed Railway, the line in this place being perfectly level; the slight curve, which was unavoidable, beautifully set out; the road-way clean, dry, and free from obstructions; and the rails firmly fixed on massive blocks of stone.' Not for nothing had the 31-mile railway absorbed upwards of £800,000 of capital.[51]

Booth could therefore show his readers a vision of the fruits of the Rainhill trials: 'The light of the moon illuminating about half the depth [of the great rock excavation forming a cutting], and casting a darker shadow on the area below – the general silence interrupted at intervals by a noise like distant thunder – presently a train of carriages, led on by an Engine of fire and steam, with her lamps like two furnaces, throwing their light onward in dazzling signal of their approach – with the strength and speed of a war-horse the Engine moves forward with its glorious cavalcade of merchandise from all countries and passengers of all nations.'[52]

In the final chapter, 'Considerations – Moral – Commercial – Economical', he explained the moral economy of the railway system. Its benevolent possibilities stood in stark contrast to the ambiguous results of recent industrialisation, especially through factory and mill. 'On the banks of the Canal a great cotton factory rears its tall sides', he told his readers as their virtual journey approached a manufacturing district, '...and the fly-wheel of its steam-engine pursuing its continuous and uniform revolutions, as if symbolical of that eternal round of labour and care, of abundant toil and scanty remuneration, of strained exertion and insufficient repose, which...have been the condition and tenure on which the existence of so large a portion of mankind has depended'.[53] Indeed, there were no guarantees that the steam engine *per se* would change that condition: 'It has frequently been a matter of regret, that in the progress of mechanical science, as applicable to trade and manufactures, the great stages of improvement are too often accompanied with severe suffering to the industrious classes of society.' Echoing Malthus, Booth inferred that the 'universal and unceasing' race of competition and the 'never-ceasing race of population against subsistence' seemed to characterise 'the present condition and aspect of society, as constituting a vast trading community'.[54]

In contrast to these double-edged features of industrialisation, then, the L&MR presented 'one great object for our admiration, almost unalloyed by any counteracting or painful consideration'. Its inauguration brought into being 'a new theatre of activity and employment...to an industrious population, with all the indications of health and energy and cheerfulness which flow from such a scene'. Henry Booth had anticipated 'the extension of Railways throughout the country, intersecting the island in every direction'. Its future replication across the land opened up a vast new source of occupation both to the labouring community and to capital. Indeed, he predicted that 'The stately Turk, with his turban and slippers, will quit his couch and his carpet, to mount his Engine of fire and speed, that he may enjoy the delight of modern locomotion.' Thus from 'west to east, and from north to south, the mechanical principle, the philosophy of the nineteenth century, will spread and extend itself'. Most strikingly, it produced a 'sudden and marvellous change...in our ideas of time and space' by enabling commercial travellers to 'live double times' and by saving on the capital deployed on rolling stock.[55]

Taken together, these results might not guarantee the visions of Utopian theorists that 'a whole community shall enjoy the pleasures and satisfactions to be expected from that happy combination of the powers and capabilities of the human race' when 'the fervour of an earnest enthusiasm – religious, moral, social – shall not be inconsistent with the calculations of the merchant, or the speculations of the political economist'. Able now to find rhetorical purchase in the name of the winner of the Rainhill trials (rather than the unqualified *Sanspareil, Perseverence* or *Novelty*), Booth noted the ambiguity of the name 'Rocket' which represented either 'the harbinger of peace and the arts, or the Engine of hostile attack and devastation'. Regardless of the progress of mechanical science, Booth concluded with the moral exhortation to his readers to make of the railway a good work: that 'though it be a futile attempt to oppose so mighty an impulse, it may not be unworthy [of] our ambition, to guide its progress and direct its course'.[56]

Such was the promise of the L&MR. But Booth's *Liverpool and Manchester Railway* had been published a couple of months *before*, not after, the grand day of opening on 15 September 1830. Booth could never have predicted that the official opening would be catastrophically overshadowed by the death of William Huskisson: the railway's chief parliamentary sponsor had been mangled in the machinery of the new railway system. Before this public relations disaster, there had been a mood of tremendous celebration.

The carriages were packed. Friends shouted greetings across the tracks. The 'highly excited populace' yelled and the once heroic Wellington, made unpopular through political office, paraded with other dignitaries in an ornamented carriage before heading off to dine with local aristocrats. Bishops were present, thus blessing the new work of man. Robert Peel, later to succeed Wellington as Tory leader, was present. Only the demise of Huskisson meant that guests partook without ceremonial of the lavish luncheon waiting at Manchester.[57]

Although the company directors had worked to get politicians, clergy and public on board for the new railway journey, the loss of such a senior figure raised doubts about the safety of rail, and the 'railway accident' was immediately a subject of concern, for medics and managers alike.[58] Opportunistically, perhaps, the gentlemen of science leapt in with solutions to shore up public confidence. At another dinner hosted a few days after the opening by one of the 'great Liverpool merchants', Babbage met 'officers of the new railway' and others, all of whom 'were more or less interested in its success'. When conversation 'very naturally turned upon the new mode of locomotion' it was dogged by talk of 'difficulties and dangers', and how to avoid 'expensive and fatal effects'. Babbage immediately came up with schemes to remove obstacles – including stray animals and people – from the railway, anxious as he was to 'diminish the dangers of this new mode of travelling'.[59]

With 'difficulties', 'dangers', unforeseen expense and embarrassing fatalities prominent in the early discourse of passenger railways there was clearly the potential for an implosion of the 'confidence' and trustworthiness which was crucial to their 'success'. Such confidence might be inspired by the words of the elite consulting engineers, around whom a star system was developing. But Babbage criticised the abuses of such figures and their 'interest with ... higher powers'.[60] Rather, Babbage saw the key to the 'security of railway travelling' as good data, mechanically produced, collected and registered reliably without the intervention of unskilled – or 'interested' – individuals: 'Even the best and most unbiassed judgement ought not to be trusted when mechanical evidence can be produced', he claimed. He therefore called for self-registering of speed throughout the journey and a dynamometer to measure the force of the engine. Such 'mechanical registrations' would become the 'unerring record of facts, the incorruptible witnesses of the immediate antecedents of any catastrophe'.[61] In short, they would manufacture confidence in the fulfilment of the promise of the new system.

Booth had championed the embryonic L&MR as representative of the 'progress of mechanical science'; Babbage saw it as a technology which might be rendered safer by 'mechanical science' instantiated in instrumentation. Invoking the name of 'mechanical science' as the guarantor of innovative engineering projects' trustworthiness would become a familiar and formidable trope in the rhetoric of the British Association. Launched within a year of the opening of the L&MR, the BAAS established its section for Mechanical Science in part as a response to the provincial phenomenon of railway construction – whilst the railways themselves would increasingly allow rapid travel to its meetings (chs 1, 3). From the earliest days of the L&MR, men of science exploited this new opportunity to expand their knowledge-making remit (not least through the geological opportunities of railway cuttings), or to assert again that practical art was truly mechanical science applied to practice.[62]

Experiments, templates and systems: the Great Western Railway

In order to cater for a market that combined freight with a mass public newly eager for mobility, the L&MR's builders had recruited and systematically combined techniques available to them from existing transport industries and elsewhere: stage-coaches already traversed the country according to an established timetable; canal builders had cut, embanked and tunnelled; the lawyers of the canal companies had grappled with compulsory land purchase to tame hostile environments, politically exerting power over landed opposition; their managers had organised teams of 'navvies' or labourers, physically engineering a way through hills and across rivers to gain maximum economy; engineers already had the skills of locomotive engineering from mines and, more recently, from the testing ground of the Stockton and Darlington Railway.

Yet in combination, these and other elements made a railway system, accessible to the analysis of Hughes (Introduction), but also the descriptive term of contemporary transport analysts (see, for example, our epigraph). It was a system of great power and it served also as a flexible template for future railways: steam traction (an option tried with limited success by the canal companies); parallel tracks allowing trains to pass simultaneously in each direction and to increase the capacity for freight; a fixed and reliable timetable; passengers divided into classes, from first to third, and, again like stage-coach passengers, paying fares calculated by the mile; purpose-built stations providing comforts and entertainment.[63] Thus historical actors and historians have presented the L&MR, unlike the Stockton and Darlington Railway, as the solution to the problem of rail transport and, indeed, the 'model' for future railways.[64]

The L&MR however did not constitute the fixed and final template for all subsequent railways. Rather, the railways were sites for continued variation, innovation – and experiment. Indeed, as David Brewster observed in 1849, the 'first of the great lines with which England is now covered was the Liverpool and Manchester Railway, which has been justly called *The Grand British Experimental Railway*'.[65] Status-conscious engineers in all fields happily described themselves as 'experimenters', with support from the technical press. Reporting Morse's telegraphic ventures (ch. 5), the

Railway Magazine applauded the United States Congress's vote of $30,000 'towards a grand *experiment* on this mode of communication'.[66] Even the Stephensons' *Rocket* was not the last word in engine design: as early as May 1833 the L&MR took delivery from Sharp, Roberts of a locomotive with numerous novelties – and aptly named *Experiment*.[67]

Engineers looking to equip Britain's new railways exploited the L&MR as a vast experiment in engineering practice and transport economy. In 1835, Peter Barlow, mathematics professor at the Royal Military Academy at Woolwich, detailed to the directors of the London and Birmingham Railway 'experiments' he had made on the L&MR.[68] De Pambour's foundational treatise on locomotive engines, brought out by the entrepreneurial engineering and architectural publisher John Weale in 1840, likewise encapsulated 'new experiments' made on the L&MR.[69] As an experiment – in the language and practice of contemporary engineers – this first major railway was the beginning and not the end of railway culture.

This climate of technological flux helps to explain why, when Brunel set about designing his own railway in the early 1830s, he could talk of making the rules from scratch, likening the process to 'nothing but the sudden adoption of a language…understood by nobody about [around] him'.[70] Brunel's sprawling technological text would be the Great Western Railway, prospected in 1833 and by the mid-1840s dominating the area to the south-west of London.[71] Even the name confirmed Brunel's love of the grandeur he had witnessed as a student in Paris.[72] Brunel treated the railway as an opportunity for exuberant innovation, so energised was he by the ethos of experiment, and with its potential payoff in bolstering the claims of engineers to high scientific status. Yet if the L&MR was a 'grand experiment', and Brunel made the GWR experimental, it was open season for the men of science. The professor and populariser Dionysius Lardner warned a BAAS audience that Brunel's audacious 2-mile Box Hill Tunnel was bound to cause asphyxiation to passengers. (One hundred men had already died in its construction.) The Oxford geologist and BAAS manager William Buckland predicted that the unlined portion of the tunnel would fall from the concussion of the trains.[73] As an experiment, then, the GWR might be ill performed according to standards laid down by authoritative scientific men.

Brunel found himself in disagreement, too, with professional engineers like Nicholas Wood, a one-time pupil and now an ally of George Stephenson. Wood had moved with astonishing rapidity to demonstrate 'with original experiments' and 'tables', the superiority of railroad to canals as an instrument of 'interior communication'. More than that, he had begun to freeze the flood of railway culture in written form.[74] It was too early to speak of railway practices as 'traditional' but they were being given a template in Wood's *Practical Treatise on Rail-Roads* (1825, with a third edition by 1838). Further, the mechanisms of the rising bureaucrats and government experts began to make them conventional. Thus the House of Commons adopted John McNeill's graphical system of displaying cuttings and gradients as a standard for all railway plans deposited for their approval.[75] Brunel routinely objected to premature attempts to formalise engineering practice. The establishment of a false system, by the invention, in print, of a tradition of practice would be detrimental to free trade in British engineering.[76]

Not for Brunel the 'standards' of railway construction laid down by others. Setting about devising his own method of laying the 'permanent way', he convinced himself and the Directors of the GWR that a gauge of seven feet – 'broad' in comparison to the 'narrow' gauge favoured by the Stephensons – would deliver greater speed and enhance the comfort of sensitive first-class passengers. The locomotives he designed were correspondingly larger than those at work on the lines in the north of England. An experimental technology like the railway might be yoked systematically to another promising experimental novelty, the electrical telegraph (ch. 5). In the autumn of 1838 a Fellow of the Royal Society suggested railway accidents might be prevented using 'the electro-telegraph of Prof. Wheatstone', whilst another correspondent of the *Railway Magazine* saw the device 'giving timely notice of whatever may be on the road' and thus auguring the solution to 'accidents of collision'.[77] Brunel quickly conferred with telegraphers about matters of safety and speed of working. All of these distinctive features testified to his enthusiasm for luxurious innovation – an enthusiasm not always approved by the GWR's shareholders (Figure 4.2).[78]

If Brunel's railway was a site for innovation and luxury, it was also to be a model for his own brand of *system*-building. A systematic approach to practical art was *philosophical*, in its contemporary sense, since its purpose was to impose order, through the exercise of the intellect, from above. Systematic design de-emphasised the roles of the contractor and craftsman. There were precedents here in the work Marc Isambard Brunel and especially his scheme for the making of ships' blocks for the British Admiralty at Portsmouth. Where previously they had been produced through the agency of skilled workmen, Brunel senior broke down the construction into a series of simple tasks, each of which could be carried out by a specialised machine. The result was systematic, organised, mass production – satisfying the Admiralty at the expense of the skilled artisan.[79] This mode of action fitted well with I.K. Brunel's increasing assumption of authority at all levels of his projects. The *Great Eastern* saw him attempting to exert supreme control over design and execution.[80]

With the GWR, Brunel also worked according to 'system' in design, construction and running. Thus he told Daniel Gooch in January 1841: 'I feel that an extraordinary degree of *system* and *method* will alone enable us to succeed.' Speaking even of webs and dropped stitches,[81] Brunel carefully designed every fragment of his railway to link coherently with every other. He began by exerting strong design control over individual components of the GWR, conceived as an 'ideal railway' to challenge the 'model' of the

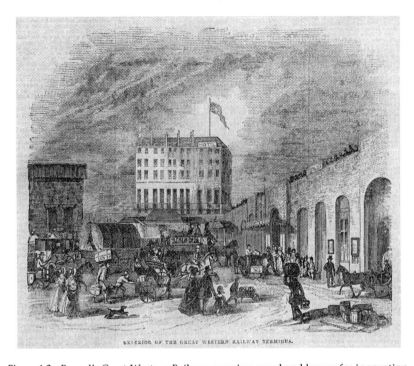

Figure 4.2   Brunel's Great Western Railway promises speed and luxury for innovation-hungry travellers.

Source: *Illustrated London News* 3 (1843), p. 52. (Courtesy of Aberdeen University Library.)

L&MR.[82] Those components, understood as simultaneously mechanical, social, economic and political, included the permanent way (in construction and gauge), the locomotives, the decoration and furnishing of the carriages, the stations, the bridges, tunnels, cuttings and embankments.

Out of these individual parts, Brunel envisioned the forging of a greater whole, the main London to Bristol line functioning as an integrated system but focused on the 'Western' city of Bristol rather than the metropolis to the east. Bristol's merchants and city fathers prided themselves on their city's culture of 'improvement', a culture within which Brunel was already hard at work.[83] The completion of the GWR between this vibrant port and the capital would therefore be seen as a major cultural 'improvement', indeed aggrandisement, for the genteel citizens of Bristol. By June 1841 the GWR was making its mark on space both architecturally and physically, and providing luxurious, reliable and swift communication for well-to-do passengers moving between Bristol and London.

Looking beyond a railway system, however, Brunel wanted a co-ordinated transport system of which the railways were one, but only one, strategic component. The most astonishing illustration of this still-grander design was the oak-built *Great Western* steamship. Proposed by Brunel to the Directors of the Great Western Railway in October 1835, launched in July 1837, and tried, in a dramatic and much-publicised competition with the Irish packet boat *Sirius*, in March and April 1838,[84] the *Great Western* would depart for America from the port of Bristol, conceived as the terminus of the GWR and as a westerly gateway to London. From the perspective of the English capital, the ship became a systematic extension westwards of the London–Bristol railway, across the Atlantic, to the commercial capital of New York and to a 'New World' which was soon to be accessible, not just by steamboats but by a distinctive American system of railroads.

The Bristol–London line was, also, effectively a 'trunk' railway with branch lines emerging from it – or feeding it. With major extensions to the west (into Wales) and south-west (towards Cornwall), the GWR conquered new territory and imposed its order on the environment with such visually stunning icons as the Saltash Bridge linking Devon and Cornwall. Above all, however, Brunel gave to the GWR system a distinctive 'culture' in its broad gauge, in its ambitious engineering, in its prestigious architectural features (notably Paddington and Bristol Temple Meads stations), and in its ethos of luxury and speed.

Although Brunel's system-building can be seen as an emphatic realisation of his 'ideal' railway, the experimental nature of his railway practices brought with it difficulties. By no means all of his innovations proved successful. Moreover, Brunel was not the only system builder. The L&MR, although not the only 'model' for railway practice, had in many aspects been replicated and extended to create a rival system which would eventually threaten the territory of the GWR and challenge its distinctive culture. Thus, by the mid-1840s the private systems of railways, associated with the various companies or allied groups of companies, were in competition for space.

The push for 'progress in mechanical science', fuelled by seemingly limitless social, economic and moral gains, saw a relaxation of investors' caution. A second surge of railway construction began in 1843 and would see Britain's railways grow from 2000 miles to 5000 miles by 1849. Railway shares seemed like a safe bet even for the clergy, whom *The Times* accused of 'forsaking scripture for scrip'.[85] The headlong rush to connect every city certainly worried one commentator who described how projectors now 'shut their eyes with a map of England before them, with the determination that whatever two points on the map their pencil should touch first should be selected'.[86] Writing in 1845, Brunel's verdict was brutal: 'Everybody around seems mad – stark staring wildly mad – the only course for a sane man is to get out of the way and keep quiet.'[87] This 'mania' matched the Victorian 'sensation' surrounding the anonymous account of cosmological and human development known as *Vestiges* (1844).[88]

With a free flow of capital, with a vast field of opportunity and the conse-quent relaxation of constraint on engineers' imaginations, 'sensational' public 'failures' were as widely publicised as 'successes'. Brunel's failures were numerous. Even his most faithful supporters believed that his prom-ises as a locomotive designer were not fulfilled. The giant locomotives he commissioned for the GWR functioned erratically.[89] His assistant Daniel Gooch stepped in with the *Firefly* (1840) and with the 4-2-2 locomotives, with massive 8 ft driving wheels (from 1846) which repaired the GWR's reputation as a rapid and stable transport system (although Brunel tended still to assume 'ownership' of, if not direct 'credit' for, Gooch's work).[90]

Furthermore, Brunel was involved in one of the most notorious cases of railway failure, the promotion of a scheme, the 'atmospheric system', described even by his closest ally Gooch as 'certainly the greatest blunder that has been made in railways'.[91] Engineers had routinely questioned the neces-sity of expending power needlessly in transporting a locomotive in addition to freight wagons or passenger-carrying carriages, when carriages alone could perhaps be propelled along the permanent way by other means. For example, Glasgow's engineering professor Lewis Gordon promoted 'the advantages of . . . locomotive carriages instead of the present expensive system of steam tugs'.[92] Engineers already employed systems whereby stationary engines pulled carriages at a distance using strong ropes; why not, then employ compressed air – indeed a 'rope of air'.[93]

The atmospheric railway, patented by gas engineer Samuel Clegg in 1838, was to be powered using a special tube running between the tracks and sucking the carriage along. Stationary pumping engines at regular intervals provided the power.[94] Brewster, who commented on its 'many advantages', recorded in 1849 that the system had been in use on the Dalkey to Kings-town line 'without any accident' from October 1843.[95] Eminent engineers had queued up to endorse an exciting new system – amongst them William Cubitt, Vignoles and indeed Brunel himself, who used it on the Exeter, Teignmouth and Newton Abbot line of the South Devon Railway. But Brunel lost money with the rapid demise of the scheme. He lost credit also, in an age where personal reputation was vital to garner funding for engin-eering projects.

The battle of the gauges

Through the 1830s, the growing power of the railway came from the con-struction and systematic arrangement of ever more lines. The London and Birmingham Railway, for example, was of immense strategic importance: 'that grand trunk', according to Booth, 'which will unite the north and the south, and bring into closer communication the Capitals of England, Scotland, and Ireland'.[96] Where the Stockton and Darlington line acted as a channel for fuel, and the L&MR offered a single trade and passenger route between a manufacturing centre and a major port, a London and Birmingham railway forged an iron chain between the nation's capital and Midlands' industry. Immediately Birmingham became a node for an emerging rail network centred at London. The line from Liverpool to Birmingham, built by Joseph Locke, who had been resident engineer on the L&MR, was a 'Grand Junction', completing the communication from the north-west port to the southern metropolis. A surge of railway building in the two years from 1835 saw no fewer than 88 new companies absorbing capital of no less than £70,000,000.[97] Engineers continued to push north into industrial centres like Sheffield. Lines to the east, south and west of London staked out the framework for total coverage.[98]

With the establishment of key 'models' of rail, chaos was avoided. Although there were thousands involved in making the railways, a relatively small group of consulting engineers designed and contracted out much of the work. Engineers with reputations for timely delivery and a high ratio of miles per pound accumulated new projects. An oligarchy of railway engineers, including Robert Stephenson, Locke and Brunel, abetted by wealthy contractors, like Thomas Brassey, ensured functional consistency from one railway to the next, rationalising production whilst also being aware of 'improvements' in practices and practical arts, learned from experience.[99] In principle, Brunel provided one model which was replicated throughout the south-west of England and might have been extended to all parts of the kingdom (and even the Empire). In practice, however, networks of lines, operated by other railway companies had already occupied much of the territory to the north and east of Great Britain; they brought with them a challenge to the territorial ambitions of the GWR.

Brunel had been highly successful in identifying the GWR with its distinctive track, so much so that one late nineteenth-century commentator even dubbed the history of this railway *The Story of the Broad Gauge*.[100] The engineer had argued that a 7 ft gauge for the track of the GWR would bring superior comfort through steadiness, luxury suited to high-class passengers, greater speed, and, at a more mundane level, more convenient access to the engine. In the mid-1830s, shareholders seemed convinced that those characteristics would augment business so much that any extra expense in land purchase, construction or maintenance would be worthwhile.[101] Yet, as Brunel's ardent supporter Babbage, himself a shareholder in the GWR, admitted, as soon as the line opened there were 'violent party movements for and against' Brunel's so-called broad gauge.[102]

The chief opponents advocated the 4 ft 8½ in. gauge, chosen (it was suggested) to conform to carriages on the common roads, used in the 'Mineral Districts' before passenger railways, and then favoured by George and later Robert Stephenson.[103] This was the system which vied most convincingly for the title of the 'standard gauge': originally the Passenger Railway Acts made the adoption of this gauge imperative but a relaxation of that rule allowed for Brunel's 7 ft gauge and indeed many other choices. With many gauges in use, and with the one deployed most extensively having been chosen, apparently, on grounds of tradition and convenience rather than being rooted in philosophical argument, it was understandable that as late as 1864 Babbage grumbled: 'The question of the *best* gauge for a system of railways is yet undecided.'[104]

When Babbage asserted in the same year that the 'question of the gauges has long been settled',[105] he could do this consistently by understanding that there were, in fact, at least four questions to be answered, intellectually and pragmatically, in the 1830s and 1840s. What would be the '*best* gauge' hypothetically in the absence of tradition, convention and any existing system? More realistically, which gauge would be 'best' assuming there was no existing system, but relying upon the experience learnt from working the existing railways? Should government, rather than the free market, exert its power and impose a single uniform gauge for future railways (in Britain and even across the Empire), compelling the conversion of existing lines to that standard? And what, if the decision were taken to exert that power in practice, would be the 'best' imposition to promote the national interest?

Babbage sought to answer the second, at least, of these important questions in 1838 and 1839 as Brunel's system came under sustained threat from GWR shareholders. After his witnessing of the L&MR opening, Babbage had fashioned himself into an expert on railway economy. By 1838, he claimed, the 'question of the gauges' had become a matter of 'considerable public importance'. In advance of the Newcastle meeting of the BAAS, the Association's Council worried privately about an impending 'battle of the gauges', especially when there were other disputes brewing about the 'question of steam navigation to America'. In Newcastle, of all places, George Stephenson was the obvious choice to chair the now-thriving Mechanical Science Section; but this chief proponent of the narrow gauge declined to be sectional president, perhaps to avoid open attack and perhaps to avoid a venue in which he might be expected to argue against Brunel. Neither unseemly option guaranteed commercial advantage and both promised social disadvantage. Reluctantly, Babbage himself, no stranger to controversy or polemic, agreed to oversee 'these dreaded discussions'. Having orchestrated a civilised debate, at least, when the much-maligned Lardner spoke on steam navigation, Babbage rejoiced: 'I have saved the British Association from a scandal.'[106]

But at another of those British Association champagne dinners, Babbage decided to determine Stephenson's 'real opinion of the gauge'. Supposing there were no railways, yet Stephenson was 'in full possession of that large amount of knowledge' he had derived from experience, suppose he was 'consulted with respect to the gauge of a system of railways about to be inaugurated'. Would he advise 4 ft 8½ in.? Babbage recalled the reply. ' "Not exactly that gauge," pronounced the creator of railroads;

"I would take a few inches more, but a very few." ' Babbage wickedly compared Stephenson to a woman who, having had a child out of wedlock, urged as extenuating circumstances, 'that it was a very small one'.[107]

When Brunel himself asked Babbage for his own opinion on the gauge question he was nimbly siphoning authority from a man celebrated as the author of the *Economy of Machinery and Manufactures* (1832). Yet Babbage and Brunel had also known each other for almost a decade by the autumn of 1838, when they collaborated to persuade the Directors of the GWR to allow the economist of machinery the access and facilities needed to generate experimental data on the various railway systems. Babbage gave Brunel information ('causes') which helped him in his 'purpose of obtaining you the means of making your experiments'; Brunel spoke to the Directors who were 'happy to comply with your request and will give directions that you may be enabled to accompany the regular trains for the purpose of making the experiments you wish'.[108] The Directors lent Babbage assistants and locomotives (including the *Atlas*). Babbage equipped a second-class carriage with measuring devices, transforming it into 'my experimental carriage'. He derided the quality of other experiments and instruments, whilst his own registered (without human agency) the force of traction, the shake of the carriage and the passage of time. Five months, £300 and a personal tour of most of the railways in Britain later, Babbage concluded, unsurprisingly, that 'the broad gauge was most convenient and safest for the public'. So remarkable were Babbage's persuasive powers – supported by experiment and data generated by self-acting instruments – that he claimed to have convinced floating voters from Liverpool, swinging the Directors back in favour of Brunel's system, and thus allowed a stay of execution for the broad gauge.[109]

Babbage's case for the convenience and safety of the broad gauge might well have convinced those who saw the GWR as an isolated line, or set of lines, temporarily or permanently unconnected with lines elsewhere. That is, Babbage's analysis of the gauge question was plausible and useful to those who accepted the claims Brunel made in December 1838. Then, he told the Great Western Railway's shareholders that he was breaking ground in an altogether new district, that the GWR would have 'no connexion with any other of the main lines (northwards)', and that he expected 'no possibility of a connexion with any other line'.[110] The GWR might be a system in itself; but it was not destined to be integrated with the railways in other districts. Here, it seems, Brunel's desire to make an ideal railway clashed with his personal ambitions for a national system. To contribute to such a national system, Brunel might conceivably have copied the form of the existing L&MR, laying down a line in preparation for a subsequent 'junction' with it; but the form of that *Grand British Experimental Railway* had *not* been ideal. Brunel had opted for a very different, and in his view superior, pattern of railway practice which might, by virtue of that superiority, ultimately displace lines of an inferior pattern.

The 'isolation' defence of the broad gauge proved a hostage to fortune. In 1845, when the GWR already dominated the south-west and south Wales, Brunel himself had tried to force a connection between his system and the rival lines to the north and east; by then commentators spoke routinely of the *national* railway system and it was impossible to see the GWR as operating in a permanently isolated district. The variety of gauges was a fact with lines having opted for 4 ft 8½ in., 5 ft and 5 ft 6 in. – in addition to Brunel's broad gauge.[111] But the growing number of junctions of railways of different gauges brought 'inconvenience, confusion, risk, loss of time, extra labor, and as a consequence, additional expense'. With a single gauge, one contemporary commentator promised, 'passengers with their luggage, gentlemen's carriages and horses, cattle, merchandise, and minerals would pass uninterruptedly throughout their journey'. That stark imagined contrast heightened debates, not over the 'best' gauge considered in isolation, but over the possibility and mechanism of creating a *uniformity* of gauge in practice when it looked impossible that there would be a spontaneous conversion by any of the rival companies to create such uniformity.[112]

Advocates and interested parties constructed a huge range of arguments pro and con. Brunel had stated the (local) advantages of the broad gauge to the GWR projectors and Directors; with Babbage's help he had defended it in 1838 and 1839 as a system independent of the other lines. But in 1845, with increasing experience of actual railways, commentators rejected the inherent superiority of Brunel's system: on the narrow gauge, trains were as fast and as steady; their engines were just as accessible. Yet it was more expensive to make and maintain the broad gauge permanent way, especially with longitudinal sleepers made of heavy timber baulks; its heavier engines were (allegedly) expensive and unreliable.[113] Overall, neither the system of narrow gauge nor the system of broad gauge was self-evidently or inherently advantageous – and, according to one author in 1845, seeking compromise and moderation in debate, even Brunel was not claiming the broad gauge was 'preferable to all others'. Engineers beginning afresh but with the experience gained in the last fifteen years might be rather more cautious – and reach altogether different solutions.[114] But the questions of whether a single gauge should be imposed, and if so which, remained.

With no clear technical choice available, and with many interested parties active, debates were protracted; many and varied rhetorical techniques came into play to secure victory. Brunel had accrued a vast stock of credit by engineering works and by well-publicised adventures. Tales of Brunel's audacious acts – and escapes – abounded: cheating death when the Thames Tunnel flooded; dangling 200 ft above the Avon in a basket suspended from a wrought iron bar during the building of the Clifton Suspension Bridge; narrowly – and through his own ingenuity – escaping death by asphyxiation in 1843 from a coin lodged in his windpipe; just prior to the maiden voyage of the *Great Western*, almost killed by fire *and* a subsequent fall (broken, thankfully, by the managing director standing below). These tales of endurance, leadership and providential escape charmed doubting publics and sceptical investors.[115]

Yet although a household name, Brunel had a reputation that was vulnerable to attack. His lack of foresight in 1838, repeatedly emphasised by critics, was just one of the factors which seriously eroded his ability to argue authoritatively.[116] Abandoning the gentlemanly decorum Babbage had attempted to maintain in the BAAS, and which was vital to sustain the disinterested 'philosophical' stance of the men of science, Brunel's opponents embarked on a vigorous and sometimes brutal pamphlet war. They dwelt on his failure to substantiate explicit technological promises: he had claimed there would be no connection between GWR and with other lines but 'the projector of the Broad Gauge' had been proved totally wrong;[117] he had argued that the GWR gauge was exclusively for flat and straight lines but had himself applied it to steep and sharply curved ones. Why trust a man who prided himself on rational action, they implied, if he was so inconsistent as to begin, after all this, to advocate an 'Atmospheric system... in all essential points... the exact converse of the Broad Gauge Locomotive System.'[118]

Equally erosive of his credibility was the opprobrium showered upon him for his (alleged) inexperience, his self-indulgent experimenting, his arrogant independence and self-generated image of 'genius', and his failure to recruit substantial allies in the community of railway engineers. '£.s.d' claimed that 'Mr Brunel has learnt to shave on the chin of the Great Western Proprietors.'[119] Henry Cole dubbed Brunel a 'railway eccentric' and quoted Bishop Hall's 'How worthy are they to smart, that marre the harmony of our peace, by the discordous jars of their new paradoxall conceits.'[120] Brunel's expensive and grandiose novelties looked bloated and gaudy against the economical and trustworthy products of a Stephenson or a Locke. These were men of engineering personas altogether humbler, more conventional, and, crucially, more 'safe'. Brunel's dangerous individualism clashed with collective action and worked to his disadvantage: only the 'eccentric genius' Brunel defended the broad gauge against 'at least *seven* first-rate safe engineers'. As for Daniel Gooch's loyal support of this errant knight of engineering: Brunel had 'not been able to find a SINGLE independent *railway* engineer to back his eccentricity, for we cannot record Mr D. Gooch of the Great Western as but Sancho to Don Quixote'.[121]

The eventual winners in disputes such as this cannot simply be understood as those with 'the best' technology according to universally agreed criteria. When the GWR ran out of virgin territory and found itself competing, on a small island, with larger rivals possessing greater territory, that competition was undoubtedly difficult. Any 'broad' track, complete with tunnels and bridges, could, with modification, house a narrow one – and the reverse was not always, or easily, the case. But with no criteria of debate accepted by all parties, and with the credibility of individuals as much as systems under attack, the final victory of the narrow gauge cannot be seen as merely the result of inevitable rational, or practical, choice. Contexts of debate included the interventions of men of science like Babbage, the rise and fall of Brunel's alternately bolstered and buffeted credibility, and the words and deeds of the advocates of multiple competing forms of railway practice (each with its own momentum, frozen in material objects and established in cultural practices). There were also 'interests' of equal power in industry and government. Only a heavy political instrument, in the form of the Gauge Act of 1846, settled the issue and solidified in law the final answer to Babbage's questions: all new lines must conform to the narrow gauge.[122]

Babbage had regarded the question of the gauges open in the 1860s; and in the 1870s, the practical outcome of the gauge-battle might still stand as an issue of national identity and of technological style. Robert F. Fairlie, for example, inventor of the Fairlie locomotive, argued in 1872 from the evidence of lines in Norway, Russia, India, Canada, Australia, New Zealand, the United States, Mexico, Peru and France, to the construction of the 'Railways of the Future' on a narrow gauge – but not the narrow gauge of George Stephenson. Fairlie wanted lines set out at 3 ft 6 in. He believed – as Babbage suggested – that it was action, not argument, which sustained the now-standard English gauge. In the United States 'opinion finds free vent' and the problem of railway progress had been earlier solved (by instituting narrow gauge lines) than if left 'in the hands of the more prejudiced and less progressive engineers' of Great Britain. Fairlie was worried about the railways of empire. Echoing Brunel's opponents, his claim was that lines built to a 'broad gauge' (of 4 ft 8½ in.) were, in India 'too magnificent' and hence too few, and in Australia part of a 'ruinous system with which the country is already burdened'.[123]

Cultural constructs: representing railways

Despite the push to standardise railway practice, the railway 'experience', as imagined by nineteenth-century cultural commentators, was not simply one of homogenisation, limitation and constraint.[124] Commentators with varied allegiances and agendas questioned engineering reputations, the trustworthiness of engineering projects, and blanket assertions of the progress of the nation and empire through 'mechanical science'. That these categories were subjected to critique is unsurprising: an early emphasis on 'novelty' and 'experiment' encouraged attention but could do little to consolidate 'confidence' through benign familiarity. To augment the familiar trustworthiness of the railways, interest groups, including gentlemen of science, entrepreneurs and engineers, and cultural commentators participated in the creation of attractive, rich – and safe – railway cultures in diverse forms. These agents dealt with visual and print culture; they imagined – and created – new social spaces, from the leisure resort to the academy. But these 'insider accounts' were not the only ones; rival, if not necessarily hostile, discourses explored alternative representations.

Visual representations of the railway destined for, or originating with, 'insiders' offered images for consumption, elucidation, comfort and familiarity.[125] T.T. Bury, Alfred B. Clayton and Isaac Shaw were only three of the artists who quickly produced saleable 'coloured views', 'views...taken on the spot' and 'views of the most interesting scenery' associated with the L&MR.[126] Fine images of key locomotives, like stallions, in boardrooms signified power, pedigree and progress to directors and financiers; pictures of track and locomotive within bucolic settings displayed the new 'naturalness' of the railway machine in the garden; images of the architecture (Classical, Moorish, Tudor, Gothic or Egyptian) clothing bridges and tunnels, or constituting the great metropolitan termini, squared the new railway machine with Western aesthetics and expressed the confidence, luxury and security of rapid travel; literary portraits of local topography and antiquities associated with railway guides made the railway a route to improving intelligence and, in parallel, an object of inspection.[127]

Certainly, new technologies of observation, representation and recording, especially photography, radically reformed the art of seeing during the nineteenth century; but the railway shows that it was large- as well as small-scale technologies that could be deployed in order to refashion Victorian visual culture. It was the visual literature of the railways, rather than untutored experience, that taught the traveller to 'see' (and describe his or her 'seeing') with a panoramic gaze. Wolfgang Schivelbusch describes this panoramic vision offered to the railway passenger, as the train carved a path rapidly with almost uniform motion through the landscape – and yet distanced from it, and cushioned from its immediate undulations.[128] Print makers silently supported the rhetoric of railways, both as experimental science writ large and as suppliers of critical, panoramic distance from the mess of experience. Bury, for example, emphasised the Euclidean nature of railways, with 'tangents and flat planes driven where surveyors and profit determined, regardless of local topography' in order to represent 'the triumph of hard engineered culture over soft undulating nature'.[129]

Controversial artists such as J.M.W. Turner must have been familiar with Bury's work and with the images of visual railway satirists like George Cruickshank, or cartoonists like John Leech who attacked the railway mania of the mid-1840s with images of locomotives named not *Experiment* or *Novelty* but *Speculation*.[130] Knowing, subverting and perhaps transcending such work, Turner depicted the raw excitement and disruptive power of *Rain, Steam and Speed: the Great Western Railway* (1844)[131] – even as Turner's tireless advocate, John Ruskin, decried the perversion of the practical arts where they denied the expression of human 'life', and where, through railway engineering did cutting violence to nature.[132] First shown at the Royal Academy in London, Turner's disturbing and technically experimental image, of a transport technology which deserved the same adjectives, divided and confused critics. The image was far more than a mere view from his first-class carriage in 1843. A down train sped away from London through the countryside of Maidenhead and over Brunel's notoriously flattened single-arch bridge. Was this a representation of the loss of 'Old England' to steam culture or, conversely, a symbol of human progress challenging – and harnessing – nature's elemental forces? The picture issued an ambivalent message 'about the casualties of progress and the impossibility of not changing'.[133] It cut a modern technological corridor through a previously unspoilt landscape replete with pastoral classical resonance – and thus signalled a culture, literally, in transition.

Travellers quickly rediscovered reading as a mechanism by which to avoid embarrassment in the new social space of the railway carriage. Although railway operators had expressed familiar cultural norms by segregating passengers into classes, sensibilities were such that there still remained a discomforting uncertainty as to the 'character' of those with whom one sat. Railway companies were happy to facilitate the production and sale of specialised literature fitted to ameliorate such predicaments – and comforting to prospective clients. Newspaper sellers, stationers and providers of cheap and popular literature, like W.H. Smith, opened stalls along the routes. A literature of travel guides, like Edward Mogg's for the GWR, took in the historic pleasures of Windsor, Bath and Bristol, and followed the new contours of Britain.[134]

Other literature pedalled a 'new history of England' along the new lines, quite literally, of the London and Birmingham Railway. *Railroadiana* (1838) had been prompted, supposedly, by a recent railway excursion when the author's time in Tring had been prematurely exhausted. New mobility brought new irritations when 'Restless as old Time himself, the driver takes his seat – the auxilium starts and the tourists remain ungratified.' The railway's annihilation of space meant there was too much to see in a day. Few doubted the railways were 'working a mighty change in the relations of different parts of the Empire' and the metropolitan 'man of pleasure' could take amusement in traversing the new routes, especially to see the towns and rural sections destined to be new 'objects of rational attraction'. For such busy hedonists, the 'literary Railroad ... [would] prove not less important than the iron one'; in this new species of tourism and history combined, *Railroadiana* discerned the 'opening of a new arena to "locomotive" minds'.[135] The railway journey provided not only an opportunity to read, but also an opportunity to rewrite the history of Britain.

Technical authors rapidly generated a specialist literature of another kind, designed to inform enthusiasts for, and those with a vested interest in consuming, railway intelligence of the state, and progress, of practice. John Herapath's *Railway Magazine, and Annals of Science*, for example, boasted 'copious accounts of all railways at home and abroad' in the second half of the 1830s – a boast which became increasingly difficult to fulfil, with the mania of 1837. On the continent parallel publications like the Paris-based *Journal des chemins de fers* echoed this trend for the technical literature which bound together a spatially dispersed republic of railway practitioners. Print was an outlet for an aspirant elite of transport philosophers like Nicholas Wood whose *Practical Treatise on Railroads* (1825 and subsequent editions) achieved the canonical status of Thomas Tredgold's definitive manuals of practical art.[136]

Antagonists undermined such attempts to project a unified, progressive and secure image of railway culture when they engaged in pamphlet wars (notably during the 'battle of the gauges') or, more rarely, despatched sonnets to their opponents. During a controversy surrounding the Kendal and Windermere Railway (in the English Lake District), Wordsworth objected to a feverish, even mesmeric, railway ideology obsessed with novelty and progress, in danger of jettisoning a more peaceful and pastoral mode of life.[137]

But the most keenly scoured railway literature was surely that produced by Bradshaw: that company's guides and timetables promoted and then reinforced 'railway time', in an astonishing attempt to standardise, discipline and regulate human behaviour. The historian E.P. Thompson famously described the enforced transition from local, ritualised time – functionally answering the needs of and responding to the contexts of the home, the domestic place of production or the field – to the march of factory time, with labour regulated according to the perceived demands of low-skill, mechanised mass production. Railway barons, linking industrial and urban centres, now wanted a further homogenisation within Britain: downgrading local times, privileging instead a single unified timing system covering the whole of Britain, centralised on Greenwich, and willing to offend the astronomers who would prefer time to follow the passage of the heavens (ch. 1).[138] Irrespective of the needs of parliament or democratic voices, these railway managers envisioned society temporally rationalised by railway track and timetable. In a strikingly concrete sense, time was now the possession – and dispensation – not of local communities and interests but of the railways. As *Bradshaw's Railway Almanack* announced in 1848:

> We...record with pleasure the decree of the railway potentates, that uniformity of time shall be observed in all places of their dominions in Britain; we hope the government or the parliament will at once sanction the decree; but whether they do or not, the railway clock and the telegraph clock will very soon become the public regulator. More especially so, if, has been proposed, the falling [time-]ball at the Greenwich Observatory should be connected with some of the principal telegraphs, and the wondrous messenger shall have tolled the mid-day bell at Edinburgh before the ball that liberated him has reached the few feet of its descent.[139]

The railway timetable emphasised the efficiency of finely regulated work and commerce, but it also enforced a distinction between work and leisure. Although the railway 'model' of L&MR linked commercial and working centres, and the railways reinvented the idea of the human commuter, travelling some distance regularly for business or employment, the GWR passed through fashionable Bath. The railways served elite pleasure spots and opened up, in some cases even creating, leisure resorts for mobile masses. Already accessible by paddle steamers, developers actively invested in coastal and estuarial resorts such as Eastbourne and Margate. The day trip to the British coast became an accepted part of life for the lower orders.[140]

Conversely, there was the practical possibility, again fostered by the railway companies, of the trip 'into town' for shopping or cultural consumption. Railways helped to create a distinction between the urban and the suburban, with the new domestic suburbs encircling the activity and bustle of the commercial and cultural centres.[141] Whether the railway should penetrate the very heart of the city, however, was a contentious issue. When Charles Fox built a line between London's Camden Town and Euston Square, he attempted to preserve order, control access, and limit pollution with trenches and surrounding walls.[142] An industrially progressive city such as Glasgow had fewer reservations. Its medieval College, sited near the Cathedral and heart of the old mercantile town, had already been replaced by a railway goods yard by the 1870s. A new Central Station, constructed in 1901–06 to incorporate the existing Caledonian Hotel as its facade, served to transpose westwards the heart of the booming Edwardian commercial city.[143]

The railway engineers carefully crafted their own institutional places and spaces, designed to embody, in external architecture and internal plan, their professional status and relationships; existing spaces might also be recoded or rejuvenated as passage points, although contested ones, for professional entrance. The growth of the railways required skilled practitioners: machine makers, locomotive builders, engine drivers, and the engineers responsible for creating and maintaining the track. During the second railway mania, the high-profile figurehead for these workers, George Stephenson, emerged as President for the northern-based Institution of Mechanical Engineers (founded 1847), which had broken away from the hitherto unified profession of the civil engineer, represented by the London-based Institution of Civil Engineers and dominated by the older generation of canal engineers.[144]

Alongside this institutional separation, the academies and universities moved to corner at least a part of the market in engineering training. In France there was a rapid response from the schools to breed up students for practical and administrative railway work. August Perdonnet at the Ecole Centrale des Arts et Manufactures added material on railways to his lectures in mining and metallurgy within two years of the opening of the L&MR; by 1837 he was delivering 'the first full course on railroads given anywhere in the world'. As part of their training, students at the Ecole Centrale inspected the Paris–Versailles line of which Perdonnet was engineer in chief. The Ecole des Ponts et Chaussées began a course on locomotive building in 1843.[145]

British colleges and universities had been capitalising on a growing market: student engineers paid for that part of their training which college tutors could reasonably annexe from more traditional forms. The University of Durham, close to Newcastle, equipped young men for the 'academical rank of Civil Engineer' from the late 1830s, with the support, patronage and adjudication of premier British railway philosopher Nicholas Wood; Wood's grander plans for the *separate* training of elite mining, civil and railway engineers, in connection with the University foundered through lack of funds. In 1841 University College London employed Vignoles, a man by then known as the engineer of the Dublin–Kingstown Railway and well connected in Britain, Ireland, Russia and France. Trinity College Dublin (TCD) had a School of Engineering by 1841. Thanks to BAAS figure Humphrey Lloyd, Dublin responded quickly to the new railways. With only a dozen miles of track in Ireland that year, the city terminus was, nevertheless, quite literally at the back gate of the college. The students' professor would be the Irishman John McNeill, trained under Telford, with links to the engineering communities of Glasgow and London, and to British government. When McNeill left the teaching to his assistants, it was to get back to consulting work on the growing Irish railway system. By 1860 Dublin was exporting significant numbers of engineers abroad, to India and the wider Empire.[146]

The academic engineers, however, worked in a dynamic and sometimes conflictual relationship with practical engineers occupying workshop spaces. Industries associated with the use and manufacture of machine tools (especially lathes and machines for boring, planing, drilling and screw-cutting) seized on the opportunities afforded by locomotive construction. Many of the engineers were associated with Henry Maudslay's Lambeth 'nursery' for mechanical engineers. The Scot James Nasmyth entered as a high-status assistant in 1829 and, having witnessed the opening of the L&MR, set up business in Manchester where he would eventually occupy vast works close to the line. Nasmyth became a multi-millionaire on the basis of locomotive manufacture, assembly-line production of tools and the 'steam hammer' which combined great power with the delicacy, allegedly, to crack an egg in a wine glass. Nasmyth wanted volume sales of reliable rather than exemplary quality both for home consumption and export. 'I do not like', he told Charles Babbage, 'to boil eggs with chronometers'.[147] Joseph Whitworth, on the other hand, although also from Maudslay's school and also working in Manchester, became a byword for precision and for standardised machine parts. Whilst Whitworth championed technical education, the pragmatic Nasmyth rejected the efforts of the academic engineers, saying: 'I have no faith in Royal "Kademys" of Music nor Colleges of Technical Education with Suitered

Professors &c &c &c Workshops are the true and only Colleges for such practical Education.'[148]

'Child of steam': the New World empires of railroad power

After mid-century, and especially after 1860, railroads expanded almost exponentially on continent-wide and empire-wide bases.[149] As one of many lines projected in the wake of the launch of the Stockton and Darlington Railway, the Baltimore & Ohio Line, opened in 1830, has come to represent the beginning of rail in the United States. Where British origin stories gravitated around the *Rocket* locomotive and the Rainhill Trials of October 1829, the Americans had their *Tom Thumb* and an alleged competition, taking place in August 1830 just *prior* to the opening of the L&MR, between this experimental locomotive and a horse. In the United States the total length of track increased from 23 miles in 1830 to approximately 3000 by 1840, after a moderate mania of building, paralleling the British case. This grew to over 30,000 miles in 1860 and to 200,000 miles by 1902. United States railroads reached an all-time peak of over 250,000 miles in 1916.[150] Writing privately to his brother Henry around 1910, the American social critic Brooks Adams conveyed the profound and panoramic image of the vast powers of nature harnessed by the modern railroads of his own nation:

> Beginning on the crest of the Rockies the tide flows down into the Mississippi valley, and then across to the eastern mountains in an ever increasing flood, with ever heightening velocity. At last you come to the lakes and Buffalo. There, I take it modern civilization reaches its focus. . . . for miles and miles we tore along with a huge train of Pullmans [the luxurious coaches carrying the hall-mark of their designer and builder George Pullman], at a rate of a mile in fifty seconds, passing lines of endless freight trains, all going fast, all headed to the Hudson, all moved by a power such as exists no where else on earth. And next year these huge engines will be antiquated, and we shall have larger engines, heavier cars, higher speed, and more mass. No one who has ever watched that torrent from its source on the divide to its discharge in New York Bay can, I think, help feeling that the hour of the old world has struck.[151]

Adams's remarks that 'next year these huge engines will be antiquated' captured something of the nineteenth- and early twentieth-century American faith in progress through invention.[152] Nineteenth-century American railroads prided themselves on an immense variety of inventions. Some, like the rail-car ferry, made possible extensions of the railroad system across major rivers such as the Susquehanna (1836) or across lakes and seas where the distance was shortened in comparison with the alternative land route. Others, like Eli H. Janney, lodged patents for automatic car couplers (1868 and 1873) which promised a saving of labour and increased speed of handling. Still others, like sleeping car patents (1856) and dining cars (1863), advertised passenger comfort over the long distances between towns and cities. Indeed, the thirst of diverse American publics for innovation arguably contributed to the relative decline of railroad systems in the mid-twentieth century: perceived as antiquated and no longer symbolic of national progress, railroads seemed to many Americans to have been superseded by the automobile and the jet aircraft, at least for passengers.[153]

At the beginning of the twentieth century, however, American railroad managers impressed the civilised world. 'What the American railroad manager does not know about advertising methods is scarcely worth knowing', wrote Archibald Williams, populariser of 'modern' invention, in his *Romance of Modern Locomotion* (1904).[154] 'If the trains are well fitted up people are more likely to travel through the country they serve; if people travel, they see what opportunities are awaiting the settler; if they settle they bring more grist to the company's mill in the shape of extra freight.' And it was freight, not passenger, receipts that brought profits to the American railroads. Special, but scarcely economic, luxuries prevailed at the turn of the century on the well-named 'Twentieth Century Limited' and the 'Pennsylvania Special' 'which make the 912 mile run from New York to Chicago in twenty hours':

> On such trains you can command a bathroom and a barber's shop. If time hangs heavily, just step into the library car, or take a stroll into the observation car. If blessed with plenty of money, you pay something extra and hire your own drawing room. For the business man's convenience in this 'hustling country', a typist and shorthand writer is part of the staff whose services are at the command of the passengers. While you whirl along you are kept in touch with the latest prices ruling on 'Change and transact your affairs as easily as in the office.

For women too the luxuries included new and highly fashionable 'electric contrivances', including those 'on which they can heat their curling-tongs' while every passenger could benefit from the provision of state-of-the-art 'electric lights...so arranged that one will be directly over the passenger's shoulder whether he is sitting in a corner seat against the window or reclining in his berth'.[155]

For turn-of-the-century America, the railroads simultaneously supported material and symbolic *power*, expressed through ever-increasing speed and ever-greater luxury. Not everyone, of course, shared in the mania for acceleration. Half a century earlier, British evangelicals such as the Free Church of Scotland geologist Hugh Miller, one-time crofter from Cromarty, felt that it was 'as if the locomotive and the railroad had been introduced into every department of human affairs, – as if the amount of change which sufficed in the past scheme of Providence for whole centuries had come to be compressed, under a different economy, within the limits of less than half a lifetime.... One seems almost justified in holding that the great machine of society is on the eve of being precipitated on some all important crisis, and that the rapidity with which the wheels revolve marks the sudden abruptness of the descent.'[156] This Victorian controversialist, then, read railroads as an indication that humankind, seduced by material wealth and power, was heading evermore rapidly towards the last days prophesied in the scriptures.

Not only evangelicals, however, read railroads as a symptom of flawed and fallen humanity, now on the edge of catastrophe. As railroads correlated with national and imperial power, the French naturalist novelist Emile Zola used the railway system, centred on Paris, as the site for a savage critique of the hypocrisy of France's Second Empire on the eve of the Franco-Prussian War of 1870. Ostensibly a major symbol, alongside the Suez Canal and the rebuilding of Paris, of the rationality, order and prestige of the French state, the railway system in Zola's *roman expérimentale* (experimental novel) represented disorder and corruption in high places. *La bête humaine* (1890) in fact deployed physiological insights to explain the criminal nature of human beings and to break down distinctions between animal and machine systems. Zola therefore replaced a theology of original sin with laboratory science to reveal to his readers the thinly veiled flaws in human beings and their creations, flaws which threatened to destroy the rhetorics and myths of human advancement through technology so loved by France's imperial masters.[157]

This naturalistic re-invention of sin found a parallel expression in the writings of Henry Adams, brother of Brooks and one-time Harvard professor of medieval history. Conscious of his and his father's heritage in New England Puritanism, with its scepticism towards human perfectibility, Adams read the history of steam power as an acceleration of both size and speed – and of concomitant concentrations of wealth and pride. Working together from the 1890s, Henry and Brooks Adams inferred that energy, especially steam power made concrete in railroad and steamship systems, equated to the economic and political power of modern nations. As Miller had done, Henry increasingly speculated in the 1900s that this progression was moving towards catastrophe and collapse.[158] It was a secular view that owed much to a life that coincided with the rise and growth of American railroads and, more specifically, with the career of another brother, Charles Francis Adams. As we shall see, Charles Francis's critical assessments of railroad power eventually led him, not in the footsteps of his grandfather or great-grandfather as United States President, but to the presidency of the first transcontinental railroad, the Union Pacific.

If French railroads had become, from the 1860s, an expression of state power, America's railroads became, after the Civil War (1861–65), the representation of the uncontrolled power of corporate capitalism driven by tycoons such as Jay Gould, Jim Fiske and Cornelius Vanderbilt.[159] In the late 1860s, critics of these builders of private empires began to surface. Charles Francis Adams, Harvard-educated and recently demobilised from the Civil War, quickly constructed for himself the role of railroad analyst. All the Adamses felt it their patrician and gentlemanly duty to uphold Enlightenment values of rationality and independent judgement. Conversely, they saw themselves as enemies of the monopoly capitalist power, financial speculation and political corruption of the new railroad 'robber barons'.

'On this continent', wrote Charles Francis in the *North American Review* (*NAR*) in 1867, 'our own country is the child of steam.... The steamboat and the railroad alone have rendered existing America possible'.[160] This article, entitled 'The railroad system', initiated a series of controversial contributions to the *NAR* on the subject of New World railroad culture. Although the *NAR* readership was small and elite – and in decline – the venerable periodical still carried authority and its reviews carried weight with editors of far more downmarket periodicals and newspapers.[161] It was from this platform that Adams launched a profound critique of the earlier promises made for the railroad system, a nuanced evaluation of its current state, and a manifesto for radical reform.

The issue at the heart of Adams's subject was the dramatically increasing power of the railroad corporations. Significantly, he chose for historical precedent not the territorial might of monarchy or aristocracy, but the temporal power of the Roman Catholic Church. Like the Popes of old, the 'ruling spirits' of the railroad corporations were not 'born to the purple... they are, usually, men who force their way up by native ability, and who rule only so long as they succeed'. As yet, the railroad corporations had not concentrated all power 'in the hands of a single individual', but since the mid-1860s there were indeed signs that different railroads were replacing competition with co-operation such that in thirty years time 'there will then exist a combination of forces, material, social, and political, united under one head, which may bear comparison even with the Church of Rome'.[162]

Adams's choice of historical precedent was doubly significant. On the one hand, the analogy with the Church of Rome highlighted his conviction that the 'material, the moral, and the political interests involved are found to be inseparably connected, and the one leads to the other, in some unexpected way, no matter from what side this railroad problem is approached'. On the other hand, the choice of analogy reflected the Adamses' inheritance of a New England Puritan culture, deeply committed to democratic and republican principles and profoundly hostile to, and suspicious of, religious and secular powers vested in institutions and organisations beyond the control of the community, state, or federal nation. As with many Bostonians, especially with Harvard links, the family had largely abandoned Calvinist religion (characteristic of the old New England Puritanism) and attended Unitarian services with their emphasis on the 'God-like in the human'. Henry and Brooks were severely sceptical of this optimistic and benevolent creed.[163]

'The railroad system' examined, from a contemporary perspective, the character of railroad power. Adams knew as well as his readers the kinds of claims made on behalf of the railroads by advocates like Booth, economists like Babbage, proponents of mechanical science like BAAS managers and a host of popular critics since the inception of railroads four decades earlier. Presenting himself as a more subtle and disinterested critic, Adams began:

> [The community] must disabuse itself of... the idea that the railroad system is nothing more than a dividend-realizing monopoly of certain great joint-stock companies.... its mere money value could not be represented in figures. The application of steam to locomotion is vulgarly [popularly] looked upon as an improvement, an advance of civilization, a great result of science, a fine investment of capital, a wonderful improver of the value of corner lots, a great time-saver, an indispensable agent for the development of new country; but it may be questioned if it is often viewed in its true magnitude, as, with perhaps two exceptions, the most tremendous and far-reaching engine of social revolution which has ever either blessed or cursed the earth.[164]

Writing with a style that resonated with biblical language, Adams wanted his (mainly New England) readers to understand that this 'engine' had recently effected some dramatic cultural transformations. Steam locomotion created a new phase of colonisation as Europeans left the Old World in search of something better. Whereas many colonies had died in infancy on account of geographical isolation, in the late 1840s whole populations rushed to the new goldfields 'and forthwith steam became their servant, and bound them closely with the older world'. Where yesterday a wilderness existed, now 'California and Australia took their places among the communities of the globe.' The 'ringing grooves of the railroad' had begun to replace a 'long, wearisome, and dangerous wagon-road, scarcely marked out across the plains' connecting 'a nomadic population of semi-barbarous, undomesticated men with a distant civilization'. Moreover, extension of 'the new process of development to Mexico and South America can only be a question of time' while 'Africa can only be accounted among the possibilities of the future.'[165]

Adams pointed out the dramatic changes, governed by universal laws of a Newtonian character, wrought by locomotion in older societies: 'From the most powerful of European monarchies to the most insignificant of New England villages, the revolution has been all-pervading.' It was the primary goal of railway cultures (paralleling telegraphic goals (ch. 5)) to promote national unity and to degrade local diversity. Thus the 'tendency of steam has universally been towards the gravitation of the parts to the centre...

Increased communication, increased activity, and increased facilities of trade destroy local interests, local dialects, and local jealousies'. After a long history of division in Italy, for example, 'every mile of completed railroad takes for Italian unity a new bond of fate, – banishes a little more of local jealousy, local interest, and local dialect. The nation sensibly gravitates.' In the Austro-Hungarian Empire, 'long made up of discordant elements, which for centuries have been retained under one head by a skilfully contrived and artificially stimulated antagonism and jealousy of forces', the 'revolutions of the steam-engine have at last rendered forever impracticable the traditional policy of the house of Hapsburg'. On the other hand, Russia possessed a 'vast, incongruous, overgrown empire' whose permanent unity appeared impossible. Yet with railroads in course of construction all over the country, 'the whole great nation will be bound together hard and fast by the iron bands'.[166]

As in peace, so in war did the railway transform the course of recent history. In the Crimean War against the Ottoman Empire and her European allies (France and Britain), Russia failed 'because she could not avail herself of the steam-engine; the Allies succeeded, because they could avail themselves of the steam-ship.... Could Russia have concentrated men and munitions with the ease and rapidity of France and England, the war must have had another close.' More recently still, during the American Civil War, European military authorities had 'told us of the vastness of the territory to be subdued, of the impossibility of sustaining our armies, of the power of a people acting on the defensive'. They were to be disabused of this view by the power of the railroad:

> The result the world knows. It saw a powerful enemy's very existence depending on a frail thread of railroad iron, with the effectual destruction of which perished all hope of resistance; it saw [General] Sherman's three hundred miles of rear, and the base and supplies of eighty thousand fighting men in security three whole days' journey by rail away from the sound of strife; it saw two whole army corps, numbering eighteen thousand men, moved, with all their munitions and a portion of their artillery, thirteen hundred miles round the circumference of a vast theatre of war, from Virginia to Tennessee, in the moment of danger, and this too in the apparently incredibly brief space of only seven days.[167]

Adams also argued that this new culture of steam locomotion was rendering nations more cosmopolitan and uniform through processes analogous to the mechanisation of print for a mass readership. In Europe, London had lost all its old quaintness of the stagecoach era. Indeed, 'London is, in all essentials but size, like Boston; Paris, like New York. Paris and London...are giving up their distinctive characteristics, to become the stereotyped railroad centres of the future.' And although Rome, 'thanks to the Papacy, has resisted the revolution a little longer... even there the shrill scream of the steam-whistle breaks the silence of the Campagna'. Close to home in New England, the 'revolutions of these few years have swept away the last vestiges of colonial thoughts and persons'. Gone were the ruling country gentry, abolished as a class together with dress, manners and houses by the railroads: 'The race of hereditary gentry has gone forever, and the race of hereditary business-men has usurped its place. Shrewd, anxious, eager, over-worked, the men of to-day will accomplish vast results, and immensely accelerate the development of the race.'[168]

The same phenomena, Adams noted in a language strongly redolent of the central planks of Newtonian philosophy in dynamics, astronomy and optics, were to be witnessed in the cultures of information where steam power 'exerts its influence with a speed and force equally accelerated'. In particular, the 'newspaper press is the great engine of modern education; and that press, obeying the laws of gravitation, is everywhere centralized, – the rays of light once scattered are concentrated into one all-powerful focus'. Thus 'To-day's metropolitan newspaper, printed by a steam-press, is whirled three hundred miles away by a steam-engine before the day's last evening edition is in the hands of the carrier.' As a consequence the metropolitan press was driving the local press out of circulation. Above all, this increase in communication from centres was generating numerous new communities with no faith in, or reverence for, the old. Thus it was now the 'old' that needed to justify itself, altering the question from that of 'Is the proposed innovation an improvement?' to that of 'Is the existing condition certainly better than that proposed?'[169]

For Adams, the locomotion revolution manifested itself most conspicuously – and bewilderingly – in the domain of trade:

> Increased communication leads to increased activity. Prices seek a level; produce is exchanged; labor goes where it is needed. England and Russia exchange bread for cotton, and Iowa and Ireland, labor for corn.... Increased activity demands new centres and channels, and these phenomena result, – those overgrown, dropsical giants of a growth that is just begun, – those portentous accumulations of the evil humors of society which men call railroad centres, and the apparition of which on the face of the earth is confounding and puzzling all thinking men.[170]

The vast increases in urban populations in European cities, Adams reflected, 'are restrained by tradition, by ignorance of their own power, and by immense exertion of force'. But in America 'these uncontrollable masses of humanity seem likely... to make their power felt'. The same steam power that had enabled the crushing of a rebellion and 'to preserve our unity as a nation' appeared simultaneously 'to be steadily sapping the very foundations of our political edifice'. In the hands of New York's populace, for example, 'the principle of self-government has become a confessed failure'. Democracy itself, once the pride and boast of the new and 'enlightened' nation, now seemed threatened.[171]

Steam, then, had 'proved itself to be not only the most obedient of slaves, but likewise the most tyrannical of masters'. Whereas European railway promoters had emphasised the positive benefits for city lives in work, leisure and consumption, Adams offered a far more disconcerting perspective. The railroad had carried to New York 'the wealth of the whole world, and makes of her an overgrown monster'. It had built up San Francisco 'like a very palace of Aladdin'. It had peopled 'Colorado, Montano [sic], and Idaho as if by magic, and transfers the seat of empire to what still appears in the maps which ornament our office walls as a wilderness.' But steam had also made 'grass grow in the once busy streets of small commercial centres, like Nantucket, Salem, and Charleston' and even 'promises to make a solitude of the wharves of Boston'. In short, it threatened irreparable damage to those very communities to which it had promised so much:

> The result of all commercial combination and concentration is necessarily individual inequality and disparity.... In these [railroad centres] we must ultimately look to see wealth enormously increased, with a population proportioned to it, with all its corresponding depths of vice, of misery, and of poverty. The fortunes of the Astors and Stewarts are but precursors. The sharp spasms of misery and poverty are not yet felt.[172]

Revealing both his patrician and puritan distastes for ostentation, Adams highlighted the consequences for morality: 'With wealth comes taste and luxury, and with it also misery and vice. Both are consolidated, both are disseminated together.... It is all the world over a race for wealth.' Unwilling to construct a 'golden age of purity and simplicity' in the past, Adams nevertheless identified the rule that 'the most democratic are the most moral communities'. In communities such as Vermont, for example, 'there are few who are very rich, and still fewer who are very poor'. From the perspective of railroad culture, they were 'good communities to be born in, provided you emigrate early'. But with 'the vast accumulation of wealth, and the consequently vast under-stratum of poverty', improved morality was not a consequence. 'Railroad morals' had become a 'proverbial expression', a 'by-word of opprobrium among man' whereby 'Robbery and gambling, under the euphonious names of "cornering" and "operating", have become pleasant matters of daily occurrence, and excite scarcely a comment.'[173]

In this forensic dissection of the railroad system for the *NAR* in 1867, Adams had already identified the key source of the problem of railroad power: that in 'their eagerness to be successful traders, they [the railroad corporations] forget that they are also trustees'.[174] This emphasis on the importance of 'holding in trust' mirrored biblical notions of stewardship in which the created world, although under man's dominion, was not to be exploited for mere personal gain at the expense of the community and future generations. To resolve, therefore, the antagonism between railroads and communities would mean introducing a principle of large receipts and small profits into railroad management. But the problem was 'not yet ripe for any satisfactory solution, through the absence of any reliable statistical tables'. Adams therefore asserted that the priority 'in any future legislation' should be

> the creation, in the various States, of bureaus of railroad statistics, under the superintendence of competent commissioners. The annual returns of Massachusetts, for instance, need to be entirely remodelled, as from them it is almost impossible to arrive at any reliable conclusions. Such bureaus should be permanent, and collect information from all civilized countries, as well as exact specific returns on all possible points from the State corporations. Knowing the peculiarities in regard to through and local travel, construction, grades, and elevations of each road, and the requirements of particular regions of the State, they could shed a flood of light on railroad legislation which will never be derived from spasmodic agitations, leading to superficial hearings before legislative committees. When such a bureau exists, and not till then, may some intelligent railroad legislation be hoped for. Until that time comes, the most important material interests of the community are in perpetual danger of experimental legislative tinkering.[175]

Within two years, Adams had followed up his opening salvo in defence of community against corporation with 'A chapter of Erie' (1869), also published in the *NAR* (with his brother Henry as editor). This, and further articles, openly challenged the allegedly corrupt practices of the railroad tycoons, notably Gould who had become notorious for diluting ('stock-watering') the value of his companies' stock by arbitrarily issuing more share certificates and increasing the percentage of his own holdings. Soon the campaign for railroad reform that the brothers had orchestrated in the *NAR* gained for Charles Francis the position he had sought as a Railroad Commissioner. Now his manifesto for railroad reform could be implemented in practice.[176]

Through a culture of scientific (statistical) reform, Adams served the Railroad Commission for some ten years. But his railroad career reached its zenith in the spring of 1884 with his appointment as President of Union Pacific, the first (and debt-ridden) transcontinental railroad. His presidency lasted until November 1890 when deteriorating financial conditions forced him to hand over to his long-time *bête noir*, Gould, whom Adams described in his private diary as the 'little wizard'. 'Gould showed me out', Charles Francis wrote of the moment of his resignation. 'As we formally shook hands, the little man seemed to look smaller, meaner, more haggard and livid in the face and more shrivelled up and ashamed of himself than usual; his clothes seemed too big for him, and, his eyes did not seek mine, but were fixed on the upper button-hole of my waistcoat.' Gould, in short, had not even the appearance of a trustworthy gentleman. Indeed, as Adams had earlier prophesied, the 'race of hereditary businessmen', degenerate in its physical form, its dress, its manners and its morals, now usurped the patrician gentleman. A few days later Adams confirmed his diagnosis: 'Today ceased to be Pres't of union Pacific and so ended my life of rail-road work ... Gould ... and the pirate band scrambling on deck at 10 [o'clock].'[177]

Interviewed on 28 November, Gould gave his own verdict on his sternest critic: 'The fact is that the [Union Pacific] road has been run on principles that have never before been carried into practice. They have appeared in books, I believe, and occasionally in poetry. The difference between the two presidents is very simple but very great. Mr Dillon is an honest, practical railroad man, while Mr Adams is a theorist.'[178] In the long-running battle between the concentrating tendencies of corporate railroad power and the bid to restore notions of trust and trusteeship on behalf of communities (local, state and federal), representatives of the former appeared to have won the day. But, as Geisst has recently argued, the Adamses' criticisms later formed a basis for the Harvard school of anti-monopoly economics. Charles Francis Adams had certainly shown the extent to which the 'material, the moral, and the political interests involved are found to be inseparably connected ... no matter from what side this railroad problem is approached'.[179]

Canadian Pacific spans the world

In the sections above we have discussed the development and representation of national (rather than simply regional) railway systems in Britain. The battle of the gauges indicated just how vital it had been to empire-builders to think beyond merely regional terms to objects of national – and imperial – importance. In the case of the United States in particular, we have seen how the Adamses recognised the nation-building potential of the American railway system, whilst remaining cautious about the kind of nation the railway might, indeed, construct: the possible benefits of dissipating petty local interest vied with the perceived evils of quasi-gravitational railway centralisation. But the very name of the Union Pacific indicated the capacity of the railway, not merely for integration, but for transit from an eastern seaboard to a western frontier eventually meeting the Pacific Ocean.

Britain's far-flung territories in India, Australia and British North America, as well as being physically remote from London, were certainly undeveloped economically (in British terms) and arguably more culturally fractured than Great Britain and even the United States. In themselves, these territories were geographically disparate, with vast regions often separated by sublime natural features: mountain ranges with their ravines and passes, deserts, and great rivers; they presented hostilities of climate and fauna (monsoons, droughts, mosquitoes and even lions) unknown in the mother country. In such conditions, imperial railways were expensive to construct but they served multiple purposes.[180]

A system of railway within such a territory might indeed create access to valued but hitherto unexploited terrain or might connect centres of administration, military power, population and commerce. Both in India and Australia regional administrators saw their territory as virgin land, and, seeking to avoid a repetition of the waste and conflicts dramatised in the 'gauge battle' of 1845 and 1846, promoted railways of a single, unified gauge. But the imposition of a standard by decree (7 ft in India, by Governor General Dalhousie) or even inter-governmental consensus (4 ft 8½ in. in the provinces of New South Wales and 'South Australia') might still fail to bring about lasting uniformity: rivalries between engineers, questions of economy, and pragmatic decisions taken in the light of difficult terrain saw India with at least three systems and Australia opting for gauges as diverse as the Irish (5 ft 3 in.) and the Queensland narrow gauge (3 ft 6 in.).[181]

But imperial railways also served to consolidate, building nations by linking regions, cultures and 'Dominions' hitherto separated. In 1878 the first Premier of the new Confederation of Canadian States, Sir John A. Macdonald, held out the promise of national unity:

> Until [the Canadian Pacific Railway] is completed our Dominion is little more than a 'geographical expression'. We have as much interest in British Columbia as in Australia, and no more. The railway once finished we become one great united country with a large inter-provincial trade and a common interest.[182]

This Canadian Pacific Railway (CPR) served as a great line of transit, in two senses: first, binding the eastern and western extremities of the Dominion one with one another; and second, making the territory into a thoroughfare for passage by an 'All-Red Route' (of steamship, railway, and steamship once again) whereby imperialists might circumnavigate the globe without departing from British territory.

'The time may come when the long steel ribbon stretching across the old territories of the painted Indian will prove the salvation of the British Empire', wrote Archibald Williams in 1904 in his account of the world's railways which drew upon information supplied directly from several of the companies concerned, including the CPR. Williams acknowledged in his account of the building of 'the Great Canadian Highway' the vulnerability of the Victorian Empire, its vast eastern territories heavily dependent upon the Suez Canal – 'easily blocked by a single hulk' – for trade and troops.

Now, however, the 'CPR system' of land and sea transport offered a much more readily defensible means of securing almost all parts of the distant Empire:

> the Atlantic tracks to Canada could be kept open by her [Britain's] powerful navy for the transports. Once on Canadian soil, the forces would be whirled by express to the Pacific coast . . . and be prepared to embark for India, or to return, in case of need, to England. In war time, too, Britain's great granary will be the Canadian provinces, now all tapped by the 10,000 miles of the C.P.R. system, as cargoes from Russia, India, New Zealand, and the United States could not be escorted.[183]

In peacetime, the CPR offered by 1903 an integrated service which not only ran from coast to coast across the North American continent but extended its lines across two oceans, making feasible fast voyages from Liverpool to Montreal (in summer) or Halifax (in winter), onwards by rail to Vancouver, and thence across the Pacific to Japan and Hong Kong, there to meet with the long-established eastbound lines of P&O or Blue Funnel. CPR not only provided what *The Railway Yearbook* described as 'an Imperial highway' bringing Canada nearer to Great Britain, in the Company's own advertising slogans between the two world wars, but it also spanned the world and urged its publics to 'Travel by the All Red Route' of Canadian Pacific steamships, trains and hotels.[184]

From its conception, CPR served political ends. In the first half of the nineteenth century, the territories of British North America looked vulnerable. Traditional loyalties to France and Britain divided the eastern states. Western and northern territories, with their formidable climate and mountain ranges, had scarcely been surveyed, let alone colonised. European hunters vied with indigenous peoples for lucrative furs. A powerful United States might well have taken an increasing share of any or all of these vast lands, but the United States itself had proved a fragile political union of states and territories in the years between Independence and Civil War. The emergence, however, of a far more powerful post-War United States, increasingly unified by its transcontinental railroads (especially Union Pacific), left Britain's isolated Pacific coast territories with an uncertain future. Although unwilling to abandon their loyalties to the British crown, the population of Columbia on the Pacific coast remained geographically separated from their fellow loyalists by the Rockies which presented a forbidding natural barrier running from north to south. Opting in 1870 to join the Confederation of Canadian States, Columbia insisted upon the completion of a railway across British North America. Acknowledging Columbia's vulnerability, Canada's Government under Macdonald agreed that a railway should be completed in ten years from 1871.[185]

For six years and for some £750,000 of Government money, surveyors mapped out a path for the 2600-mile railway. In charge of these surveys was Sandford Fleming, Scottish-born enthusiast for standard time (ch. 1) and CPR engineer-in-chief. In his Boy's Own adventure account, Williams later dramatised for his readers the sublime challenges faced by these teams in their wrestle with nature's most awesome powers: 'Engineers explored the mountain passes in all directions, experiencing the hardships and dangers inseparable from travel in wild, icebound country, fissured by huge chasms, along the side of which they had to creep, and flanked by towering peaks that hurled down devastating avalanches.'[186]

Only in 1880–81, however, did the private CPR Company take control of the project. George Stephen, who had emigrated from the Scottish Highlands and who had made his fortune by turning around a bankrupt Minnesota railroad, led the syndicate that formed the CPR. Indeed, Scottish influence – political, engineering and economic – was immensely strong throughout. Under the terms of the contract negotiated with the Government, CPR would receive not only a large cash injection but also 25 million acres of land in the so-called Fertile Belt traversed by the railway. CPR also received a 'free gift' of 710 miles of track still under construction by the Government. For its part, the Company undertook to lay the remaining 1900 miles, including the crossing of the Rockies, by the spring of 1891 and to the same standard of construction as that of the Union Pacific.[187]

Beginning work as CPR general manager early in 1882, William Van Horne – experienced in every aspect of railroading since an early age and lately general superintendent of the Chicago, Milwaukee & St Paul Railroad – accelerated the pace of construction through carefully planned organisation in a manner reminiscent of Thomas Brassey. From an average of 2.5 miles of track per day in 1882, the rate climbed to 3.5 miles a day in 1883. Twenty miles of track were laid in one three-day period. By the end of the year, the Rocky Mountain summit was just 4 miles distant.[188] The journal *Engineering* gave a vivid, first-hand account of the certain fulfilment of Macdonald's promise:

There were three hundred men with thirty-five teams in this [track-laying] gang. Moving along slowly but with admirable precision, it was beautiful to watch them gradually coming near, everything moving like clockwork, each man in his place, knowing exactly his work and doing it at the right time and in the right way. Onward they come, pass on, and leave the wondering spectator slowly behind... The returning locomotive, with her long string of empty cars rushing past him, awakens him from his reverie, and another, pushing before her more slowly her heavy load and taking them up to the front, shows him that where an hour before there was nothing but the upturned sod, two ditches, and a low embankment, there is now a finished working railway, and that the great Pacific highway is a fixed fact before his eyes.[189]

Waxing in a visionary way, *Engineering* asserted that the 'fixed fact' of the 'great Pacific highway' transformed yesterday's 'hunting-ground of the Indian and the home of the buffalo' into '[the land of] Britain today, not in name only but in use, and will probably be occupied within a week by some hopeful and happy British family, who in another season or two will make it a smiling home, and the abode of lasting comfort and prosperity'. It was, moreover, 'a sight that hundreds came to see' – 'a miracle of progress, the visible growth of an empire, the practical realisation of the dream of centuries, as the highway was gradually being laid down destined to conduct the commerce of Europe to that wonderful Orient where a prodigal Nature pours out her riches to supply the wants and luxuries of the world'. It was, in short, the accomplishment 'by that motley gang and those few locomotives' of the long-dreamed of north-west passage to Asia.[190]

The 'fixed fact' of the 'great Pacific highway' served the new nation near the beginning of 1885. Already experienced in moving men during the American Civil War, Van Horne organised the rapid shipment of troops to quell an indigenous revolt in the Northwest – along a railway fragmented by over 100 miles of gaps. Thus even before its completion, the CPR had demonstrated its practical value. Then, towards the end of the year, lines from west and east met in the Eagle Pass of the Gold Range of mountains and, with little ceremony, no ostentation and no gold spike as in the Union Pacific, joined to form the world's longest railway of the time, more than five-and-a-half years ahead of the original completion date.[191] In 1907 the *Montreal Gazette* considered that the 'building of the CPR gave us an undiscovered Empire. It opened up million upon million of waste acres – it preceded population and drew in its wake immigration with great industrial awakenings and it breathed life itself into the nostrils of Canada.'[192]

Only a year after the completion of the CPR, its President, George Stephen, wrote to Prime Minister Macdonald to inform him of a meeting with one of the founders (all Scots–Canadians) of the Allan Line which since 1856 had provided the principal mail service from Liverpool and Glasgow to St Lawrence ports in summer and to Portland, Maine, in winter:

I had a visit from Andrew Allan this morning and gave him roughly my ideas as to what would be necessary to *perfect* the Liverpool end of the C.P.R. At first he was startled, but breathed easier before he left me and after he understood me better. He now knows that nothing but the very best and fastest ships will be of any use to us, and that whoever owns them the C.P.R. must have a substantial control over them so as to ensure a unity of action.... He admits Halifax could be made in 5 days from Liverpool; so could Quebec barring fog or ice or both. I hope you will take time to send Tupper [Minister for Railways and former Premier of Nova Scotia] a line now and then to keep him hot on the Pacific line.[193]

In order to avoid dependence on New York or Boston, CPR, through
Stephen, recognised its 'enormous indirect interest in the Canadian Service
on the Atlantic being as good as any'. He therefore urged his Prime Minister
to 'advertize for tenders for a fast mail service', possibly from Southampton

PUNCH, OR THE LONDON CHARIVARI.—October 15, 1887.

THE NEW NORTH-WEST PASSAGE.

BRITANNIA. "NOW, FROM MY WESTERN CLIFFS THAT FRONT THE DEEP
TO WHERE THE WARM PACIFIC WATERS SWEEP
AROUND CATHAY AND OLD ZIPANGU'S SHORE,
MY COURSE IS CLEAR. WHAT CAN I WISH FOR MORE?"

Figure 4.3 *Punch* playfully represents the modern realisation of an old vision:
a north-west passage to the Orient.

Source: *Punch* 93 (October 1887), p. 175. (Courtesy of Aberdeen University Library.)

or Plymouth rather than Liverpool, with the option of calling at a Continental port. 'If you stipulate for the proper service', he told Macdonald, 'I do not care who gets the contract, certainly the C.P.R. will not undertake the service.'[194] Despite CPR's encouragement of rival operations on the Atlantic, the Allan Line continued to dominate the direct route from Britain to Canada until CPR inaugurated its own transatlantic *Empresses* in 1906 and secretly bought out the Allan Line three years later.[195]

CPR's early priority, however, lay in establishing a Pacific service: 'We *must* undertake that service – if it is to be done at all', Stephen told Macdonald in September 1886.[196] Already in May, Stephen had informed his shareholders that 'The negotiations with the Imperial Government for the establishment of a first-class line of steamships between the Pacific terminus of the railway [Vancouver] and Japan and China are still pending' and that 'The question of connections with the Australasian Colonies is also receiving the attention of the Directors, and they are also looking towards such steamship service on the Atlantic as will fully meet the requirements of the Company.'[197] British officials at the time, such as Sir Andrew Clarke, inspector-general of fortifications, understood the significance of such a service in securing the imperial network: 'Once a regular steamship service with the East by way of Canada and the Pacific is established, the whole Empire will be firmly knit together and the chain of communications between British stations will literally girdle the world.' And the Earl of Harrowby told the House of Lords in 1887 that 'This great Canadian Pacific Railway was, perhaps, the greatest revolution in the condition of the British Empire that had occurred in our time (Figure 4.3).'[198]

Three former Cunard ships, chartered from the Clyde shipbuilder and naval architect Sir William Pearce (successor at John Elder's Fairfield yard), entered the transpacific service in the same year. It was not until 1890, however, that the CPR reached agreement on a £60,000 per annum, ten-year mail contract. With carefully chosen names – *Empress of India*, *Empress of China* and *Empress of Japan* – three purpose-built liners entered service in 1891, the year that George Stephen was elevated to the British peerage as Lord Mount Stephen. Many more famous Pacific empresses followed in the years up to the Second World War, but that conflict brought the service to an abrupt and permanent end. It was a conflict that also marked the beginning of the end for the British Empire as a whole.[199]

5

'The most gigantic electrical experiment': The Trials of Telegraphy

After the harassments and disappointments of a year, when wealth and labour, care and anxiety, skill and invention might appear to have been absolutely thrown away, and to have gone to swell the vast amount of profitless labour which is done under the sun, it is no small solace to meet with such sympathy as you now manifest.... [By] your presence here this evening you show an interest in the great undertaking, animated by a conviction that the foundation of a real and lasting success is securely laid upon the ruins which alone are apparent as the results of the work hitherto accomplished. (Cheers.)... What has been done can be done again... improbable, impossible as it seemed only six months ago – chimerical and merely visionary as such a project seemed ten short years earlier – instantaneous communication between the Old and New Worlds is now a fact. It has been attained. What has been done will be done again. *The loss of a position gained is an event unknown in the history of man's struggle with the forces of inanimate nature.*

> – *Professor William Thomson rebuilds confidence in the recently failed transatlantic telegraph cable in a speech to an audience of Glasgow's leading citizens reported in the* Glasgow Herald *on 21 January 1859*[1]

In the frantic debates of the spring and early summer of 1858, prior to a renewed attempt to establish a submarine telegraph line between Britain and North America, *Scientific American* told its readers that once the cable was 'successfully laid down' it would 'remain the most gigantic electrical experiment ever made'.[2] The ambivalence of that statement – the telegraph as a robust product of electrical science, or as vast but fragile innovation – seemed to be justified by the events of the late summer. On 6 August 1858 Ireland and Newfoundland were in electrical communication; by 20 October 1858 telegraphic communication between Britain and her North American empire totally ceased.

In a little over two months barely more than 730 messages had been transmitted, most of them in the first three weeks of operation. Vast capital investment, much of it raised by Liverpool merchants eager for direct news of cotton prices in the Southern States of the USA, lay 3 miles beneath the waves of the Atlantic. Yet William Thomson quickly recast this spectacular failure as a strategic position soon to be made secure in the inevitable march of telegraphy against the hostile forces of nature. Riding on a wave of civic and academic pride, Glasgow University's professor of natural philosophy affirmed that the cable had made the vision of instantaneous communication across the ocean an irreversible historical fact which only required an unshakeable trust from the investing publics to overcome all setbacks. For Thomson, that unshakeable trust would be built upon the foundation of laboratory-based experimental science.

In this chapter, our interest is in the 'reputation' of the telegraph, as much as in the reputation of its advocates; that is, we want to understand the simultaneous creation of a stable role (or roles) for the telegraph and an authoritative role (or roles) for its advocates. We recall that Watt deliberately made himself into a new kind of philosophical steam engineer, as part of a co-ordinated programme of social and technical actions designed to generate that confidence in the commercial 'reputation' of the steam engine demanded by Boulton (ch. 2). A useful framework here is that of the fashioning of engineers and their artefacts: assembling, primarily from existing social tropes, a new and valued identity; simultaneously assembling, from a set of existing represented 'needs' and 'uses', a trustworthy and reliable technology with a well-defined meaning.

We develop this analogy of fashioning in the case of the telegraph. It is very clear that the telegraphs were not 'self-fashioning' (a technology can hardly speak for itself); but they had powerful advocates in their 'inventors'. The originator of a system of semaphore communication in late eighteenth-century France, Claude Chappe, fashioned himself as 'telegraphic engineer'; W.F. Cooke remade himself as electrical telegrapher and fought vehemently to secure an answer, in his favour, of who was 'inventor of the telegraph'; the Society of Telegraph Engineers, looking for a 'founding father' of electrical engineering, hit upon Francis Ronalds, who had proposed an electrostatic system of signalling in the 1810s; in America, Samuel Morse too refashioned himself from artist to economist of intelligence; William Thomson made a transition from Cambridge mathematical wrangler to scientific expert in all matters of 'electrical engineering'. In all these cases, we see individuals, representing social 'types' (inventor, originators, disinterested commentator), refashioned as trustworthy, credible and expert advocates for telegraphy.

These advocates issued 'manifestos of promise' about their own roles and especially about the prospective 'uses' of their technologies. For natural philosophers concerned with electricity and magnetism, manifestos often incorporated claims regarding the illustration and elucidation of the regularities of an ordered natural – and moral – universe; but electrical and magnetic 'toys' could also be projected as filling niches in spaces of exhibition and popular amusement, in elite sites of science like observatories, or in aristocratic contexts. Telegraphic projectors effected transitions of such toys from the places of natural philosophy to local, national and even international arenas of large-scale commerce, where they were transformed into telegraphic technologies validated as trustworthy and worthy of confidence. As with Booth and his railway prophecies (ch. 4), in attempts to specify the future meanings of these practical commodities, projectors made diverse, opportunistic – and indeed 'visionary' – claims.

Advocates agreed that telegraphs promised rapid – even 'instantaneous' – communication at a distance.[3] Elaborated in their manifestos, however, were the specific utilities of such intelligence and the conditions placed upon it. For Chappe, the visual telegraph would send instructions and edicts, enhancing surveillance and centralised control, aiding military campaigns, and building a Napoleonic empire. For Ronalds, mimicking Chappe, an electrostatic telegraph would strengthen monarchy, entrap criminals and keep friends in touch. For Cooke, an electro-magnetic machine would preserve public order, or economise steam engines' use for the business entrepreneurs of the new railway system. In America, Morse made telegraphy an instrument of government, territorial consolidation and the spirit of the frontier. In the 1840s, the brothers Brett projected a telegraph which would communicate orders to empire, bring imperial intelligence to the centre, facilitate local government, strengthen the grip of army, navy and the new police force, direct such objects of national importance as the lighthouse service, maintain national boundaries – and trounce smugglers. International and transatlantic projectors constructed telegraphy as a technology made trustworthy in an alliance of electrical physics and practical skill, and coded as an imperial asset, a harbinger of peace, a decisive military weapon, a mercurial messenger for commodity brokers, – or as a stentorian auxiliary to the ship's captain.[4]

The proposed identities of the telegraph (or other technologies) were not simply 'accepted' by relevant social groups. In issuing manifestos of promise for a technology, there were various strategies to generate acceptance or, indeed, credibility. Some manifestos were extravagant; but such talk eroded the confidence of sober backers and, moreover, could easily be 'discredited' by critics or vested interests (including those committed to existing communication services placed under threat). To work from deeds alone was possible only where capital was already in place. To push forward a single interpretation of a technology was risky but indicated stability of purpose. A nuanced account, projecting difficulties and sketching solutions to them (as Booth tried to do) expressed caution and in so doing augured reliability. Another possibility, in such a manifesto, was to offer a view purporting to stand above and beyond any particular interest.

Who, then, could count as the 'disinterested' evaluator of the telegraph? From the late 1850s, William Thomson would increasingly present himself as an *independent* expert in virtue of the authority of electrical science; in this respect, Thomson's role was akin to that of C.F. Adams, who also had, as it were, an interest in disinterest with respect to American railroads (ch. 4). Both could claim to be taking the 'view from nowhere' in order better to understand new systems and networks, unencumbered by traditional practice and sceptical of the many divergent promises once made, not all of which had been fulfilled. Both, of course, used disinterest to win key roles in their respective industries; in both cases, communal, quasi-legislative instruments gave authority to their roles – intellectually and pragmatically. In the immediate aftermath of Thomson's comments above, the Joint Committee of 1859–61 linking government, commercial interest, electrical science and engineering practice provided the forum to accredit the independent scientific expert.

Claude Chappe and the revolutionary telegraph

The French Revolution of 1789 and the wars that followed provided an opportunity for telegraphic projectors, like the four affluent and well-educated Chappe brothers. Claude Chappe, a well-paid priest living in Paris, indulged his interest in electrical science until, in November 1789, as a result of the Revolution, he lost his position. Returning to the provinces, where his brothers were also unemployed, he began to consider the possibilities for the new regime of rapidly sending messages over great distances – using apparatus eventually known as a 'telegraph'.[5] Despite being demonstrated before the dauphin and the Duchess of Orleans at the time of Louis XIV, visual telegraphs by the natural philosopher Guillaume Amontons did not win government interest: such devices were considered mere toys. But in the days of Revolution, the Chappe brothers promoted their telegraphs as crucial tools in a war effort.[6]

With the large private capital available to them, they developed the visual telegraph in three different forms: a 'synchronised system' using clocks, rotating disks and a sounding gong or electrical signal (1791); a 'shutter system' (1791); and a 'semaphore system' (1792) since elongated bodies proved easier to see.[7] Each of these forms was intended as a means of communication in support of the regime; and the Chappes attempted to win support through carefully staged demonstrations in evermore elaborate – and public – circumstances. First, they experimented privately, sending messages between stations separated by almost half a kilometre. By 1791, in a public demonstration in Chappe's native Sarthe, they succeeded in communicating over a distance of 15 kilometres. They next went to Paris, the centre of political influence and finance, and planned further shows, with stations set up on the Champs Élysées and in private parks. But the interpretation given to these projects was crushingly negative: hostile parties destroyed their apparatus; inflamed observers were convinced of the meaning of their telegraph as a means of communication between royalist supporters or even with Louis XVI then incarcerated in Paris.[8] Such were the potency of the meanings of telegraphic communication that experimentation required police protection, sought and won from the Legislative Assembly – on which Claude Chappe's brother Ignace, as a deputy, had influence.[9]

The turbulent and unstable political environment offered opportunities for presentation but it was hard to sustain promises when the nature of the audience, and the legitimacy of its authority, frequently changed. In March 1792 Chappe presented his 'tachygraphe' – or 'rapid writer' – to the French Legislative Assembly with promises of speedy transmission of messages at day and night; he prophesied a system of lines connecting Paris to its frontiers and seaports. The opportunity of the committee of public instruction to examine its merits foundered, however, when the Assembly itself was abolished the following September – to be replaced by the Convention which was too busy to deal with one of very many such nascent projects.[10]

A new opportunity came when on 1 April 1793 Gilbert Romme reported favourably on its prospects to the National Convention. Romme elevated it above several other contenders before a joint committee of war and public instruction. He recommended a test so that accredited assessors could decide on the likelihood that it would substantiate the promises indicated to the committee: its potential usefulness in the war, especially speeding the exchange of dispatches between government and armies. The earlier demonstrations in the department of Sarthe, now retrospectively judged satisfactory, added authority to Chappe's claims; Chappe himself had the opportunity to speak in favour of his telegraph, tripping up only when it came to the question of how it might work in fog. The powerful advocate Romme foresaw difficulties (how might messages be kept secret?) and proposed solutions (codes, known only to station agents, might be used). The Convention then ordered and funded a commission of inquiry.[11]

On 12 July 1793, a committee of men of science and government officials combined to elaborate the promises of the telegraph in a large-scale experiment, with these specific aims in mind, using two stations 33 kilometres apart and a third between them. Members (including revolutionaries Cambon and Monot) disagreed about the results of those tests; but one, Joseph Lakanal assumed the role of Chappe's sponsor.[12] His rhetorical claim that a normal military despatch would reach Paris from French Flanders in under fourteen minutes stirred the Convention to applause – before any line was built. During 1793 and 1794 France was under attack from the allied forces of Austria, England, Prussia, Hanover, Hesse, Spain and the states of Italy, Marseilles and Lyon had broken with the French government, and the English (in August 1793) had taken Toulon. Little surprise that Chappe was reimbursed for his efforts, given the rank and pay of military lieutenant, and awarded a special title to go with his new role: *ingénieur-télégraphe* (telegraph engineer) – the name 'tachygraphe' having been jettisoned after a discussion between Ignace Chappe and the head of the department of the interior.[13] Claude Chappe's title was redolent of contemporary moves to create a corps of public engineers on military lines, focused especially in one Parisian institution, the École Polytechnique (established in 1794), as a militaristic school promoting, also, elite science and engineering – in a time of war.[14]

Promising rapid communication of intelligence between vast but unco-ordinated forces fighting on many fronts in post-Revolutionary France, the Chappe telegraph flourished. Although no businessman, Chappe was instructed to construct a line of 230 kilometres from a central terminus in Paris outwards to Lille and the Army of the North. The system epitomised the centralisation of government and the bureaucracy of the state. An ord-nance, which made it the duty of the Garde National to protect the apparatus, reduced the possibility of sabotage from men unwilling to see Chappe's promises fulfilled.[15] Public funds, privileged access to essential materials, and permission to remove obstacles in the line of sight saw the route completed, largely through the efforts of the brothers who promoted it, by August 1794. That date proved timely since it allowed the telegraph to be used to inform the administrators in Paris of the recapture of the fortress town of Quesnoy, just north of Lille, from Austria, only an hour after French troops had moved in.[16] The event, however, illustrated the continued dependence of the telegraph on old forms of communication: a courier had still been required to bring the message to the station. Media management by interested parties was essential to win support for the lines and to publicise their fulfilment of promise. A member of the Convention cited the news as an example of science serving liberty.[17] The subsequent recapture of Condé in French Flanders from the Austrians, and the defeat of their 80,000-strong army, at the end of August 1794 provided another media opportunity, in early September. Then, Lazare Carnot, minister of war, member of the Committee of Public Safety, and father of Sadi Carnot announced: 'Citizens, here is the news just received this instant by the telegraph that you have established.' 'Indescribable enthusiasm' in the Convention duly followed.[18]

News of victories, once arrived in Paris, became opportunities to applaud Chappe.[19] Such events clarified an iconic history of the telegraph as military weapon and, more immediately, provided convincing arguments for the development of additional lines. They would be popularised from the mid-nineteenth century by Louis Figuier in his various *merveilles* (i.e. marvels) of industry and electricity. In a pattern that would be repeated with the electric telegraph from the 1840s, the drama of these events, and the speed with which news of them was communicated, delivered on promised actions and helped to secure the future of the technology, at least in the short term.

The system grew under war conditions. The Convention wanted lines from Lille to Ostend, and to Landau in Bavaria via Metz (under the charge of Chappe, who also looked after the factory making the equipment) – although with no money for workers and a shortage of metal and wood that was easier said than done. With the new Directory in power from October 1795, support came for a line to Strasbourg: this, finished in 1798, had 46 towers (closer together than those of the first lines, which had been too far apart). The Lille line was extended to Dunkirk; there was a new line from Paris to Brest and another from Paris to Lyon. But the expense of these ventures was vast, annual maintenance and service far outweighing actual construction and no income from public use to defray costs. At first in favour of the enterprise, Napoleon choked at the huge expense. Refusing to see the telegraph as the 'military solution', he cut the budget, and allowed lines (including the Paris to Lyons) to be abandoned.[20]

Chappe had several times attempted to revalue the telegraph, in 1795 suggesting it would be serviceable as a means of transmitting weather reports; but there was no such service until 1856.[21] After the cut in funding he suggested it might be used, not as a military messenger, but as a business machine, able at the right price to deal with stock quotations, to announce ships' arrivals, to provide information for newspapers – and, especially, to solve a problem concerning the national lottery. The results of the lottery were first announced in Paris; and it was widely known that there were fraudsters who would sell tickets in the provinces after the result was known in Paris – but before it reached the towns. The lottery management proved to be the major source of private funding for the telegraph and thus re-created Chappe – as the saviour of the lottery.[22]

When Chappe died in 1805 – by suicide – it was, according to one account, precipitated by anxiety over challenges to his priority claim as the telegraph's inventor.[23] His tombstone, emphatically, bore a carving of a semaphore tower and the words, lest they forget, of 'Reprise de Condé'. Thus man, machine and military campaign (but not lottery) were forever identified. His brother Ignace produced an *Histoire de la télégraphe* (1824), all carefully charting Claude's role. A further identification, with the mail service, came in 1859 after which date the tombstone, sliced into two to make front and back visible, graced the entrance to the headquarters of the Administration of Posts and Telegraphs.[24] Much had been done to perpetuate the name 'Chappe' and to consolidate the connection of 'Chappe telegraph'. After his death, the Chappe brothers Ignace and Pierre, long involved in the administration of the telegraph, took control of the lines as business concerns. Another brother, Abraham, worked to solve problems resistant to repeated attacks, like that of working in fog.[25]

The material and social technologies involved drew upon existing forms. Chappe advocated the construction of a series of semaphore towers, each about ten feet high, and with no more than 10 miles between each other.[26]

The stone towers were similar to those used for windmills; where materials were short existing buildings served – including Strasbourg Cathedral and the Louvre in Paris – and from 1793 Chappe had been permitted to situate his apparatus in any belfry he chose;[27] the mechanism used cranks, ropes and pulleys familiar to millwrights.[28] To fulfil the promise of secrecy made by Romme on Chappe's behalf, he had developed his first codes for the system in correspondence with a relative, Léon Delaunay, who, had worked as a consul in Lisbon. Indeed the basis of the first code was that used by Delaunay routinely for his diplomatic correspondence. Ultimately there were three code-books of 92 words on each of 92 pages, giving words, phrases and places.[29]

But there were also new material and social forms. Skilled trained operators were required to work the semaphore system, recognising symbols and transmitting quickly with military precision; mistakes were punished by a station director with the power of instant dismissal or imprisonment; the menial and low-paid 'stationnaires' were selected (rather like the human calculators at Airy's Greenwich Observatory) for their lack of ambition and for their dull but reliable intellect (ch. 1).[30] Mounted on the top of each tower was a beam (the 'regulator') which could assume various positions; pivoted at the ends of the regulator were two wooden arms (or 'indicators') taking the form of windmill sails, each of which could take several different positions. The regulator and indicator could be moved by wires and levers into any one of 196 distinct and therefore meaningful configurations, all using a mechanism mimicking the position in miniature. In the hours of darkness, lanterns at the ends of the various limbs, and at the pivots, helped to make the positions clear. Chappe's demonstration device of 1793 had drawn upon the clockmaking skills of Abraham Breguet redeployed; without the achromatic telescopes of the late nineteenth century, Chappe's towers would have been too close together for effective benefits in times of transmission. Thus the telegraph could only be successful with the skills of all its makers and of its human operators.[31]

The efforts of Chappe's brothers and their successors, and the disciplined working practices of the operators, sustained a system of 4000 kilometres, with 500 stations, in France and with additional lines in places of territorial ambition (like Algeria and Egypt). The network remained viable well into the nineteenth century, only gradually being displaced by the opposition of the electrical telegraph. Indeed there were new lines built as late as 1846 (as the electric telegraph began to dominate Britain).[32] This is less surprising when we understand that to displace the Chappe system, the advocates of an electric telegraph would likely need to be able to promise to match the times it had achieved, as a smoothly working system of man and machine, from Paris to Calais (3 minutes), to Lille (7 minutes), to Brest (8 minutes), or to Toulon (20 minutes).[33] Even in 1894, British periodicals claimed that the electric telegraph was slower than Chappe's. Certainly, the promises made by the earliest advocates of electrical telegraphy for 'instantaneous communication' had not been fulfilled.[34]

Although the Chappe system flourished in France – proposed as a rapid form of communication, advocated as a military machine and then as a lottery safeguard – similar optical systems had their projectors in the varied contexts of Sweden (1794) and the United States (1800). In Ireland, it was the threat of invasion by France in 1797 and 1798 that gave renewed meaning to optical systems developed in the 1760s by James Watt's Lunar Society friend Richard Lovell Edgeworth; as that threat had subsided, so the credibility of the system, seen as a response to it, dwindled – and it fell into disuse.[35]

In England, aspirant optical telegraphers gained support from the Admiralty. Military officers had witnessed Chappe's system in France; from 1796 an imitation, with shutters, instead of semaphores, adapted to the distinctive British climate, stretched to the channel ports where – so its promoters claimed – it would help to repulse Napoleon's invading forces, should they appear.[36] In this race of military technologies, the British would counter French might with copies of mechanisms for intelligence collection and distribution. A visual telegraph linked London with naval bases such as Portsmouth to the south and Plymouth to the south-west. A line to Deal on the strategic and vulnerable south-east coast of England (little more than 20 miles from France) could receive a message from London in a minute's time. Towers dominated the landscape as symbols of power inscribed in geographical space – and often recorded on maps as 'Telegraph Hill'.

The credibility and stability of those lines depended, as in France, on the intensity of fighting and, moreover, on the level of perceived threat which gave them meaning as military technologies. In times of peace that meaning evaporated and without a new and different promise, fulfilled or not, they were vulnerable to closure. John Macdonald, Royal Artillery Lieutenant-Colonel, one of the fiercest advocates of telegraphy in Britain, argued from 1797 until the 1810s for the development of a science of telegraphy, put together a telegraphic dictionary, and prepared a transmission code – ready for Admiralty use. But Macdonald's unrestrained manifesto of promise looked beyond the military context to visions of 'approximation of time and space' that would reform the intelligence infrastructure of Britain for a far wider range of purposes:

> It is unnecessary to dwell...on the incalculable benefit that would arise to commerce, to public revenue, to private convenience, and to public safety and security, by establishing a ramified telegraphic system, extending from the metropolis to the principal seaport towns, inclusive of a methodized intercourse with the principal [nearby] cities...Such an undertaking would be a sublime attempt at an approximation of time and space; and would be truly worthy of the high character of our

mighty nation. I...prophesy that future ages will see this magnificent idea fully realized.... Man is a progressive animal.[37]

Nevertheless, the last British optical telegraph line closed in 1847 – after, that is, the promises of the rival electric telegraph (first voiced in the 1830s) had been, in part, fulfilled.[38]

Demonstrating the electric telegraph: philosophical toys and commercial promises

When Francis Ronalds approached the British Admiralty with a new system of telegraphy, developed in 1816, he was famously rebuffed. With an existing shutter system working to fulfil naval and military objectives, Ronalds's proposal was just another unnecessary irritation.[39] Writing in 1823, Ronalds asked pompously: 'why should not our kings hold Council at Brighton with their ministers in London? Why should not our government govern at Portsmouth almost as promptly as at Downing Street? Why should our defaulters escape by default of our foggy climate?...Let us have *electrical conversazione offices*, communicating with each other all over the kingdom, *if we can*.'[40]

The gentlemanly Ronalds had 'amused' himself in the intervening period 'wasting', he said, time and money, 'in trying to prove, by experiments' that electricity could be used as a 'most accurate and practicable means of conveying intelligence' – or, as he put it, 'so *diligent* a courier', since he was convinced, by experiments in his Hammersmith garden, of the '*instantaneous* transmission of electric signs'. It was Ronalds's scientific friends that proposed to him his discoveries might 'actually be employed for a more practically useful purpose than the gratification of the philosopher's inquisitive research, the schoolboy's idle amusement, or the physician's *tool*'.[41]

Originating in contexts of philosophical experimentation, electric telegraphs initially competed for public attention alongside the shocks and sparks associated with the demonstrations and displays much beloved by the popular lecturers and showmen of electrical science.[42] The eighteenth century was the scene for the creation of 'Newtonian' sciences, including a science of electricity, emphasising the law-abiding phenomena of electrical repulsion and attraction, and harnessing those phenomena for astonishing and amusing electrical spectacle – sometimes with the gleeful electrification of gendarmes and boys.[43] With static electricity, such schemes promised popular or courtly entertainment rather than bureaucratic and commercial reform. An anonymous correspondent of the *Scots Magazine* suggested, in 1753, the transmission of messages using a wire activating a pith ball for each letter of the alphabet. In Spain, Francisco Salvá courted the royal family as patrons of his system with offers of an electrical connection from Madrid to their spring residence at Aranjuez. Here not a pith ball but a shocked human operator registered the receipt of the message.[44] Gentlemanly enthusiasts like Ronalds, and philosophical chemists like Joseph Priestley, deployed silken cords, pith ball electrometers, Leyden jars (as stores of electricity), glassware and other readily available apparatus in experimental philosophy. From 1800 the 'voltaic pile' became a reliable and convenient source of current electricity, vital to those like Humphry Davy who appropriated electricity to differentiate new chemical elements.[45]

Current electricity offered greater scope for a practical system of general utility. As Davy showed his gentrified audience at the Royal Institution, philosophical explorations could be seen to serve useful ends. In 1809, the Kassel professor of anatomy and inventor, S.T. von Soemmering, impressed by Chappe's telegraph in ousting the Austrians and informed about Salvá's courtly experiments, created a system using the current of the voltaic pile, passing through an acid solution, to release visible bubbles of gas. He claimed that by using a separate wire for each letter of the alphabet he could transmit messages over a distance of 2 miles, despite the difficulties experienced in 'reading' the bubbling signals. To substantiate that promise in 1810 he orchestrated a show before Baron Paul Schilling, a German attaché at the Russian Legation in Munich. In this diplomatic context, von Soemmering won the support of an influential patron; but the special circumstances of war that specified a meaning for Chappe's telegraph did not exist for von Soemmering.[46] He could not effectively counter charges of tardy transmission and, with his numerous conducting wires, excessive expense.

A very different kind of practical demonstration in 1820 provided natural philosophers with new arguments in favour of the unity of natural forces whilst also offering an opportunity for the creation of a rival social group with different practical agendas. A Danish experimental philosopher, Hans Christian Oersted, observed that, when a wire was connected to the two poles of a voltaic pile, the current passing through the wire caused the deflection of the needle of a magnetic compass.[47] In the same year André-Marie Ampère invented the 'galvanometer' which exploited this phenomenon as an indication – and ultimately as a measure – of electrical current. From the mid-1820s, natural philosophers (like Davy's protégé Michael Faraday) and an emerging group of practical electricians (personified in William Sturgeon) disputed the nature of the electrical world, with the former speaking of the harnessing of natural powers, like electricity, in practical machines, and the latter celebrating their own elaborate technical apparatus of electrical exhibition – of which a telegraphic system was simply one part.[48]

Ampère had suggested that a galvanometer needle might be used in an electric telegraph. But it was von Soemmering's patron – and then disciple – Schilling who, from 1822, experimented with these electro-magnetic receivers, constructed a code to reduce the number of costly wires and, in 1835, promoted his electro-magnetic telegraph in public. To associate himself with the authority of science, he staged a demonstration before a group of natural philosophers in Bonn. The German natural philosophers were already projecting roles for the electro-magnetic telegraph which placed it at the heart of specialist sites of science. In 1833, the magnetic expert and mathematician J.C.F. Gauss and the natural philosopher W.E. Weber had created a telegraphic link, with their own code for efficient working, and acting over a kilometre between Gauss's University in Göttingen and the Magnetic Observatory. In November of that year, Gerard Moll, director of the Utrecht Observatory, told Faraday about this 'very pretty' and 'very curious' apparatus with its wires communicating 'through the open air over roofs and steeples'.[49] Steinheil, likewise, set up a line starting from his observatory at Zerchenstrass, passing through the Academy of Sciences at (Schilling's) Munich, and thence heading off to the Royal Observatory at Bogenhausen – having traversed a mountain, a river and a busy suburb.[50] Devoid of informants on lottery numbers or troop movements, these lines rather linked elite sites of science, patronised by government and monarchy, by using the very electro-magnetic phenomena which philosophers were attempting to fathom.

In this competitive European scientific context, it was Charles Wheatstone and William F. Cooke who made the electrical telegraph an attractive proposition beyond the showroom, the observatory, and the university lecture theatre.[51] An endlessly inventive scientific (and musical) instrument maker – one of his more bizarre creations was the hybrid 'concertina violin' – Wheatstone began in 1836 to construct electric telegraphs of different designs. He was also a professor of experimental natural philosophy at the young King's College London; although the rival University College London tried to cater for the new professions ignored in Oxford and Cambridge, it was King's that became the key English centre for the training of professional engineers, reconceived as men of practice informed by academic science. Amongst those teaching the engineers from the late 1830s, apart from Wheatstone himself, was his colleague Frederic Daniell, inventor in 1836 of the 'Daniell cell'.[52]

The relationship of Cooke and Wheatstone was complex, complementary and ultimately fraught. Cooke provided an injection of commercial *savoir faire* whilst trying to refashion himself from military officer (from 1825 to 1833 he held a commission in the East India Company's army) and anatomical model-maker to scientifically accredited projector in the practical electrical arts. Anatomical modelling had taken Cooke to Heidelberg, where in 1836 he witnessed a telegraph of Schilling's design in action. Cooke's father, a professor of anatomy at Durham, had been a friend of Ronalds. Now Cooke quickly began to re-imagine the commercial possibilities of an electrical telegraph in the England of the 1830s and sketched a prospectus designed to garner patronage from 'Government and the mercantile potentates'. In general matters, he claimed, the telegraph – with a system of lines buried and inaccessible beneath the common roads, with trusted 'confidential clerks' recruited perhaps from a community of deaf-mutes skilled at abbreviating messages – would be cheaper, more efficient, and more secret than any of the many existing forms of communication.[53]

Specifically he looked to government, commerce and individuals for support. As a government machine, the telegraph could provide information about impending riots (in an age of Chartism); government agents would have instruments dedicated to them, and the power to cut off the public network. Thus the telegraph was a secret agent of suppression and surveillance. For the commercial world he held out the prospect of provinces in communication with metropolis, and thus placed on an equal footing with London with regard to the daily state of the markets. As for individuals, they would be furnished with 'security and confidence', and insulated from the tricks of swindlers and forgers, thanks to this new trustworthy mechanism of distributing intelligence; friends could learn quickly of the state of the sick; and in total, the telegraph would contribute to the 'aggregate of comfort and happiness to the nation' (for Benthamite readers). In just one further example of opportunism, Cooke saw the new railways as a venue for his projected commercial telegraph: might the stationary steam engines used to provide power on steep inclines, and usually kept permanently fired, be more economically deployed if a telegraph could signal ahead to warn them of impending use?[54]

Cooke was aware of the utopian promises made on behalf of the railways – especially by Dionysius Lardner – but in 1836 he appears to have seen roads, not railways, as the key protected routes for a telegraphic system (ch. 2). At that date, the railway system was barely more than a few scattered lines, despite the plans of many more routes to come very soon. Rather, Cooke tried to substantiate the claim that an electric telegraph would save time and fuel for a stationary 'winding' engine employed to haul freight wagons and passenger carriages up steep inclines. Through his father's connection Joshua Walker, Cooke negotiated a meeting with the Liverpool & Manchester Railway Directors in January 1837. His proposal now was to set up a telegraph for signalling through a 1-mile tunnel leading from the main Liverpool terminus and serviced by a winding engine. The Directors, however, had already decided to use a simple pneumatic whistle for the job and were unconvinced that Cooke's more versatile 'sixty signal' telegraph was necessary. Although the directors gave Cooke a space for continued experiment, they left him still in search of the means of fulfilling his telegraphic promises.[55]

From late 1836, Cooke had been attempting to solicit the explicit and timely approval of powerful men of science (like P.M. Roget, Secretary of the Royal Society) for his projects whilst undertaking practical experiments in private over a mile of wire. At the end of the year, an interview with Faraday elicited an authorisation of the correctness of the general principles of his machine, although no view on its efficacy over a great distance. (Faraday was also circumspect about Cooke's *'perpetuum mobile* principle' wheeled out at the same time.) In February 1837 Cooke called on Wheatstone, only to discover to his dismay that the experimental philosopher had been working for some time on his own telegraphic instruments, knowing of Ronalds's efforts but basing his own on Schilling's designs, and with plans to use them practically.[56] The two men entered into a business partnership. Working in a highly competitive environment of electrical practice, Cooke and Wheatstone petitioned for a patent for a 'five-needle' electric telegraph in May 1837 (although the illness of William IV delayed its confirmation till June).[57]

Contemporaneously with their efforts to patent the telegraph, Cooke and Wheatstone continued to imagine possible demands for it, in order to give it a stable meaning, fitted to diverse spaces and multiple markets. Thus in May 1837, again through Joshua Walker, they approached the Directors of the London and Birmingham Railway offering a means of communicating at a distance, again for a mile and a half, between the stationary engine at Camden Town and the terminus at Euston. Whilst in some degree repeating what Cooke had done for the L&MR, Cooke and Wheatstone used the opportunity offered them not only to meet a specific demand (signalling a short distance to a stationary engine) but also to carry out an experiment on a much larger scale designed to persuade onlookers of the practicality of signalling over distances of up to 19 miles. (Wheatstone later explained how earlier experimenters had 'failed to produce any well-grounded belief in the practicability' of a telegraph.)[58] By doing that they responded to the questions Faraday had raised the previous year, and – augmenting Cooke's original perceptions of the limited railway 'role' of the telegraph – made plausible projections of a telegraphic network imprinted on that of the railway system.[59]

Cooke and Wheatstone prepared carefully for a demonstration of their five-needle telegraph scheduled to take place before the Company Directors on 4 July 1837. Unlike the Rainhill trials, there were no competitors here; but like those trials, and the large-scale authorised trial of the Chappe telegraph under Lakanal, it was essential to convince powerful advocates that the telegraph had 'practicability'. The ambition, as well as pleasing the Directors, was to build confidence, convincing railway managers that electricity could be transmitted, not only over the short distance of an incline or tunnel, but for long distances, equivalent to those between stations, thus scaling upwards – and moving outwards – from the philosophical toys of the natural philosopher to the commercial installations of business. Robert Stephenson, the railway's consulting engineer, witnessed early trials of the apparatus, declared himself, in Cooke's words, 'a convert to our system' and urged careful rehearsals for the public show so that 'all may go off in "good style"'.[60]

After a further and more private show in Cooke's 'den' at the Camden Town station during August, when an hour of constant working passed without 'mistakes or blunders of any kind', Stephenson admitted he was 'convinced of its practicability' and began to talk of more elaborate means to protect a system of telegraphic lines projected to run all the way to Liverpool – with a total cost (to shareholders) of £100,000.[61] Now, with the authority of the builder of the *Rocket* behind him, Cooke developed plans with the company Directors for just such a telegraph system, matching a planned railway system yet to be constructed. There would be lines connecting London, Liverpool, Manchester – and Holyhead, assuming a railway to Holyhead would eventually be completed. In this vision of a combined railway and telegraphic system of conveyance and intelligence, for government and commerce, Cooke looked to the knitting together of the United Kingdom of Great Britain and Ireland; messages reaching Liverpool from America could quickly be communicated to the capital. Astonishingly, an imagined project of empire-building thus lay in front of his audience before any system of railways had been laid down and before any practical telegraph of more than a few miles had escaped from the natural philosophy classrooms. The refusal, in December 1837, of the Directors to countenance the extension of Cooke's visionary world beyond Camden showed only too clearly its fragility.

Cooke began to cultivate Brunel in September 1837, after the London and Birmingham trial of July, but before the negative decision of the London and Birmingham Railway Directors was known.[62] There was serious competition in prospect. Edward Davy exhibited a rival apparatus at the Exeter Hall in London in January 1838; he contacted Brunel, and indeed many of the railway companies, hoping to get them to adopt his telegraph. By June 1838 he had a patent for his apparatus, removing Brunel's and others' doubts concerning the legitimacy of his actions; but legal battles in connection with Davy's estranged wife saw him flee to Australia – leading to a collapse of his prospects.[63]

Soon after their initial contact, Brunel took Cooke to Maidenhead to view the ongoing railway works.[64] It seems clear that the plan was merely to establish the telegraph along the route as a trial of a long-distance communication system – not specifically, or primarily, to work the line by regulating the trains' movements. Also, there was no formal discussion of the adoption of the telegraph at shareholders' and directors' meetings: rather, the arrangements were at first undertaken privately between Cooke, Brunel and a few interested Directors. Eventually, the company set up a committee to formalise the commercial agreement with Cooke. The chairman of that committee announced, according to Cooke, that 'they must have a telegraph and they believed ours to be the best, an opinion strongly supported by their Engineer [i.e. Brunel]'.[65] Brunel's carefully cultivated support was vital in giving Cooke any chance of winning over the Company Directors, just as Stephenson's had been with the abortive London and Birmingham plan.

Early in 1838, Cooke described a series of encounters with the Great Western Railway Company. After protracted discussions, a memorandum of agreement was close; then, unilaterally, the Company proposed dramatic changes. At this point, Cooke took a gamble. It was one which would secure both the reputation of the telegraph (seen as a means of communication to be tried along the new lines) and his own reputation as an independent, credible, disinterested, gentleman (who could not simply be dictated to by the Company). Returning the altered document unmarked, and writing that he would withdraw from discussions, Cooke found himself suddenly treated with cordiality and respect. He was presented with an agreement that carried 'flattering circumstances to myself, and the most propitious to the invention'. Later he met Brunel and, describing the encounter to his mother (to whom he appears to have confided every detail of his ventures in marketing and self-presentation), he wrote:

> I had a long interview (from 10pm till past midnight) on Monday with Brunel... He let me have a peep behind the scenes, and it appears that the Company had never before been treated so coolly, and were absolutely pleased with the novelty of the occurrence... The agreement is again in my hands for alteration. If we come to terms, they will now be satisfactory, though not very lucrative, yet still handsome; but once started under Brunel I may make money elsewhere. I think I have secured my being treated as a gentleman by everyone connected with the Company. Pray keep all that I have written to yourselves [i.e. his mother and family] as I would not for the world that anyone should know that I am so gratified with what has occurred or had so fully calculated upon the result of a bold step.[66]

In such deliberate (and deliberated upon) actions, Cooke fashioned himself as independent, dispassionate and aware of his (and his machine's) value; it was equally important that those assembled characteristics, and bold actions, were not seen as actively produced, cynically calculated or productive, in him, of a secret, ungentlemanly, jubilation.

The adoption of the telegraph, and the accreditation of its inventor, by the GWR took place in a context not only of intense railway building, but also of intense competition amongst the railways and of careful marketing of the young technology (ch. 4). Even though the L&MR had not adopted Cooke's device, it had allowed him to continue his experiments; the London and Birmingham had apparently funded Cooke and Wheatstone's demonstration experiments and then adopted the apparatus, albeit on a small scale. In this context, it is unsurprising that the GWR should have joined the current fashion of the railway. Moreover, it was characteristic of the GWR, built for, and marketed to, elite passengers for speed and luxury and noted for its 'novelty', vested in the person of Brunel, that the expensive and innovative electrical telegraph should be adopted. When Cooke quoted a figure for the means of protecting the cable running beside the tracks (by wood), it was Brunel that proposed a more expensive, and more lasting, solution (using an iron tube).[67] The adoption of the telegraph also reiterates Brunel's aims as a builder of systems (ch. 4). We know that by the autumn of 1837 Cooke had presented extended views of a system of telegraphic communication on the railways controlled by Stephenson and his allies. On the GWR, telegraphic communication would consolidate the imagined system stretching from the metropolis of London to the city and port of Bristol and then across the Atlantic to New York: on 8 April 1838 the *Great Western* began her maiden voyage out of Bristol.

Work on the 13-mile experimental line from the London terminus at Paddington to West Drayton was ready to proceed by May 1838. Once again Cooke worked to ensure that the line was laid down without a hitch, carefully training his men, out of the view of GWR engineers, before starting work on the railway itself.[68] Later, Brunel himself, now an interested party in seeing the telegraphic projections fulfilled, arranged for the visible support – through direct 'witnessing' – of the elite of science. He recruited Babbage, a man whose computing 'engines' threatened to displace unreliable human computers and who was now willing to support a practical mechanism for removing skills conventionally required in the transmission of 'intelligence'.[69] (Wheatstone's alphabet telegraphs, pointing out letters rather than using codes, reduced the need for skilled human operators – like those by Chappe.) By 14 October 1838, Brunel was telling Charles Babbage: 'I am to be at Paddington at 3 o'clock to try a few miles of the Elect. Mag. Telegraph[.] I will call on you at 2½ . . . should you be disposed to witness our experiment.'[70] This 'experiment' took place just before Babbage used GWR facilities, and an 'experimental carriage', to demonstrate the superiority of Brunel's broad gauge (ch. 4).

The telegraph on the stretch between Paddington and West Drayton was operational by July 1839; but the GWR Directors refused to pay for its extension (at a time of economic downturn). Cooke negotiated a deal in which he could extend the line at his own expense if he paid a small rent and transmitted railway messages gratis. He went ahead. Unable to afford the iron tubes Brunel had wanted, Cooke suspended his wires from iron posts. Whilst mocking his earlier promises of invulnerability, secrecy and safety, these telegraph poles became a ubiquitous marker of 'progress' in the electrical arts where other innovations lurked in exhibition halls or out of public view.[71] By 1843 Cooke's line had reached the station at Slough close to Windsor Castle and to the Queen: the inventor wondered about the possibility of a direct link between Victoria and her government in London.[72] The same year Albert the Prince Consort inspected in person the galvanic telegraph (Figure 5.1).

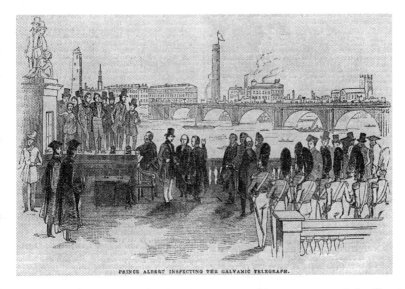

Figure 5.1 Albert the Prince Consort witnesses a public experiment of the electric telegraph in 1843.

Source: *Illustrated London News* 3 (1843), p. 5. (Courtesy of Aberdeen University Library.)

The GWR line, constructed at great personal expense and licensed to an entrepreneur who had opened it up to novelty seekers at a shilling a time, provided publicity useable in the campaign to solidify the telegraph as more than a mere novelty (just as the Chappe line had been linked to the dramatic Condé announcement). Two sensations – a royal birth and the capture of a murderer – achieved wide publicity in contemporary newspapers, and in accounts of railway culture, like the best-selling account of the London and North-Western Railway, *Stokers and Pokers* (1849).[73] In August 1844 news of the birth of Victoria's second son, Alfred Ernest, at Windsor Castle, reached London by Cooke's telegraph (with the help of a royal messenger on a thoroughbred steed traversing the journey from Windsor to Slough station in some 8 minutes). *The Times* commented on the 'extraordinary power of the Electro-Magnetic Telegraph'.[74] Newspapers also widely reported the 'capture by telegraph' on New Year's Day 1845, of the poisoner John Tawell in transit from Slough to Paddington and ill-advisedly dressed in distinctive Quaker's garb (giving pause for thought to the Q-less alphabetic telegraph).[75] Tawell's case radically increased interest in the invention, enriching Cooke's licensee, but despite these sporadic uses, and others like signalling ahead to prepare horse-drawn carriages at Paddington, the GWR telegraph line fell into disuse.[76] In 1849 the GWR removed it – allegedly because of the collapse of one of Cooke's telegraph poles on to a locomotive.[77]

The telegraph had not, then, delivered safety and speed to railway working, at least on the GWR. In fact, in response to the bad publicity of accidents which dogged the companies, even since the opening of the L&MR, Wheatstone and Cooke cannily envisaged – and actively promoted – the telegraph as a means of regulating railway traffic more reliably than human operators, and perhaps even allowing economies in construction for financially pressed shareholders. In 1842, for example, they advertised a 'self-acting telegraph' (Babbage would have approved) to give notice of approach of trains towards tunnels, level crossings, stations and inclined planes worked by stationary or assistant engines. Cooke had attempted to make the case for cheaper railways with single lines and passing loops – but (and because) these could only be operated with 'perfect safety' by means of electric telegraph.[78] It was only later that the railway companies endorsed the particular meaning of the telegraph as regulator of railway traffic – rather than rapid messenger boy for the elite.[79]

The lack of enthusiasm from the railway companies left Cooke gloomy in 1843; perhaps the key event in the revival of his spirits and of the invention's reputation came in 1844 with the chance to fulfil a promise made by Ronalds in 1823: that of linking London with Portsmouth by electrical signal. Where hitherto the railway companies had been unwilling to adopt extended lines, the London and South Western Railway worked with the Admiralty to construct a line along the Railway's route, with separate wires for Admiralty, Railway and public use. The line was effectively guaranteed by the Admiralty with a royalty of £1500 per year for twenty years. Such a contract provided Cooke not only with cash but, as he had always wanted, with confidence in the telegraph, considered as a conduit of intelligence for both commerce and government.[80]

By 1845 Cooke and a group of influential railway entrepreneurs purchased Wheatstone's royalty rights on the British use of the telegraph for £30,000. Built upon these foundations, the Electric Telegraph Company (ETC) (1846) aimed, according to its prospectus, 'to establish a complete system of Telegraphic communication, connecting the Metropolis with the different Ports and Cities of the Kingdom'.[81] Railways by then offered an extensive network of 'ribbons of protected land' upon which – or above which – lines, relatively immune from tampering, could be established. The ETC quickly did this. These lines catered for a growing public thirst, manufactured in part by the companies, to communicate electrically. They were used for signalling and 'safety', at a time when a variety of methods – using flags, bells and hand signals remained in use.[82] With time-balls above their offices in London's Strand, at the top of the Nelson monument in Edinburgh, on the tower of the Sailors' Home in Glasgow and on the ETC's offices in Liverpool, the ETC from 1852 distributed Greenwich time in a radical show of the new alliance of railway, telegraphic commerce and government (ch. 1).[83]

The ETC installed 4000 miles of telegraph in just six years from its foundation: exploiting the confidence in 'practicability' engendered by demonstration to authoritative engineers, the protected routes available in the railway infrastructure, the capital available in the mania of the mid-1840s, and the model of a viable commercial and government information service in the London–Portsmouth line. Thus Cooke began to fulfil the promise of a telegraphic *system*; but that fulfilment, in enhancing the reputation of the inland telegraph, made its 'authorship' a matter of vicious dispute. Such had been the case with Watt's engine (ch. 2). Blast and counterblasts in print, appeals to fathers of the engineering profession like Marc Isambard Brunel, and legal manoeuvres now worked to elaborate extensive histories of the genesis of the 'electric telegraph', and to determine the answer to Cooke's burning question: *The electric telegraph: was it invented by Professor Wheatstone?*[84] Cooke's fashioning as gentleman and electrical entrepreneur, and his precise role as telegraphic projector – when the purpose and function of the telegraph was itself so unstable – had been undermined almost from the start by intense rivalries with Wheatstone.

By 1857 most large towns in Britain were linked.[85] In the mid-1850s, communication between city telegraph offices was routine. But to initiate that communication meant a visit to the office of the dominant ETC, to the office of its rival the English and Irish Magnetic Telegraphic Company (established in 1851 to connect Great Britain and Ireland), or, later, to the office of the Magnetic Telegraph Company (MTC) which resulted from their merger in 1857.[86] Co-operating with the MTC, the London District Telegraph Company (LDT), established in 1859, offered local, rather than inter-city, services. With its main office at Charing Cross, the LDT satellites spanned 4 miles outwards from a central co-ordinating hub at which messages were coded, decoded, or relayed not by self-acting instruments but by human operators.[87] In an attempt to generalise the LDT's plan, and again to cut out unreliable human intermediaries, the Universal Private Telegraph Company (1860) offered connections between business offices in centres of commerce including Glasgow, 'second city of the empire', the manufacturing hub of Manchester, the port of Liverpool, the textiles and finance centre of Leeds, and Newcastle, nexus of coal and heavy industry.[88]

New World electric telegraphy

Entrepreneurs in countries other than the United Kingdom projected telegraphic systems as indicators of national improvement, progress and pride. In the United States the key player was Samuel Morse: painter and first President of the American Academy of Design. Joseph Henry, head of the Smithsonian Institution in Washington, insisted that Morse made not a single discovery in electricity or magnetism applicable to the invention of the telegraph. Rather, Morse's case shows how the creative, even ad hoc, assembly of common mechanical parts (from clocks) and readily available materials (from the printing trade) might launch a new technology and that no technology can succeed without effective marketing.[89]

Whilst in Europe from 1830 until 1832, Morse had inspected the Chappe semaphore system; he seems also to have been aware, like Cooke and Wheatstone, of the philosophical demonstrations of Schilling, Gauss and Weber and he began his own experiments by 1832. No practical electrician, by 1835 Morse had nevertheless assembled an electro-magnetic telegraph.[90] Where Cooke looked to national opportunities, Morse, by 1836, was talking of a global reach of telegraphic signalling. Morse patented his electrical telegraph, like Cooke and Wheatstone, in 1837. Where Wheatstone had emphasised 'alphabetic' telegraphs which could be 'read' by any literate individual, Morse used from 1838 the so-called 'Morse code' of dots and dashes. The code, developed in collaboration with his partner, from 1837, the ironmaster Alfred Vail and borrowing features of typesetting practice, was originally tapped out by skilled operators and then, after transmission, printed out on strips of paper to be translated; later Morse found that skilled telegraphic operators (differentiated on gender lines) preferred to translate direct from sounds rather than printed marks.[91]

Like Wheatstone and Cooke, Morse promoted the telegraph through demonstrations from 1837 first in private, and then in public. The near simultaneity of these American demonstrations with those in Britain emphasised national rivalries over the authorship of the telegraph. Where Cooke and Wheatstone looked to railway business and government patronage to adopt the telegraph in Britain, Morse used strategies of seeking patronage, from the American and other national governments, by demonstrations in elite contexts and by offering gifts of instruments to potential patrons as far afield as the Ottoman court.[92] In the United States, the 'Morse Bill' of 1843 empowered the painter turned entrepreneur to begin a system of lines.[93] Railway infrastructure and government demand (as with the London–Portsmouth telegraph) provided the context for Morse's first line – along the railway route between Baltimore and Washington, DC in 1844; if careful media management focused attention on Cooke's GWR line, this could not compete with Morse's carefully chosen – and soon iconic – first telegraph message: 'what hath God wrought'.[94] The completion of the line was timely, coinciding with the eve of the Democratic Convention in Washington and the selection of the party's presidential candidate.[95]

The growth of the British lines through the ETC was echoed by the rapid implementation of a telegraph network in the United States: by 1848, Florida was the only state east of the Mississippi not connected; by 1852, 18,000 miles of telegraphic wire criss-crossed the Eastern third of the USA in a pattern dominated by small providers. Only the creation of the Western Union Telegraph Company in 1856 ended this intense competition. The Company won government support to 'Go West' – pushing the telegraphic 'frontier' to the Pacific coast, which it reached by 1861, some eight years ahead of the Union Pacific Railroad project (ch. 4).[96] Despite such integrative nation-building, worries mounted about a surfeit of undifferentiated, contradictory, irrelevant, incomprehensible or sensitive telegraphic news causing havoc to an over-responsive government, trade and public; and, during the Civil War, for the first time in the history of the United States, the federal government established an apparatus of censorship never before possible.[97] After the Civil War, the growing scale of inland operations dwarfed those in the United Kingdom. Western Union had 2250 offices and 100,000 miles of line by 1866.[98] In their lines of conveyance transmitting information and material commodities alike, the 'Western Union' and Union Pacific began to fulfil Morse's ambitions with systems spanning the North American continent. Meeting Morse's expectations for a global system, however, required a new kind of telegraphic technology.

Submarine telegraphy: connecting an empire

When the first commercial telegraphic land-lines came into use in Britain, postal systems and patterns of trade and commerce had already spread far beyond British shores. Mail contracts with new steamship lines from the late 1830s exemplified the nature and importance of imperial communications between land masses separated by water (ch. 3). The question telegraphic projectors asked, again from the 1830s, was how it might be possible to use their new means of communication across rivers, seas and even oceans.

Wheatstone's definitive statement came in February 1840 before the House of Commons Select Committee on Railway Communication. When the Chairman, Lord Seymour, inquired 'Could you communicate from Dover to Calais in that way [i.e. by telegraph]?', Wheatstone replied, 'I think it perfectly practicable.'[99] Wheatstone was unclear about how to insulate such a cable, although he considered using rubber. But by October *The Times* reported that 'Professor Wheatstone conceives that it is possible to communicate with his apparatus between Dover and Calais.'[100] By 1841, the utility of such a projected line would be enhanced, ironically, with a new optical link planned from London to Dover – and an existing Chappe telegraph taking over from Calais to Paris.[101] In 1843 Wheatstone had been careful to ensure that his rights to the telegraph, although in part signed over to Cooke, still allowed him to establish such an international link for his exclusive profit. By 1844 he was engaged in secret experiments in Swansea Bay.[102]

In July 1845, as Cooke prepared to launch the inland networks of the ETC, two brothers, Jacob and John Watkins Brett, wrote to the Tory Prime Minister Robert Peel, 'to submit a plan for general communication by means of oceanic and subterranean inland electric telegraphs'. In a remarkable rhetorical flourish, the Bretts printed their petition using their own apparatus before delivering it to Peel and eventually putting it into the public domain in a pamphlet. Their printing telegraphs augured many advantages, amongst which they listed:

I. THE IMMEDIATE COMMUNICATION OF GOVERNMENT ORDERS AND DESPATCHES TO ALL PARTS OF THE EMPIRE AND THE INSTANT RETURN OF ANSWERS TO THE SAME. FROM THE SEATS OF LOCAL GOVERNMENT. ETC. ALL DELIVERED IN AN UNERRING AND PRINTED FORM.

II. A GENERAL TELEGRAPHIC POST OFFICE SYSTEM UNITING THE CHIEF AND BRANCH OFFICES IN LONDON IN CONNECTION WITH ALL THE OFFICES THROUGHOUT. THE KINGDOM. FOR TRANSMITTING MESSAGES OF BUSINESS. ETC. FROM MERCHANTS BROKERS. TRADESMEN. AND PRIVATE PERSONS. AT A FIXED RATE OF CHARGE. THESE COMMUNICATIONS WOULD BE PRINTED ON PAPER AND ALL ENCLOSED IN SEALED ENVELOPES. AND ADDRESSED BY CONFIDENTIAL CLERKS. AND ISSUED BY SPECIAL MESSENGERS OR THE USUAL POST OFFICE DELIVERY.

III. THE ADVANTAGES OF THIS PLAN. APPLIED TO POLICE ARRANGEMENTS THROUGHOUT THE UNITED KINGDOM. AND TO THE ARMY AND NAVY DEPARTMENTS MUST BE AT ONCE OBVIOUS TO THE GOVERNMENT. BY IT INSTRUCTIONS MIGHT BE CONVEYED INSTANTANEOUSLY. AND THE MOVEMENTS OF THE FORCES SO REGULATED THAT ANY AVAILABLE NUMBER OF THEM MAY BE BROUGHT TOGETHER AT ANY GIVEN POINT. IN THE SHORTEST POSSIBLE TIME NECESSARY FOR THEIR CONVEYANCE.

THESE ARE SOME OF THE ADVANTAGES. OTHERS READILY SUGGEST THEMSELVES. NAMELY. GENERAL COMMUNICATION BETWEEN STATIONS ON THE COAST. SUCH AS LIGHT-HOUSES CHANNEL ISLANDS. ETC. SO THAT A GENERAL SUPERVISION OF THE COAST MIGHT BE OBTAINED FOR THE USE OF THE NAVY. LLOYDS. AND FOR THE PREVENTION OF SMUGGLING. ETC.[103]

Many such claims had been made piecemeal by earlier telegraphic projectors with diverse audiences. But for Peel the Bretts laid out an integrated and imperial vision of order, governance, record and registration. This complex vision thus encompassed an appeal to centralised government control, a recognition of empire, the creation of confidence and confidentiality through self-acting instruments, the co-ordination of forces of law at home and of colonial military power abroad, and the surveillance of coasts – in objects of national importance (ch. 1).

By 1846 Europe had an extensive and growing telegraph network; along a busy and commercially important railway line, the telegraph linked London with Dover. A submarine cable connecting Dover with Calais would now be one connecting the growing system of the ETC with the networks of continental Europe. To implement that project, however, required a submarine cable of about 20 miles, which was greater than the width of any European river. It also required the approval, not simply of the British government but also the French. In 1847 the Bretts obtained a concession from Louis Phillippe to lay a cable but they could not garner sufficient public support to fulfil this ambition; during the revolutions in Europe in 1848 there was little opportunity to renew such a project. By February 1849 *The Times* was also talking of a Dublin and Holyhead submarine telegraph, allowing a direct connection between two capital cities of the United Kingdom. The Chester and Holyhead Railway, with its spanning of the Menai Straits by the spectacular tubular Britannia Bridge, provided the route for a telegraphic line installed by the ETC.[104]

In amongst these speculations, experience of electrical telegraphy under rivers, across estuaries, and around dockyards demonstrated not the easy completion of plans for systematic connection – but rather the ever-present spectre of failure, with telegraphic ventures revalued not as objects worthy of national support but as speculations and swindles. On 7 February 1850, for example, *The Times* noted: 'The people of this country have been recently edified by an American project for connecting New York and the Isle of Wight by a sub-marine electric telegraph. Perhaps some of our readers experienced a little jealousy at this signal display of enterprise and daring on the part of our brethren in the States'; happily *The Times* could report: 'a well-considered and elaborate prospectus now lying before us [for] a simple line of railway...to connect...the two stations of Calais and Mooltan [in India. Thus] we have at last fairly beaten the Americans in comprehensive surveys and audacious speculation'.[105]

The Bretts had gained a new concession from the French government on 10 August 1849 for ten years, on the understanding that communication would be effected by 1 September 1850. They could not simply re-use the uninsulated overhead cables familiar to land-line engineers. Wheatstone himself had been unsure about how to insulate a submarine telegraph line but by 1843 new candidates were becoming available. In that year the multi-purpose rubber-like substance 'gutta percha', derived from a tropical tree, had been 'discovered' by Westerners and presented to the Royal Asiatic Society. Karl Wilhelm Siemens came across samples in London and sent them to his brother Werner in Berlin: by 1846 he was using gutta percha as an insulator for lines laid underground (to secure them against the possible attacks of a rebellious populace) (Figure 5.2).[106] Apparently independently, Faraday had published his observations on the electrical insulating properties of this substance early in 1848.[107] The Bretts now, in 1850, took a single copper wire, insulated it with a half-inch layer of gutta percha, and prepared to weight it with lead in order to construct their international line. After

Figure 5.2   Collection of the juice of the *Isonandra gutta* tree, source of the best quality gutta percha as insulator for Imperial submarine telegraph cables.

Source: Charles Bright, *Submarine Telegraphs. Their History, Construction, and Working*. London: Crosby Lockwood, 1898, facing p. 256. (Courtesy of University of Kent Library.)

intensive work, reported in detail in *The Times* from April 1850, they used the steamship *Goliath* to lay out a cable from Shakespeare's Cliff at Dover and the opposite chalk headland of Cape Grisnez on the French coast, midway between Calais and Boulogne, on 23 August. It appears to have been working by 28 August only days before the deadline. *The Times*, which had followed events closely from the beginning, devoted an editorial of 31 August to the cable, claiming now of this scientific miracle: 'The jest of the scheme of yesterday has become the fact of to-day. The wildest exaggeration of an Arabian tale has been outdone by the simple achievement of modern times.'[108]

The line ruptured – allegedly having been scooped up by a zealous French fisherman – and after less than a week the transmission of messages between nations had utterly ceased. 'It was, in fact, looked upon as a mad freak – and even as a gigantic swindle – indulged in only by wild minds', recalled Charles Bright in 1898.[109] Not surprisingly there was considerable difficulty in raising the capital to fund another attempt, despite subsequent efforts to revalue the cable as a necessary experimental trial in the science of sub-marine telegraphy, prefacing the establishment of durable links between Britain and France.[110] At least in 1851 engineers could speak more easily of success, thanks to a new operation which was socially robust and a cable which had been materially fortified to withstand severe mechanical strain and the hostilities of the underwater environment. Four copper wires replaced a single fragile strand; the manufacture reinforced an insulated sheath of gutta percha with a covering of spun yarn saturated with tar; and this sheath, in turn, was armoured externally with a strong wrapping of rope made from ten galvanised iron wires. The technique of wire-rope making, rather like the substance gutta percha, had been imported. In the 1830s Lewis Gordon saw this technique being used extensively in Central European mines. Although much contested, it was probably Gordon's partner, R.S. Newall, who had the idea of using this product to make a durable submarine electric cable.[111]

Like its failed predecessor, the 1851 line was a very public affair. This was absolutely the choice of the project's sponsors. Thomas Crampton, engineer of the London, Chatham and Dover railway, was himself a key shareholder of the Submarine Telegraph Company which established the line. Having volunteered to execute the work and, more importantly, to provide half the capital for it, it was Crampton who, on 25 September 1851, announced the connection in the Crystal Palace, just as Victoria, surrounded by a group of men of science, formally drew the Great Exhibition to a close.[112] Later commentators would interpret Crampton's work as 'the first step in submarine telegraphy ... [from which] have sprung the prodigious developments of the system which now connects all parts of the civilized globe in a network of electrical wires'. Merging the reputation of the man with machine, Crampton must therefore be 'the father of submarine telegraphy'.[113] Other dignitaries had been involved in a more spectacular demonstration of international co-operation:

> It was a singular coincidence, that the day chosen for the opening of the Submarine Telegraph was that on which the Duke of Wellington attended in person to close the Harbour Sessions, and it was arranged that his Grace on leaving Dover by the 2 o'clock train for London should be saluted by a gun fired by the transmission of a current from Calais.

As the train started a signal was passed, and instantly a loud report rever-
berated on the water and shook the ground – a 32-pounder loaded with
12 lbs. of powder had been fired by the current. The report had scarcely
ceased ere it was taken up from the heights, the military, as usual, saluting
the departure of the Duke with a round of artillery. Guns were then fired
successively on both coasts, Calais firing the gun at Dover, and Dover
returning the compliment to Calais.[114]

Anticipating the fulfilment of the line's promise, Julius Reuter had moved
to London in June 1851 where he quickly set up his 'Submarine Telegraph'
office in the Royal Exchange Buildings in the City in order to distribute
news and commercial information.[115] From the time of Chappe there had
been much news *about* the telegraph – and indeed, the opening of the
Dover–Calais cable for public use in November was the opportunity for
cartoonists to caricature John Bull at Dover and the French Emperor at
Calais as Siamese twins linked by 'electric wire'.[116] High-profile events made
known *by* telegraph had assisted in winning support for it as a carrier of
intelligence; and from the 1840s, a group of New York newspapers, banded
together as the Associated Press to share the costs of telegraphic newsgathering,
ready to distribute 'electric news' like that of sporting events.[117] Expanding
networks in the United States, Britain and continental Europe radically
increased the capacity of the telegraph to *distribute* news, although that
news continued to be gathered by conventional means, like mail, acting in
newly negotiated roles complementing the electric telegraph.[118] Entrepreneurs
like Reuter invented telegraphic news as a valuable information commodity.
Already experienced in the handling of news and market information by
land telegraphs on the Continent, Reuter now used the new cross-channel
line to orchestrate the exchange of stock market prices between Paris and
London brokers. Soon he was also in the business of supplying information
on grain prices in the Baltic region.[119]

Like Reuters, the Gutta Percha Company quickly exploited the cable,
placing adverts in *Bradshaw's General Shareholders' Guide, Manual, and Railways
Directory* in an attempt to promote confidence in further submarine
projects. An image of the cable itself accompanied a detailed description of
its construction. Under the heading 'Perfect Insulation', the Company
expressed its readiness, 'having completed the Covering of the Insulated
Wires for the Submarine Telegraph Communication between France and
England . . . to undertake contracts on favourable terms. By means of their
improved machinery', they claimed, 'they are enabled to insure perfect
insulation'.[120]

Within five years projectors sank short submarine cables in relatively
shallow waters to tie Great Britain more securely to Ireland: a line from
Portpatrick (in Scotland) to Donaghadee (Co. Down) strengthened the
existing mail and steamer route; another from Holyhead to Howth (north of
Dublin) making concrete earlier visions of direct communication between

London and Dublin. Britain was spliced also to Belgium (Ostend), Holland (the Hague) and the Channel Islands. Even by 1853, the Submarine Telegraph Company could list, in the pages of *Bradshaw's*, more than 200 European 'Cities and Towns…in Electrical Communication with Great Britain.'[121] Although a major hub for the telegraphic network, Britain was not the only hub. There were also submarine cables between Denmark and Sweden, between Corsica and Italy, and, in British North America, a line between Prince Edward Island and New Brunswick in Canada. In 1855 the British military made the submarine telegraph a conduit of intelligence under the Black Sea from the Crimean theatre of war – connecting Varna to Balaclava at a distance of 300 miles.[122]

'Hitherto', wrote Charles Bright in 1898, 'the efforts of the early projectors of submarine telegraphy had been confined to the work of connecting countries divided only by narrow seas, or establishing communication between points on the same seaboard'. In 1852 George Wilson spoke of 'schemes for telegraphing across the Atlantic and the Pacific [which] have been triumphantly expounded to the wonder-loving public'.[123] Telegraph projectors and engineers hoped to cash in on a stock of credibility made available in the short-sea cables of the early 1850s, forgetting the preliminary – and even subsequent – failures. Advocates claimed that a range of commercial, political and social benefits would justify the investment of capital, administrative work, diplomatic endeavour, international co-operation, mechanical and electrical engineering skill, and individual perseverance. They insisted that steamships with the capacity to carry a 2000-mile cable could weather the storms of the unpredictable North Atlantic Ocean, that a cable could be constructed to withstand the mechanical strains placed upon it by the heaving swells, and longer by an order of magnitude than anything yet in place.[124]

But by 1853 the engineers had already begun to experience alarming problems with the apparent retardation of the electric current in long lines – in other words, far from being an instantaneous communication, the electric signals appeared to move evermore slowly as the length of the line increased.[125] In a presentation to the BAAS at Liverpool in 1854, Edward Bright, manager of the English and Irish Magnetic Telegraphic Company, argued that the speed of transmission in subterranean conductors (insulated by gutta percha) was less than 1000 miles per second compared to about 16,000 miles per second on bare overhead wires. This reduction in signalling speed presented a serious obstacle for those who wished to argue that a direct Atlantic cable would be viable.[126] Recognising this, engineers tried to win support for a route subdividing the Atlantic into sections of no longer than 670 miles. In 1854 the King of Denmark granted a concession to use his territories for a 'North-about route' from Scotland to Labrador via the Faröe Isles, Iceland and Greenland. Using self-acting relays, London could 'communicate direct with America' just as it could with other European cities; for those who had no trust in such relays, a 'staff of telegraph officials' might be established at each telegraphic terminus.[127]

The argument of retardation amongst telegraphic advocates continued. In 1855 and 1856 the surgeon and electrician Wildman Whitehouse flourished details of experiments – using a continuous circuit of 2000 miles of underground wire – which countered Edward Bright's depressing prognosis. Whitehouse concluded that, with a clever use of alternating current directions, 'signals were clearly and satisfactorily transmitted over this vast length... with a facility that would answer every commercial requirement'.[128] Although William Thomson used mathematical theories inaccessible to practical electricians to raise further doubts about the economy of long-distance cables, he pointed to a solution: larger cross-sections (of copper conductor and gutta percha) than those already being proposed for a direct Atlantic cable would reduce the retardation effect and increase the rate of signalling.[129] Investor confidence in the 'direct' project therefore rested heavily on the optimistic forecasts of Whitehouse and Thomson.[130]

In 1856 the dynamic American entrepreneur and industrialist Cyrus West Field looked for support for a direct route between Newfoundland and Ireland. He met the engineer of the Prince Edward Island to New Brunswick cable, twice sounded the Atlantic ocean floor over the proposed route and organised the Atlantic Telegraph Company. The company was based not in America but England, the dominant centre of expertise in submarine telegraphy and a potential source of capital. So kind was the geology and oceanographic circumstance of a gently undulating 'Telegraphic Plateau' between Valentia in south-west Ireland and Trinity Bay in Newfoundland, in contrast to the much greater depths further south, that proponents such as United States Navy Lt M.F. Maury spoke of the plateau as seeming 'to have been placed there especially for the purpose of holding a submarine telegraph'.[131]

As early as 1845 the Brett brothers had registered their own 'General Oceanic Telegraph Company' with a view to substantiating the grand project outlined to Peel. Already acquainted with Cyrus Field, John Brett joined with him and the young telegraph engineer Charles Tilston Bright in an agreement of 29 September 1856 that, as the projectors, they would exert themselves 'with the view and for the purpose of forming a company for the establishment and working of electric telegraphic communication between Newfoundland and Ireland, to be called the "Atlantic Telegraph Company"...' Government guarantees, through the Admiralty, had secured the future of the London–Portsmouth telegraphic line in 1844. Now, the British Government, convinced of the imperial worth of the Atlantic Cable in linking London, not with the United States, but first with British 'dominions' in North America, provided a guarantee of £14,000 per annum during the working of the cable and promised naval vessels to assist in the laying.[132]

The correspondent of *The Times* W.H. Russell wrote in 1865 of the need for capital 'which the *confidence* of moneyed men in the United States did not induce them to supply'.[133] Capital came not from the Americans, or from a general public, but overwhelmingly from groups like Liverpool's tightly knit mercantile communities consisting largely of merchants and ship-owners. Liverpool was also, from 1857, the headquarters of the newly merged MTC under Edward Bright as manager and his brother Charles Tilston Bright as engineer. Although presented in the broadsheets as a national, or imperial, good it can be seen very much as a project of private empire-building, in which relatively closed groups of entrepreneurs, each interested in complex ways with the various shipping, railway, telegraphic and commercial concerns, took on trust the risks – and the potential benefits.

Babbage had written in the 1830s of the great benefits to 'English' trade of 'confidence' in commercial interaction, even to the extent of removing the need for written agreements. Although a prospectus was issued in November 1856, and Edward Bright held meetings of Liverpool worthies, the £350,000 necessary for the work was 'obtained in an absolutely unprecedented manner. There was no promotion money, no advertisements, no brokers, no commissions; neither was there at that time any Board of Directors.' Election of a Board would be decided at a meeting of shareholders and remuneration to the projectors would be dependent upon a return of 10 per cent to shareholders, after which the projectors would divide the surplus. The capital was quickly raised, principally from shareholders in the MTC and their friends, in 350 shares of £1000 each. Directors elected included John Brett, John Pender, and William Thomson. Charles Tilston Bright became chief engineer and Whitehouse, the electrician.[134]

In April 1857, the *New York Herald* boasted that this would be 'the great work of the age'. Projectors and publics alike hailed a 'gigantic step' which was worth the risk of sinking vast amounts of capital to depths of more than 2 miles.[135] Representing the practical professionals, *The Engineer* identified *with* the Atlantic telegraph project, and also identified it *as* a significant event in the 'march of human progress', advancing inevitably like a series of waves 'driven onwards by the same resistless force'. Accordingly, all kingdoms of Europe had linked themselves together by the electric telegraph, and America was 'in a high state of electric tension'. With continental systems of communication in place, all that was needed was to complete the transatlantic conductor and thus provide yet another link in the 'electric nerve' system of the world.[136]

The Company ordered some 2500 miles of cable. Two separate contractors – Glass, Elliot & Co. and R.S. Newall & Co. – each constructed, insulated and armoured half of the cable (which was made in short pieces 2 miles in length spliced together). There were two principal ships, representing the two participant nations and indicating international co-operation. Half of the cable went with the American frigate *Niagara* (the United States's largest warship); the other half went to the British man-of-war HMS *Agamemnon*. In August 1857 both steamships moved out from the eastern terminus in Ireland and headed west, gradually extending the telegraphic frontier of Europe towards the New World. Yet the cable ignominiously broke after only 300 miles, due to an accident with the paying out machinery, and the ships returned to port.[137] With no fait accompli to present, commentators targeted Bright who had been responsible for managing the workforce. *Scientific American* easily claimed that 'Mr Bright [was] evidently not bright enough to lay a telegraph cable', with gusto insisting that it was 'palpable that the laying of the cable...more resembled a lottery than an important scientific, engineering and nautical operation'.[138]

Engineers and entrepreneurs, however, were practised at turning such disasters to rhetorical advantage. Field rhapsodised: 'The successful laying down of the Atlantic Telegraph Cable is put off for a short time, but its final triumph has been fully proved [sic], by the experience we have had since we left Valentia. My confidence was never so strong as at the present time, and I feel sure that with God's blessing, we shall connect Europe and America with the electric cord.' Astonishingly, Field insisted that the 'accident' which had jettisoned the cable, would be 'of great advantage' to the Company.[139] With a slump in confidence, *pace* Field, the Company nevertheless replaced the lost cable and devised a new paying out machine. The two ships now rendezvoused in June 1858, each carrying a cable of over 1000 miles. Having spliced the cable together they moved outwards from the middle of the Atlantic in opposite directions, with the *Niagara* heading for Trinity Bay and the *Agamemnon* to Valentia. The cable broke; they tried again; it broke again. But on the third attempt, in mid-July 1858, despite problems with the insulation, and being buffeted by prolonged storms, the two ships reached their ports. On 5 August 1858 the New and Old Worlds were electrically linked.[140]

Earlier fiascos forgotten, there were riotous celebrations both in Britain and in America. This 'vast enlargement...to the sphere of human activity', wrote *The Times* on 6 August, was second only to Columbus's 'discovery' of America.[141] On the other side of the ocean, fireworks set fire to the cupola of New York's City Hall. *Scientific American*, forgetting its earlier snipes at Bright, effused: 'All hail to Anglo-Saxon genius! and two nations' heartfelt thanks to the noble, aye and mighty men of science, capital and energy whose uniting zeal and indomitable perseverance have linked the hemispheres with the electric cord!' Here was an opportunity for ceremony not to be missed: sanctioning, once again, the electric telegraph, Queen Victoria offered personal greetings to President James Buchanan and a knighthood to Charles Bright. Amidst civil celebrations, journalists, once again, predicted a new era in international relations: the British government did indeed use the cable to communicate to Canada that a peace had been concluded with China; an order for two regiments of soldiers to return to England for service in India was countermanded, allegedly saving £50,000.[142] Such exchanges of intelligence appeared to provide a tangible demonstration of the Bretts' vision of 1845.

Julius Reuter had already set up deals which included arrangements with the ATC (costing £5000 per annum), the establishment of a central bureau in New York with subsidiaries in all the main cities in the United States for the collection of despatches and especially money market information, and an agreement with *The Times* to supply messages at five shillings per word, reduced by half if Reuters were credited.[143] Entrepreneurs and news brokers, like Reuter, then, watched as on 17 August ATC officials transmitted a message which escaped the sphere of celebration and the realm of government and military edict to intervene in arenas of commerce, safety and confidence: 'ward Whitehouse Mr Cunard wishes telegraph McIver Europa Collision Arabia put into St Johns No lives lost Will you do it stop anxiety non arrival De Sauty'. The cryptic 'telegram', hand-written at Valentia on a prepared printed sheet and taking twenty-five minutes to send, routed via ATC staff a request from the owner (Samuel Cunard) of two steamers, *Europa* and *Arabia*, that had collided off Cape Race (at the southern tip of Newfoundland), to report no loss of life to the Liverpool partner Charles McIver.[144]

Despite talk of the durability and technical robustness of the technology, however, the 1858 cable – like the 1850 Dover–Calais cable – was embarrassingly short-lived. From the beginning, transmission, by Morse's system, was painfully slow. Newly invigorated advocates of the 'North-about route' gleefully reported the ATC electrician De Sauty's admission that the message of 99 words and 509 letters sent by Queen Victoria took from 10.50 a.m. on 16 August 1858 to 4.30 a.m. the next day – a time of two minutes for each letter. Even kinder estimates of four seconds per character undermined the possibility that the cable could be commercially viable since slow traffic meant poor returns for the owners, or else luxury use only.[145] Men like Cromwell Fleetwood Varley, chief electrician of the merged ETC and International Telegraph Company, struggled to rebuild public confidence in two ridiculed practices – spiritualism and long-distance electrical telegraphy – using the authority of experiment.[146] Whitehouse's (and Bright's) communications of promised actions suddenly appeared extravagant and speculative, those terms of abuse heaped on failed Victorian businessmen of all kinds. Confidence in the telegraph, and in the men who staked their reputations on its trustworthiness, began to crumble.

Practical science had not after all, it seemed, provided reliable predictions. A shot of 2000 volts, designed to counteract this sluggishness, and administered by an increasingly desperate Whitehouse a month after the first use of the cable, simply rendered the cable defunct. As news spread, the company's share prices on the London Stock Market plummeted. By 20 October 1858 the cable failed completely, having transmitted, on the most optimistic report, only 732 messages. At the end of the year Field's office and warehouse in

New York mysteriously burned down and two years later his private firm admitted bankruptcy.[147]

Failure of the 1858 cable breathed new life into proponents of rival schemes. To 'work a cable of that length for commercial telegraphy', argued the devotees of the 'North-about route', 'is a problem that requires new discoveries in science to solve', since the retardation of the signal, now patently demonstrated in experience, was also 'in obedience to the laws of the propagation of electricity' as expressed by Thomson. Although Thomson asserted in January 1859 (see epigraph) that the 1858 cable showed a direct route to be a position gained – and soon to be recaptured – in the onward march of practical telegraphy, opponents countered that *they* had witnessed a demonstration 'by experiment, that a direct line could not be expected to succeed as a commercial telegraph'. This had been a period of trial, delivering 'a rich store of facts and experience' (on the best mechanical form of a deep-sea cable and the appropriate 'electrical conditions' for it) and focusing the attention, and the debate, of men of science. The next generation of telegraphers would not have to 'grope their way along untrodden paths without a chart'. In May 1860 an 'influential deputation' lobbied Lord Palmerston – and the Admiralty provided the paddle steamer *Bulldog* to survey the route once again. But their desire was clear: to construct a new form of cable cut into manageable elements to give it 'greater speaking power'.[148]

In the years after the demise of the 1858 cable the world's telegraph lines continued to grow so that by 1862 they stretched some 150,000 miles, of which 15,000 were British, 80,000 covered continental Europe and 48,000 spanned America. With the growth especially of land networks, interested parties came increasingly to argue that seas and oceans represented congested information bottlenecks, not simply in the case of the Atlantic. The British had been attempting to establish other networks to consolidate their imperial power, in Australia and India.

Key to the introduction of telegraphy in India were Governor-General Dalhousie, central advocate of the railway system there (ch. 4), and surgeon and electrical lecturer William O'Shaughnessy. In 1849 the Court of Directors of the East India Company asked Dalhousie to investigate the building of telegraphic lines; India did not have the ribbons of protected land associated with the railways in Britain and so valuable to the ETC, but by 1852 O'Shaughnessy had completed a line of 130 km from Calcutta (on the east) to Kedgeree on the Bay of Bengal, primarily to give advance notice of the arrival of ships to the British community. The same year, the Company provided a generous budget for a 6000 km telegraphic network, designed, in Dalhousie's terms, to expedite 'transactions of Indian business' (both in government and commerce). O'Shaughnessy toured England and the continent to garner expertise; he requisitioned materials and recruited a troop of sixty soldiers dedicated to the scheme. He chose simple instruments easy to make and repair in India and prepared to counter the local difficulties of insects, monkeys and elephants. Lines linked Calcutta, the eastern port and centre of administration, with Agra (by 1853) and by 1855 there was a skeleton of main lines taking in the port of Madras on the east and Bombay on the west.[149]

With a network of 6840 km and 46 offices in place by 1856 there was a telegraphic deficit between India and London. After the 'Mutiny' in the summer of 1857 (which precipitated the demise of the East India Company), the British government attempted to exert greater political control over India. Advocates, like Lionel Gisborne, raised the prospect of a cable linking London and the Indian administrators – and then argued about how best to fill what now, with extensive lines constructed or constructing across and along the Mediterranean, came to be seen as 'gaps' in a route to India. In 1857 and 1858 there were two main contenders. Although passing through territories in the Middle East beyond London's control, the first route promised ease of access and the reliability of landlines proven throughout Europe and America. The telegraph line would pass through Ottoman Turkey, along the Tigris and Euphrates Valley and then via the Persian Gulf to India.

The alternative was a cable originating in European territory, passing beneath the Mediterranean from Malta to Alexandria, then to Suez, and then along the Red Sea, via Aden, to Karachi (now in Pakistan) on the west of the territories historically dominated by the East India Company. The Red Sea submarine portion raised questions concerning practicability, especially in the light of the difficulties experienced in 1857 by the Atlantic cable; yet with a route much of which was underwater and little of which traversed lands beyond British influence it was considered more secure. There was also a complementary scheme to have a 'direct line' via Gibraltar avoiding any commitment at all to other European powers. The British government guaranteed £36,000 per annum to the Red Sea and Indian Telegraph Company.[150]

Yet the joint Board of Trade/Atlantic Telegraph Company Committee set up in 1859 and reporting in 1861 claimed that of over 11,000 miles of submarine cables laid since the Dover–Calais line, only 3000 miles were actually working, and most of them were identified as 'shallow-water cables' down to about 100 fathoms (600 feet). Deep-sea lines not working but laid included the Atlantic (2200 miles), the very Red Sea and India cable (3500 miles) which had promised direct communication with London, the Sardinia, Malta and Corfu (700 miles) and the Singapore and Batavia (550 miles).[151] The first cable to India cost £800,000 but, even with the 'perfect insulation' of gutta percha, proved vulnerable to marine bore worms – and the line transmitted not a single message.[152]

Established to investigate the state of the submarine telegraph industry, the joint Committee demonstrated the close links between Government (represented by Captain Douglas Galton of the Royal Engineers), mechanical science (represented by William Fairbairn, BAAS President in 1861), electrical science (represented by Charles Wheatstone and the telegraph engineers Latimer and Edwin Clark), the ATC (represented by its Secretary George Saward) and general engineering (represented by G.P. Bidder and telegraph engineer Cromwell Varley).[153] The status of such bodies was itself contested: in the aftermath of the Royal Commission investigating the collapse of Robert Stephenson's iron bridge across the Dee in 1847, Isambard Kingdom Brunel had raged at the 'despots' who would 'lay down, or at least suggest, "rules" and "conditions to be (hereafter) observed" . . . or, in other words, embarrass and shackle the progress of improvement to-morrow by recording and registering as law the prejudices and errors of to-day'. Thus Brunel implied that these assemblies, of government experts, drawn from academia as well as practice, served merely to atrophy engineering practice, preventing the progress that Brunel believed came naturally as a result of competition in the engineering market place.[154]

The establishment of such a committee in 1859 indicates the intense *national* and indeed *imperial* importance afforded the cable industry – at a time when it was subjected to perhaps its harshest criticism. Ad hoc statements by proponents – Whitehouse, Thomson, Field and so on – had attempted to defuse, in a piecemeal way, the apparently disastrous evidence of repeated failure, as cables snapped, sank, were consumed or ceased to respond. Rival accounts of differently interested parties countered their statements. The mechanism of the committee provided a space in which the dirty linen of telegraphy could be fully aired – then expertly laundered. It was a space, combining so many interests as to appear disinterested, in which sporadic, unexpected and decisive failure could be revalued as essential and extensive experience laid down and collated; it was, also, an arena of judgement, in which scapegoats (Whitehouse) could be singled out and saviours (Thomson) identified, and of progress, in which mechanisms of therapy could be sanctioned.[155] A consensus of interested groups fabricated disinterest, and constructed confidence in the reliability of telegraphic visions.

The Committee concluded, publicly and conveniently, that most of the technical problems of submarine telegraphy had, either by scientific investigation or through painful trial and error, been solved. The failure of the Red Sea line was 'attributable to the cable having been designed without regard to the conditions of the climate or the character of the bottom of the sea over which it had to be laid; and to the insufficiency of the agreement with the contractor for securing effectual supervision during manufacture and control of the manner of laying'.[156] The experts also looked at the way in which the old Atlantic cable had been constructed and laid – as a prelude to claiming, without originality, that better ways would be available for doing both. The Committee proposed new schemes of management, re-examined issues of construction (the cable's insulation and mechanical strength), addressed questions concerning the paying-out apparatus and the signal-receiving mechanisms, and, perhaps most important of all, recommended the standardisation and surveillance of cable manufacture by experts.[157]

The 1859–61 report concluded bombastically that oceanic telegraphy would in future 'prove as successful as it had hitherto been disastrous'.[158] Following the publication of this report, the ATC appointed a 'consulting scientific committee' consisting of Galton, Fairbairn, Thomson, Wheatstone, and Joseph Whitworth – the embodiment of standardisation in mechanical engineering. The committee began to lead the project in a context of shaken confidence, when American investors were occupied with the events of the Civil War. Their task was to examine the samples of cables from various manufacturers – fulfilling Brunel's prediction, they refused to countenance an untried material (India rubber) as insulator, and by 1863 had accepted the tender of Glass, Elliot & Co. who were now deemed to have a 'successful and varied experience in the manufacture and submergence of cables in different parts of the world'. They also agreed to undertake the work at cost and subscribed heavily in the ATC. By spring 1864 the new Telegraph Construction and Maintenance Company (TCMC) had been constituted, with an authorised capital of £1 million and with the financier John Pender as Chairman, to take over the merged Gutta Percha Company and Glass, Elliot & Co. as well as to purchase more than half of the ATC's stock.[159]

In the renewed Atlantic attempts, Brunel and Scott Russell's *Great Eastern* (ch. 3) played a vital role. Guided by the 'consulting scientific committee', the TCMC constructed a new cable of 2300 miles, in a single piece to avoid any problems associated with splicing mid-ocean – but raising the issue of how to carry a cargo of 5000 tons (with another 2000 tons of water for storage). The 1859–61 Committee had suggested that the only vessel then large enough was the *Great Eastern*, launched in 1858 and having made her maiden voyage in 1860. A combination of traditional paddles and innovative screw propeller made the 'Leviathan' extremely manoeuvrable – and promised to revive her fortunes as a commercial cable-laying vessel.[160] By 1865 a syndicate of three businessmen, including Daniel Gooch and Thomas Brassey, had bought the white elephant at a price which astonished even them – just £25,000 – and they were immediately in a position to charter her to the TCMC for £50,000.[161] Gooch and Brassey also became Directors of the TCMC: the experiments of the *Great Eastern*, the Atlantic Cable, and railway enterprise were thus intimately linked.[162]

A new attempt to link Ireland and Newfoundland began in Valentia on 23 July 1865. That campaign would be documented in fine detail by W.H. Russell, *The Times*'s reporter and war correspondent famous for his despatches from the Crimea between 1854 and 1856. The account of this 'historian of the enterprise' was an authorised version, dedicated by 'special permission' to the Prince of Wales. Thanks to the professional artist (Robert

Dudley) that travelled with him, the book, published before the end of the year, was lavishly illustrated (adopting pictures already reproduced in the *Illustrated London News*). Such carefully orchestrated publicity, channelled through 'The Thunderer' – and complemented by reports filed in *Blackwood's*, the *Cornhill* and *Macmillan's Magazine* – guaranteed the veracity of the project and also served the function of creating additional witnesses to a gigantic electrical experiment (Figure 5.3).[163]

After 1200 miles, only 600 miles from the destination of Newfoundland, the cable snapped and sank 2100 fathoms to the bottom of the Atlantic. Attempts to grapple it failed, the ship returned to port – and now the ATC itself failed, although a successor – the Anglo-American Telegraph Company (AATC) – emerged. Those who had been on board the *Great Eastern* during the attempt, including Captain James Anderson, Daniel Gooch, William Thomson and Cromwell Varley once again frantically tried to bolster credibility for – and close the chapter of – the Atlantic cable, issuing a collective testimony to no fewer than twelve points by now incontrovertibly demonstrated: 1858 had proved messages could be transmitted; 1865 showed the *Great Eastern* was the right ship for the job; the cable could be grappled from great depths (even though it had not been recovered); the paying out machine worked perfectly (although the cable had snapped); good signalling speeds could be obtained (from ship to shore, at least); the cable was strong enough to withstand strains far greater than it had experienced (even though it broke); the 1865 cable had an insulation more perfect than the 'then perfect' insulation of the 1858 cable; electrical testing had taken place with 'unerring accuracy' (enabling immediate fault location); and a steam engine attached to the paying out machinery would enable the recovery of the 1865 cable – for subsequent repair.[164]

Forgetting the inevitable success of the previous cable – vouched for by the scientific committee – the TCMC constructed another in a robust new design, and modified, yet again, the paying-out apparatus. The ship left Valentia, once again, in June 1866 and after an uneventful crossing, landed the cable at Heart's Content Bay on 27 July. The engineers succeeded in grappling the 1865 cable, spliced it to a new cable and landed that, also, at Newfoundland on 8 September. Two working Atlantic cables were quickly connected to the existing telegraphic systems of the Old and New Worlds. In a now well-rehearsed pattern, compliments passed between the appropriate dignitaries.[165]

More than that, the fixing, in several senses, of the telegraph as a position secured in the inexorable advance of electrical communication for empire coincided with the social refashioning of half a dozen men, of science, business and engineering practice, who had acted – in consort, if not always in harmony – to fix that practical fact. In recognition of their contributions to the prestige of the British Empire, Queen Victoria conferred titles on six of the leading figures in both the 1865 and 1866 enterprises at Windsor Castle on 10 November 1866. Daniel Gooch (AATC vice-chairman) and Curtis

215

Figure 5.3 Satirists speculate on the true cause of the failure of the Atlantic Submarine Telegraph Cable in 1865.

Source: *Punch* 49 (August 1865), p. 47. (Courtesy of Aberdeen University Library.)

Lampson (ATC vice-chairman) became baronets while Thomson, Samuel Canning (chief engineer), Captain Anderson and Richard Glass (managing director of Glass, Elliot & Co.) received knighthoods.[166]

For Edward Bright, writing in January 1867, the telegraph was now a demonstrated fact: 'It may now be said that wherever civilisation exists, there telegraphic communication has penetrated linking up continents, countries, and races.' The global civilising reach of British telegraph stood in stark contrast to its origins as a mere philosophical plaything. Bright invited his readers to reflect on the momentous transformation which had been effected as telegraphy left infancy and at last reached its maturity:

> This means of conveying thought in a moment between distant points, which twenty years ago had scarcely emerged in a practical form from the philosopher's studio, now covers the world with its vast network of communicating fibres.[167]

Telegraphic and electrical empire-building

By the 1860s, men of science, acting collectively in a 'consulting scientific committee', promoted themselves to the role of disinterested evaluators of submarine telegraphic practice, independent, they would claim, of particular interests and with an authority underwritten by eternal scientific verities. Thus, Edward Bright asserted in 1867: 'The whole system of electric communication has...become a recognised science in its various branches.'[168] Although this campaign had a particular object – the successful outcome of the 'most gigantic electrical experiment' – it was indicative of wider ambitions to see a babble of localised practical skills reformed through new practices focused in laboratory science. This campaign would be expressed not in merely intellectual arenas but in pragmatic action. That pragmatic action, rooted in missions for universalisation, allowed a new, more subtle, form of imperialism.

Disseminating British instruments, standards, and practices throughout the Empire sustained and empowered that empire. The British dominated the submarine cable industry globally until the Great War – through expertise, cable-laying ships, manufacturing and constituent materials.[169] The very instruments which allowed submarine telegraphic communication, developed and calibrated within the laboratory, would punctuate the lines according to a global standard. Standards of measure, developed in a context of quality control and a culture of precision, would be aggressively exported as 'imperial' – or even 'absolute' – units, by which imperial lives would be guided. Standards of professional electrical engineering, regulated in teaching regimes, institutional structures and patterns, and printed journals accompanied the import of human, as well as material, commodities – and the export of disciplined experts whether to India, Australia or Japan.

It was William Thomson, very much the 'saviour' of the Atlantic to Whitehouse's 'scapegoat' in the verdict of the 1859–61 Committee, who worked in a context of telegraphic engineering and academic physics laboratory to provide a solution to the problem of economical (and rapid) signalling on long-distance submarine lines. Inland telegraph lines in Europe and America displayed an extensive array of signalling, receiving and testing instruments with those of Morse dominating those of the United States and Europe. Both practice and physics indicated that for a long submarine cable, messages were best sent with small currents. Thomson's sensitive 'marine-mirror galvanometer' replaced the Morse sounder for receiving messages on such lines.[170] He patented this instrument in February 1858, after the failure of the first Atlantic attempt, but just ahead of the second and also coinciding with a major expansion of his physical laboratory at the University of Glasgow. Thomson received £500 from the ATC to develop his signalling apparatus and invested heavily himself in telegraphic instrument design and trials.[171]

The activity in Thomson's laboratory and the directives of the Committee of 1859–61 were mutually reinforcing. Thomson's expanding laboratory empire set standards of production and surveillance for a business of telegraphic engineering under investigation. By the 1860s the laboratory's functions in relation to telegraphy were three: inventing new instruments, testing and calibrating such instruments, and using 'volunteer' students to gain hands-on experience in associated manipulations and measurements. A former student explained in 1870 that the 'success of the Atlantic Cable is in great measure the result of years of patient work in the Glasgow Laboratory, – experimenting on the strength of batteries, the tenacity and electric conductivity of wires, and the capacity of different substances for resisting the action of water, and testing and perfecting the numerous exquisite instruments of Sir William Thomson's invention'. Furthermore, 'the excellent electrometers turned out by the Glasgow makers owe much of their value to the fact that each one has been carefully tested and regulated in the University Laboratory before it is sent out for service'. Finally, there was the training of students for home and foreign telegraph service: 'at the laying of the French Atlantic Cable [in 1869, by the *Great Eastern*] two of the best practical and scientific electricians were young men selected from the Glasgow class'.[172]

In 1883 Thomson claimed that from 1866 all signalling on ocean telegraphs was carried out with his instruments. His personal financial returns on the design, construction and marketing of these instruments were considerable. In 1881, for example, the marine-mirror galvanometer, the quadrant electrometer, and the siphon recorder – first patented in 1870 as a device to economise on the human labour at the receiving end – were generating through patents (taken out with various partners, notably Fleeming Jenkin and Cromwell Varley) over £1500 from the Anglo-American Telegraph Company and no less than £5000 from the Eastern Telegraph Company (controlling the route, now built, from London to India via Gibraltar). Knighted in 1866, as we have seen, and made wealthy from his share in the telegraph and other patents, Sir William increasingly fashioned himself as an aristocrat of science, complete with schooner yacht (serving both as social space and floating laboratory), country seat at Netherhall overlooking the Firth of Clyde, and, in 1892, as Baron Kelvin of Largs, the first scientist to be elevated from commoner to the House of Lords.[173]

There was also a wider professional dimension. The Committee of 1859–61 had concluded that it was necessary first, to standardise the *quality* and production of materials; second, to standardise *measures* of resistance and other electrical quantities; and third, to introduce regimes of *precise* measurement, in part as a way of standardising the training of an emerging discipline of electrical engineers engaged in an 'engineering science'.[174] On the first question, Thomson had claimed that the reason for the failure of the 1858 Atlantic cable had been inconsistencies in the resistance and chemical properties of the cable; a standardised cable, not brute electrical force, promised success.[175]

On the second question, there were many rival measures to choose from, often linked to local materials, practices – or heroes. At the meeting of the BAAS in Manchester (1861) telegraph engineers Charles Bright and Latimer Clark had tabled a paper appealing for a standardised system of electrical units and proposed a unit of resistance to be called the 'ohmad', after Georg Simon Ohm's law of electrical resistance. Thomson, unable to attend in person, had already been preparing his own way for the establishment of a BAAS Committee 'On Standards of Electrical Resistance'. A potentially damaging clash between practical telegraphers (represented by Bright and Clark) and natural philosophers (represented, in Thomson's absence, by his protégé in telegraphic science Fleeming Jenkin) was avoided. Jenkin reported back to Thomson that he had got hold of Bright and Clark to tell them that the Committee had already been fixed. At this news 'Latimer Clark looked delighted and is eager to have it all explained – they laid their paper on the table without reading it, I understand.'[176]

The BAAS Committee included the venerable Wheatstone, as well as Thomson and Jenkin. Initially, scientific men dominated but from 1863 three practical electricians, also committed to electrical science, were added (Bright, Varley and C.W. Siemens). The Committee's task was to decide how best to fix such a unit of resistance. Its solution was one long advocated by Thomson: to express units for current, electromotive force, and resistance not in British Imperial units typical of practice but in the French metric units of mass, length, and time ('absolute' or 'mechanical' units) typical of recent electrical science. Unlike the 'absolute units' for electrical quantities of the German physicist Wilhelm Weber, however, the new system of units linked not to abstract 'force' but to the engineering unit of 'work' and through that to the new science of energy. The key link operated through

J.P. Joule's law of electrical heating relating resistance, current and heat, with heat correlated to mechanical work through Joule's celebrated 'mechanical equivalent of heat'. In the 1880s a series of international commissions, paralleling international exhibitions, amidst competing claims of national prestige, eventually agreed to standardised names and methods for electrical measurements.[177]

The organisers designed these congresses to fix units of scientific measure. It was at such venues that the once-local names of Faraday, Kelvin, Joule and Watt entered into a 'universal' language of science (ch. 2). The distinction between scientific and commercial measure, however, was not clear-cut. Controversy raged at the local congress of the BAAS meeting at Bath in 1864 over units of length: Rankine dissented from his scientific peers who had 'recommended the abandonment of British units of measure' in favour of the metre; he insisted that the inch was a quantity enshrined in law and custom, and 'used for practical purposes, in regions inhabited by one fourth of mankind'. In such circumstances, science and commerce could not – and should not – be separated.[178]

On the third question, precise measurement would become an essential feature of a regrouping of electrical telegraphic practice into a laboratory-trained electrical engineering profession. This profession would be engaged in promoting clearly distinct skilled actions including – but not limited to – the telegraph business. It would be organised as a group to represent that specialism. This was true not only of Britain but also, for example, of France, a country with its own imperial ambitions and in which electrical engineering institutions were very closely associated with a State Telegraph Administration (successor to Chappe's administration).[179]

The Society of Telegraph Engineers (STE) was founded in 1871.[180] It was based, like most of the national British societies, in London. It took as its model the Institution of Civil Engineers (founded in 1818), the meeting rooms of which it used. The STE provided a home for the electrical technicians who had moved from the 'experiments' of the 1830s into the 1860s' and 1870s' service of the public or of private industry on a large scale; it was designed to bring about the advancement of electrical and telegraphic science and promoted a gentlemanly ethos allegedly beyond that of the mechanical engineers.[181] From 1872 C.W. Siemens was the first president of the STE: he used his opening address self-consciously to define the institution's role in regulating professional activities. Within a few years Siemens made the STE a present of a bust of father-figure Francis Ronalds.[182] Such figures populated the history of the STE with accredited founding fathers.

The transformation of the specialist telegraph engineer into the 'electrical engineer' was signalled by the change of name of the STE to the Institution of Electrical Engineers (IEE) in 1889. The creation, in the IEE, of a new association of engineers beyond the parent body of the ICE was part of a pattern of professional fragmentation where each fragment represented a new division of labour (and society). IEE membership mushroomed from 110 in 1872 to 2100 in 1890. The electrical engineers were key to the new industries of electrical power and electrical lighting from the 1880s.[183]

The case of William Ayrton illustrates the intimate connections between Empire, physics, telegraphy and standardisation. Ayrton trained in physics at University College London from 1864 (before the 1865 Atlantic attempt); he was sent by the Indian Government Telegraph Service (successor to the projects of Dalhousie and O'Shaughnessy) to study in William Thomson's laboratory in 1867 and 1868, by which time transatlantic signalling was commonplace. He then worked with the TCMC (responsible for the 1866 cable). From September 1868 he was in Bombay where he stayed until 1872 as electrical superintendent in charge of telegraph offices in Bombay and Calcutta. On his return to Britain he worked with Jenkin and Thomson, and for the India Office. This imperialist then, in 1873, set off for the new Imperial Engineering College in Japan – set up by Rankine's former student at Glasgow Henry Dyer and with a staff dominated by British pedagogues – where he took up a professorship which combined 'natural philosophy and telegraphy', occupying the post until 1878. Ayrton taught laboratory courses in electrical engineering at the City and Guilds Institute in London from 1879; and by the time he arrived at Finsbury Technical College in 1884 he had an 'orderly regime' for electrical engineering.[184]

Ayrton's connections with the STE (and its successor in the IEE) were very strong. In 1892 he was President of the IEE; but he had been the STE's honorary secretary for Japan since arriving in Tokyo, indicating the imperialisation of a Society uniquely able to call upon international means of communication, and with an unusually dispersed potential membership. Back in England from 1878 Ayrton edited the STE's journal. His career thus illustrated, simultaneously, the mobility and export of technical expertise throughout the Empire (formal and informal); the institutionalisation in teaching and organisation of electrical engineering as a subdiscipline of experimental physics; the emphasis on laboratory measurement and electrical standards; and the imperial context.[185]

Implementing empires: a 'vast network of communicating fibres'

Having secured the Atlantic route, British telegraphic projectors and their supporters in government continued in their project to implement empire. Edward Bright asserted, again in 1867, that the 'mission of the sentient wire is a noble one: to bind together nations, and colonies to their mother countries. Great Britain is now joined to India on the one hand, and to Canada, the United States and even Vancouver's Island, on the other; our cousins at the antipodes in Australia will shortly be added.'[186] By 1853, Ireland and Great Britain were indeed bound together and Ireland would become a key passage point for telegraphic messages to North America; Britain and India were in telegraphic contact by 1864 via the Persian Gulf; the Atlantic cable linked Britain and Canada in 1866; the nationalisation of the British inland telegraphic networks in 1868 would free up capital for investment elsewhere – so that by 1871 lines would link London, via India, with Singapore, Japan and China.[187]

Although telegraphers from Chappe, Cooke and the Bretts on had made startling claims for the political utility of telegraphy, those claims had little recognised the complexity of political action and the diverse forms of power telegraphic communication might support in the practice of empire. Further, in speaking of the telegraph as a means of expediting 'transactions of Indian business', for example, Dalhousie had not – and perhaps could not – separate the intelligent business of government from that of commerce.

In July 1887 the electric telegraph celebrated its golden jubilee. That knight of the Atlantic telegraph, Sir William Thomson, used the occasion to mock the efforts of his fellow Irishmen, Home Rulers, who sought a degree of political autonomy through a separate Parliament for their native island:

I must say there is some little political importance in the fact that Dublin can now communicate its requests, its complaints, and its gratitudes to London at the rate of 500 words per minute. It seems to me an ample demonstration of the utter scientific absurdity of any sentimental need for a separate Parliament in Ireland. I should have failed in my duty in speaking for science if I had omitted to point this out, which seems to me a great contribution to the political welfare of the world.[188]

For a Liberal Unionist and imperialist like Sir William, then, it was science implemented in the telegraph that demonstrated the absurdity of Home Rule (let alone full independence) – which did not mean liberation but a return to internal strife, rural barbarism, and historic sectarian and tribal disputes which would threaten the progress of civilisation. The telegraph might drive politics, rather than simply being its instrument.

Whereas Ireland's Home Rulers attempted to distance the island politically and culturally from Great Britain, more distant territories sought closer unity with the mother country. The electrical connection of Valentia and Trinity Bay in September 1858 had been but one link in a chain all too briefly filled between Sir Walter Carden, Lord Mayor of London and the humble folk of Sackville, New Brunswick (eastern Canada) who no longer felt 'as distant colonists, but that we actually form a part of the glorious British Empire – God save the Queen!'[189] The more durable connection, of 1866, between Valentia and Heart's Content Bay in Newfoundland quickly provided the opportunity for the loyal Governor-General Viscount Monck stationed at Ottawa to inform the Queen, by way of the Earl of Carnarvon, 'that her Majesty's gratification at the additional strength which the completion of the Atlantic telegraph will give to the unity of her empire, is shared by all her subjects in British North America'. The yet more remote inhabitants of Vancouver's Island, not to be outdone, went to considerable lengths to communicate through their Mayor the message: 'The infant colony Vancouver, 8000 miles distant, sends telegraphic cordial greetings to Mother England.' It had taken not seconds but three days to arrive, crossing North America by land-line, followed by a seven-hour journey by steamer from Cape Breton Island to Newfoundland, the local cable being out of order.[190]

Far from being a distant outpost of telegraphic communication, Vancouver would become a ganglion in a great nervous circuit with the creation of an astonishing 'All-Red Line Around the World' passing only through British territories. From London, cables passed to the Bermudas, Barbados, Ascension Island, Capetown, over land to Durban, Mauritius, Cocos Islands, Perth, overland to Brisbane, then Norfolk Island, Fiji, Fanning Island, Vancouver, thence across Canada – and to London under the Atlantic.[191] The transpacific cable (completed in 1902) was no less than 3500 miles long, almost twice the length of the Atlantic, spanning Vancouver, via Fanning and Norfolk islands, to Auckland, on the North Island of New Zealand.[192] Driven by Canadian projectors, most notably Sandford Fleming of the Canadian Pacific Railway, the line matched the ambitions of British ship-owners like Alfred Holt who in the same year started to span the Pacific with services eastwards from China and Japan to the developing West Coast of North America. Indeed, since the late 1880s the Canadian Pacific Railway not only crossed Canada but had won mail contracts on the transpacific routes from British Columbia to the Far East, especially Hong Kong. Other British services began to link Australasia to the Pacific coast of North America. With the opening of the Panama Canal (1914) in prospect, shipping lines could readily encircle the globe while Canadian Pacific's integrated Atlantic, trans-Canadian, and Pacific sea and rail systems produced similar claims to offer an all-red route (ch. 4) (Figure 5.4).[193]

Part of the 'all-red' telegraphic network, of course, traversed Australia from Perth to Brisbane. Even by the early 1850s, steamship companies were bidding for mail contracts to the antipodes – and if Brunel's claims for the *Great Eastern* had been realised, Australia would have been connected to England by the largest ocean steamship afloat. But in the 1850s 'Australia' was a vast domain fragmented by great distances into largely autonomous provinces; the history of telegraphy in Australia was one of dynamic competition of private (and indigenous) enterprise, state government action, inter-state rivalry, and direction – ultimately – from the heart of scientific regulation in the Greenwich Royal Observatory.

Samuel McGowan, associate of Morse and Ezra Cornell (key to telegraphy in America), emigrated to Melbourne for gold but carried with him Morse equipment; he tried to set up a company working lines around Melbourne (in Victoria), Sydney (to the north and east in New South Wales), Adelaide

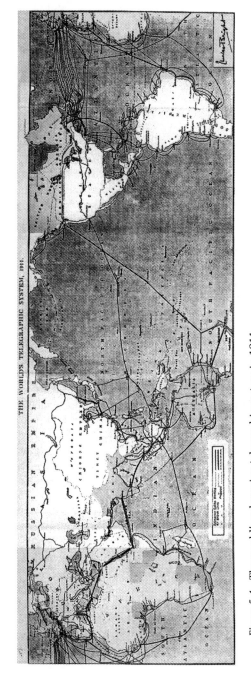

Figure 5.4 The world's submarine telegraphic system in 1911.

Source: Charles Bright, *Imperial Telegraphic Communication*. London: P.S. King, 1911, facing p. 212. (Courtesy of University of Kent Library.)

(to the west in South Australia) and the goldfields. Demonstrations of his equipment in Melbourne helped to persuade the government of Victoria to award the contract for a short Melbourne–Williamstown line, which was operational by 1854. Within three years there was a web of lines in Victoria using Morse instruments; New South Wales had lines radiating from Sydney; there were lines in Tasmania and from Adelaide (South Australia) to its port (1857); Brisbane (in the north-eastern state of Queensland) had lines from 1861; and Western Australia (outwards from Perth) by 1869.

Key for the imperial cause was Charles Todd, appointed in 1855 as superintendent of telegraphs, and observer (astronomer) of South Australia. He represented British expertise, having been a human 'computer' at Greenwich, an assistant to the Cambridge professor of astronomy James Challis, and had also been involved in the electrical time service at Greenwich (1854). Airy supported his appointment and prepared him for the task of setting out a system of telegraphy from Adelaide – even though the Astronomer Royal worried that he lacked the 'boldness and independence of character' needed to thrive in the raw Australian territory he entered. But Todd's plans were grand: he looked forward to 'the time when the telegraph system will be extended to join the several seats of commerce in Australia and also, it is no idle dream in the present age of wonders, when I shall be able to meet Dr O'Shaughnessy by connecting Asia by submarine cable thence via Calcutta to London'.[194] From Adelaide, Todd at first implemented lines on a British (underground) model until he found it too costly and unfitted to the environment; he linked his telegraph system with that of Victoria using McGowan's over-ground system; links from Melbourne to Sydney (1858) and Sydney to Brisbane (1861) followed. Two Oxford-educated brothers, one, E.C. Cracknell, Todd's assistant, took over as superintendents for New South Wales and Queensland in a process of joining the string of capitals.

The telegraphs brought in good dividends for sponsoring state governments and were used extensively to communicate sentences of law, track criminals, make official appointments, control railways and distribute commercial information. Two audacious schemes then connected all parts of the continent: the first, an East–West line, linked Albany (on the south-western tip) along the coast to Adelaide. The second anticipated the establishment in 1872 of a submarine cable from Java to Darwin (in the far north) and raised intense rivalries amongst the states about which would monopolise connections with the outside world. Ultimately, the system-builder Todd would spend two years overseeing the construction of a 2000-mile Overland Telegraph (1871–72) between Darwin and Adelaide (via Port Augusta) – using the surveying skills of Benjamin Babbage (Charles's son) and three thousand wrought iron poles imported especially from England (wooden ones being susceptible to white ants). Airy no doubt approved that for Todd the Australian telegraph became a meteorological resource.[195]

REACHING THE OVERLAND TELEGRAPH LINE.

Figure 5.5 Explorer John Forrest reaches the overland telegraph line between Adelaide and Port Darwin, symbol of home-coming and civilisation.

Source: John Forrest, *Explorations in Australia*. London: Sampson Low, 1875, facing p. 258. (Courtesy of Aberdeen University Library.)

For the Fellow of the Royal Geographical Society and explorer of south-western Australia John Forrest, cables stood in 1875 as emblems of civilisation in a vast cultural wilderness only just beginning to be tamed. Forrest wrote in his journal, after trekking halfway across the continent from the western coast: 'reached the telegraph line between Adelaide and Port Darwin, and camped. Long and continued cheers came from our little band as they beheld at last the goal to which we have been travelling for so long. I felt rejoiced and relieved from anxiety; and... very thankful to that good Providence that had guarded and guided us so safely through it.' He recorded how he was 'in high glee at the prospect of meeting civilised habitations again' once he had met the officer in charge of the telegraph station at Peake (one Mr Blood) and tucked into 'roast beef and plum pudding' (Figure 5.5).[196]

Conclusion: Cultures of Technological Expertise

Preceding chapters have explored steam power, steamship, railway and tele-graph technologies in the making (chs 2–5). These chapters have focused on the cultural construction of empires: usually built as personal and business empires but often closely interconnected one with another and in relation to the larger geo-political empires of the nineteenth century. In this concluding chapter we expand upon several cultures of technological expertise, already touched upon in our specific investigations of the new technologies. These cultures, we suggest, played fundamental roles in generating, promoting and sustaining many of the engineering projects with which we have been concerned.

First, we turn to the question of 'technological tourism' and ask: why did engineers and those other historical actors interested in cultures of practical arts travel so widely and what opportunities for inspection were available to them? We consider, second, the related issue of 'technological exhibition': how did cultures of display inform technological change and consumption? Engineers have often learnt through experience of such displays: but what was the complete context of 'education for the engineers'? The period we have focused upon in this book saw radical changes in the ways in which engineers and those active in technological arenas were trained, both for the domestic market and for the trials of empire. And third, we ask how in our period cultures of reading and of writing altered – and more specifically, how they affected technological and engineering outcomes, opportunities, recep-tions and meanings.

A cultural history of technology will also take into account three further issues. First, engineering had its distinctive visual culture; second, in justifying and motivating engineering practice, actors looked to the context of religion – so vibrant and yet so fractured during our period, and third, engineers worried about how they themselves were 'fashioned into 'gentlemen engineers', with status, decorum, respectability and trustworthiness.

Technological tourism

From the final third of the eighteenth century, manufacturers understood that the new factory sites of production made excellent advertisements for their wares. We have seen, for example, how important it was that James Watt visited Matthew Boulton's works near Birmingham (ch. 2). Wedgwood's works, similarly, doubled as place of production and visitor attraction. But one of the most important mid-nineteenth-century shrines of 'technological tourism' was the notorious Thames Tunnel between Wapping and Rotherhithe, built by the Brunels in consort with an array of resident engineers and unnamed workers. The Brunels achieved this astonishing feat by using a complex, purpose-built 'shield' built by Maudslays, and allegedly modelled on the ship-worm, which made it possible to burrow through the clay and soft mud immediately beneath the River Thames, leaving a consolidated brick tube wide enough for the passage of pedestrians and carriages (Figure C.1).[1]

It was a high-profile media event as soon as tunnelling began. Brunel senior was used to public attention: the Portsmouth dockyards had been a favoured site for visitors keen to see his system of block manufacture at first hand (ch. 4). In London, the younger Brunel emphasised the drama of technological progress (an entirely new form of tunnelling) with human drama. Injured and nearly drowned, as six died around him when water suddenly flooded into the tunnel in January 1828, Brunel provided ample copy for the London papers – even as they began to speak of the 'triumphant bore', with its prospects of completion ever receding.[2]

The Thames Tunnel was the occasion for a plethora of cheap memorabilia. Mugs, peepshows, thimbles and gin flasks serviced the visitors who came on day trips, or for longer periods from Europe and beyond, to witness the tortuously slow excavation and to experience the thrill of standing beneath the Thames. Banquets and brass bands made the tunnel a site of leisure and pleasure – when its promised utilities seemed hopelessly distant; they kept the tunnel in the news – as did the Tunnel's very own newspaper, where most publicity was good publicity, and they gave the directors a small income when all else was expenditure. Engineers knew that their technological ventures were public – given meaning by publics, financed and supported by them, perhaps hindered by them and certainly conditioned by them.

There were echoes of the Thames Tunnel in the 1850s when the sponsors of Brunel's *Great Eastern* looked to generate fluid cash by staging the launch of the vessel as public show – with consequences well-known.[3] Similarly adverse publicity resulted when the completed ship anchored in the Mersey in June 1861. The local press eagerly anticipated 'the magnificence and matchlessness of the *Great Eastern* steamship' through vivid descriptions that 'contributed materially to the success of the big ship's exhibition'. But when the ship opened to the public, a different picture emerged. 'The dirty decks and untidy cabins might with little trouble have been rendered somewhat

Figure C.1 Celebrating the completion in 1843 of Marc Isambard Brunel's Thames Tunnel: 'triumphant bore' and metropolitan tourist attraction.

Source: *Illustrated London News* 2 (1843), p. 227. (Courtesy of Aberdeen University Library.)

ship-shape, and those remarkably unclean officials and smutty-faced individuals, who were always in everybody's way, might have been compelled to wash their hands and faces, and put on more respectable attire, in honour of their Liverpool visitors', commented *The Porcupine* indignantly. 'But no such preparations were made, and the manner in which the greatest ship in the world has been shown to the people of this town is a positive disgrace to captain, agents, and all who are in any way responsible for its management.' Worse, indeed, was to come for the estimated seven or eight thousand persons on board at any one time:

> only *one obscure outlet* was provided; and to reach this, a stifling, terrific, and dangerous struggle, varying from one to three hours for each individual, had to be endured. Ladies were borne along, twisted round corners, jammed against iron walls; young children were torn from their parents; there were faintings and hysterics; there were oaths from uncivil officials, and curses both loud and deep from the surging multitude; there were men whose faces had turned to the red hue of a newly-boiled lobster, and fair ladies who looked as if they had been spending about half-an-hour in a steam boiler; but there were no responsible persons in sight; no captain, agent, or prominent officer to quell the storm or see to the safety of the struggling multitude.[4]

Both the *Great Eastern* and the Thames Tunnel were unusual attractions; more usual were the descendents of Boulton's Soho works – those factories that doubled as show-spaces for the practical works (chemical, mechanical) that fuelled the British economy through export. It was places of production like these that more philosophically minded tourists of industry visited from the early nineteenth century: men like John Robison, later Sir John Robison, knighted, in part, for his conspicuous and assiduous services to science by committee and council, in Scotland and in the cabals of the BAAS. That Robison's father, also John, had been professor of natural philosophy in Edinburgh and friend and publicist of Watt, meant that few doors were barred for the younger John – a fact that led to any number of urbane notes on the latest technological 'improvements' at home and abroad – from steam navigation on the Forth and Clyde Canal to the new 'daguerreotype': 'I can hardly imagine', waxed Robison, 'anything to be superior to some of the specimens which M. Daguerre showed me in Paris'.[5]

If the reports of Robison could be ad hoc, William Farish, Jacksonian Professor at Cambridge University and Robison's contemporary, claimed to have surveyed 'almost everything curious in the manufactures of the Kingdom' in preparation for classes delivered during the first decades of the nineteenth century to well-to-do youths. When his student and eventual successor, Robert Willis, likewise planned an encyclopaedia of British industry in the early 1830s for the Society for the Diffusion of Useful Knowledge, he worked to make this venture up-to-date and its results authoritative by planning a tour of the industrial centres of London, Nottingham, Derby, Manchester, Liverpool, Birmingham, Leeds and Sheffield. Experience at first hand would be recorded by an accompanying draughtsman. Both Farish and Willis then puzzled to see how the apparently irreducible scale and complexity of diverse and dispersed mechanical endeavour might be reduced to classroom size – a puzzle each solved, in his own way, with models encoding the analysable, transferable, and easily reproducible elements of British industry.[6]

That kind of technological tourism had its parallel in the peregrinations of Willis's correspondent Charles Babbage. We have already seen how quickly Babbage became enthused by the possibilities – and problems – of the new railway transport (ch. 4). But even at the beginning of the railway age, when the coverage of the network was just a fraction of what it would become, Babbage had visited the places and spaces of British (and European) science and industry, collecting, collating and tabulating data of potential value on the dynamic interconnected systems of means of production and things made. The literary product of his technological tourism would be the *Economy of Machinery and Manufactures*.[7]

Babbage's pontifications brought to him a status equal to that of the advocate of radical automation, Scottish chemist Andrew Ure. It made Babbage one of the country's chief 'philosophers of manufacture' and commentators on the 'economy' of a factory system which was simultaneously scientific, moral and commercial.[8] But there was an alternative to this legitimate insiders' view of philosophical technological tourism: this was the despicable – although not always despised – practice of industrial espionage. Showing and reporting shaded into concealment and selective revelation: they were curiously ambivalent technological practices. We have already seen Boulton and Watt's paranoia about the theft of ideas from the workplace, their pushing for legislative responses – and yet their happiness to use their sons and assistants to communicate competitors' techniques (ch. 2).

When the British copied techniques learnt abroad or exploited valuable commodities there, few in Britain complained of intellectual or practical theft. Lewis Gordon's travels in Europe brought back knowledge of wire-rope manufacture vital to submarine telegraphy and key to his personal fortune. 'Discovered' by Westerners in about 1843 as part of a programme of 'economic botany', the natural plastic 'gutta percha' was available only from Malaya, Borneo and Sumatra but it proved indispensable, from 1848, as an insulator in the same technology. When practised by 'foreigners', however, whether non-British Europeans or North Americans, who toured Britain in search of valuable and reproducible technique, tourism could indeed be revalued as despicable.[9]

In many respects, then, technological tourism was a national and indeed imperial question, a point further illustrated by the case of Karl Wilhelm/ Charles William Siemens.[10] A visit to London and Birmingham in 1843 to market a new electroplating process was just the first of a series of visits through which he learned about British technologies which, unprotected in the absence of international patent law, might well be exploited by his many brothers, notably Werner Siemens, back in Prussia. One fruit of Karl's intermittent tourism was to learn of the curious air engine promoted with dramatic effect by James Stirling in Dundee (ch. 2). Werner Siemens, in the know thanks to his brother, was quickly extolling the virtues of 'one of the most important inventions of our time...especially in Germany, where its use is not restricted by patent'.[11] National issues remained paramount. Yet ironically, when in 1865 and 1866 disputes over whether to deploy 'absolute' or more practical 'mercury' standards of electrical resistance reached their peak, the two brothers found themselves allied to different national groups, with Charles closer to the universalising British, represented by the BAAS Electrical Standards Committee, than to his brother and the Prussians – rivals in both science and industry (ch. 5).[12]

In a Britain increasingly mobilised by railways, technological tourism had always been welcome in the meetings of the BAAS – wandering as it did, year by year, from one (non-metropolitan) city of science or centre of practical art to the next (ch. 1). As well as providing an opportunity for delegates to converge and to wonder at the local natural sublime, these meetings were also carefully orchestrated to show off local technological products, albeit universalised by a discourse of mechanical science orated by BAAS managers. They made the local, as it were, into part of a national British possession. Although the first ports of call of the BAAS were in England, its directors enthusiastically responded to the entreaties of men of science and commerce that they visit Edinburgh and Dublin; eventually they would use the new steamships to reach Canada, South Africa and Australia, equating British science with the science, and technology, of the British Empire.[13]

At the same time, as Michael Adas has pointed out, the very display of 'superior' technologies carried with it a degree of moral subjugation for those cowed by their exposure to it.[14] He cites the example of one of John Ruskin's correspondents who, touring China, witnessed the new steamship technology in a show of 'civilising' force. The correspondent identified three different kinds of 'civilisers'. First, the Philistine militarist who 'out here is a perfect Goliath' and whose remedy for civilisation took the form: '"Let us have a war, and give the Chinese a good licking...everything else will follow."' Second, the missionaries 'are working to civilize the people here ...by the usual plan of tracts and preaching; but their system is not much in favour, for they make such very small progress among the 360,000,000'.[15] But there was a far more fashionable third way to civilisation:

We were sailing on the river in a steam launch, which was making the air impure with its smoke, snorting in a high-pressure way, and whistling as steam launches are wont to do. The scene was appropriate to the conversation, for we were among a forest of great junks – most quaint and picturesque they looked – so old-fashioned they seemed, that Noah's Ark, had it been there, would have had a much more modern look about it. My friend, who is in the machinery line himself, gave his opinion. He began by giving a significant movement of his head in the direction of the uncouth-looking junks, and then pointing to his own craft with its engine, said 'he did not believe in war, and the missionaries were not of much account. This is the thing to do it', he added, pointing to the launch; 'let us get at them with this sort of article, and steam at sixty pounds in the square inch; that would soon do it: that's the thing to civilize them – sixty pounds on the square inch.[16]

Exhibiting technology

The Thames Tunnel was only one of the many shows and exhibitions of 'scientific London' and further afield.[17] Rather than wait for seekers after intelligence and modern wonders to travel to technological installations, educators and entrepreneurs found ways of bringing practical innovation to accessible centres of display. Scottish lecturers of chemistry, natural philosophy, mechanics and natural history, and purveyors of voluntary (non-compulsory) science lectures in Cambridge, for example, furnished rather different student bodies with the excitement of technology. To do so meant investing in the material capital of exhibition with artefacts often modelled specially, or created for obvious visible effect.[18]

Iwan Morus has written extensively about technological show-spaces outside the university arena and keenly in touch with practical artisan culture; he has examined their interconnections with cultures of commodity and consumption. Such spaces were the forums for working out questions of skill and its value and for sorting out personal and national rivalries, not least those revolving around questions of priority – and ownership – of invention.[19] In celebrating diversity, practice and manual skill, such places vied with the attempt to impose upon practice the 'philosophical' authority and expertise of Cambridge-trained mathematicians.

The Glasgow meeting of the BAAS provided the opportunity to reconcile these two cultures of elite men of science and practical artisans. James Thomson, engineering enthusiast and ambitious older brother of William Thomson, orchestrated an elaborate exhibition of models illustrating the diversity and excellence of 'mechanical science' (for Section G) in and beyond the city. The models, transported free of charge by Glasgow steam packet companies, exemplify the themes of this book. A death mask of Newton provided an emblem of mathematics and physics (for Section A) and a morbid token of the Association's hierarchy. Tidal sections and charts of the city's harbour reminded visitors of Glasgow's growing prestige as a port of empire. The Newcomen model took pride of place as a hallowed relic of the sacred Watt. Models of vessels deposited in a water tank by the BAAS committee testified to Section G's commitment to bring naval architecture into the domain of science. A chart of the Atlantic 'exhibiting the connection between Europe and America, by steam – the Archimedes, &c' held out the promise of new screw-propelled steamships. A marine engine, exhibited by David Napier, testified that claims for such a transformation were not mere chimera, whilst the engine of the *Comet*, restored at the considerable cost of £120, gave Glasgow marine engineering a secure, iconic, and progressive history. Railway engineer Charles Blacker Vignoles provided a model of his Ribble Bridge. 'Alexander's Electro Galvanic Telegraph' projected the prospects of telegraphic communication. Whitworth screws, meanwhile, suggested the benefits of precision engineering and national – perhaps international – standardisation.[20]

Although this temporary exhibition accumulated and displayed practical innovations from all the provinces of Britain, in order to enhance Glasgow's civic pride, it could not hope to compete with the shows of the London metropolis for variety and enduring international impact.[21] The best known but in many ways the least typical of these concentrations of material culture was the 'Great Exhibition of the Works of Industry of all Nations' master-minded by two dynamic members of the Royal Society of Arts, Henry Cole and Albert, the Prince Consort. It was held in London's Hyde Park in 1851. Exhibits arranged within the cathedral-like transepts and naves of Paxton's iron-and-glass Crystal Palace celebrated the continued fertility of British technical innovation, contrasted it favourably with the simpler crafts of other lands, and established beyond reasonable doubt the stunning fulfilment of prophecies made in the decade since the Glasgow meeting. Amongst the clutter of this exuberant Victorian collection, visitors of all classes and both genders witnessed steam usefully applied even to agriculture, a giant lighthouse reflector symbolising safety and commercial security, and telegraph engineer T.R. Crampton's express locomotive 'Folkestone'. Steam power, steamships, railways and international telegraphic communication (announced at the exhibition) gave practical proof of 'progress' and engendered trust in future imperial prospects.[22]

London, 1851, also provided the stage for tableaux of an American industrial revolution. The US Commissioner to the London event, Edward Riddle, returned home and immediately began to garner support, from Wall Street and Washington, for New York's own 'Crystal Palace Exhibition' which opened in 1853. This, on the face of it, was a better venue for a young nation, seeking to forge its own integrity by means of telegraphic, steamboat and rail communication. It was the opportunity to contrast a fabled Yankee ingenuity with the alleged extravagance of the Old World; it was the venue for a Fourth of July oration on 'the American idea'.[23] Yet despite point-by-point imitation of the Great Exhibition, with guides, catalogues, and a

Crystal Palace unashamedly copying Paxton's, the venture did not reproduce the critical and commercial success of the London show.[24] Although once again lighthouse equipment, models of ships, railroads and telegraphs (including Morse's) were on display, the exquisite Alabama-made *Southern Belle* steam engine promised far more than it ever delivered and recent innovations were thin on the ground.

When, as here, other nations began to display their wares independently, it was often the British that came to judge, and to report back for government, professionals and public. Twenty years after Babbage had dissected British industry, Joseph Whitworth, now the doyen of precision engineering and himself an exhibitor of machine tools in New York, analysed the 'American System of Manufactures', with its focus on mass production of identical machine-made commodities. For Charles Lyell, another reporter, the great scientific triumph of New York was the large exhibit of surveys, reports, maps and charts of the United States Coastal Survey under Alexander Dallas Bache: this, for Lyell, was 'the most important scientific work now in progress in the United States'. It was truly, we might add, an object of American national importance; the surveys had achieved a standard of accuracy greater than that of the British survey in India. Yet an *English* theodolite was the primary tool of triangulation.[25] Lyell watched as the United States, although still dependent on British instruments, flexed its muscles as a new maritime empire.

The exhibitions of 1851 and 1853 demonstrated British ingenuity, power and judgement against a backdrop provided by the industry of 'all Nations' and even of the vibrant young nation of the United States. Yet paralleling this rhetoric of technological, moral and, perhaps, philosophical superiority were suggestions that Britain was in decline – and indeed, had been so for two decades before the staging of the Great Exhibition.[26] In 1830 Babbage and others had attacked the government and centres of traditional learning for their (supposed) failure to support science in Britain (ch. 1).[27] Sniping of this kind in part reflected the divisions of unreformed British science between Oxbridge elites and men of practice.[28] In 1851 Babbage chastised the organisers of the 'Exposition of 1851', of which he was not one, for their astonishing lack of foresight (not least, in his view, for failing to exhibit his difference engine).[29] And although other commentators announced that the London exhibition of 1862 continued to show British pre-eminence, especially in the new chemical industries, the Paris exhibition of 1867 both showed off – and showed up – British industry. One of the more dramatic verdicts was that Britain was 'falling behind'; or, failing that, that its progress relative to other countries was diminishing. These verdicts gave Lyon Playfair and others the opportunity to launch inquiries into, and campaigns in favour of, technical education. In the view of such critics, the solution to Britain's perceived problem would be education for the industrial classes and for the engineers of empire.[30]

Cultivated engineers

Viewed as both makers and interpreters of systems of technology, engineers often addressed questions of education. In the late eighteenth century, James Watt well summed up the mutual dependence of an engineering system and the human specialists who built and maintained it. Watt admitted that, in inventing the steam engine with separate condenser, he needed also to re-create the steam engineers who would construct and control the new powers: a new engine meant (and required) craftsmen trained in new skills. In 1797 his collaborator John Robison (senior) had talked of a need for a 'Collegium or Corporation of Scientific Engineers, with three degrees of Bachelor Master and Doctor – not merely academical honours, of no more value than the offices of a Mason Lodge, but to have Civil Consequences.'[31] No doubt the impoverished Professor Robison was thinking of good, fee-paying consumers of his lectures and purchasers of the books he would write to accompany them. But Watt, with his 'carefully cultivated inventiveness' (in Smiles's words) probably had something else in mind: the firm of Boulton and Watt had been a great training ground for steam engineers, notably Murdock and Southern – not to mention the sometimes rebellious but ultimately malleable sons of Boulton and Watt (ch. 2).

This last kind of training – proprietary, in-house, and linked directly to the exact nature of the business carried out – seemed the ultimate justification of the British economic preference for free enterprise, or laissez faire, unhindered by state intervention; what need for a training, and especially a training in science, for engineers apparently so successful without it? Watt himself was, on some accounts, a case in point: allegedly untrained, too busy to attend the lectures of Joseph Black in chemistry – which (Black and Robison seemed to claim) expressed the principles upon which Watt's engine was based, this man, nevertheless, had invented the power that transformed the nation – and fuelled an explosion in imperial expansion (ch. 2).

Civil engineers in our period frequently demanded to be taken seriously as – and to have the authority of – men of science, or represented themselves as engaged in great 'experiments' (chs 2–5). This rhetoric has often led historians to assume that engineers, especially English ones, 'needed' a greater dose of college education in book science; yet the identity of the engineer was entangled with contrasting notions of masculine independence and authority gained through personal experience rather than the consumption of ready-packaged theoretical education. Samuel Smiles captured this when, in his chapter on 'self-culture', he quoted Sir John Robison's great friend Walter Scott: 'The best part of every man's education is that which he gives to himself.'[32]

It is not then surprising that when, in the late 1830s and early 1840s, academic courses for engineers began in Britain (specifically in Durham, London, Edinburgh, Glasgow and Dublin), their reception was complex. So many courses were there, by February 1841, that James Walker as President of the Institution of Civil Engineers worried about overproduction. In the aftermath of railway mania and with Britain in recession (since 1837), the Glasgow University alumnus and Durham external examiner wondered whether there would be jobs for all these college-trained engineers:[33]

> Is then the demand for professional gentlemen likely to *increase*? Is it not likely rather to *decrease*? Now certainly the number of Engineers or Students for Engineering is increasing. If we look at the number of students in classes for Civil Engineering at the different Universities and Academies...we are led to ask, will this country find employment for all these? I freely confess that I doubt it.[34]

Walker implied that such men might find work abroad. Clearly the making of the empire meant not just the export of material objects, or the establishment of secure means of communication for goods and intelligence. It also meant the production and export of human specialists, including engineers, who, as experts in an age of expertise, could extend the scope of technological systems beyond British shores.

British science professors wanted to sell their skills as lecturers in natural knowledge, to enhance, as they saw it, untutored and unprofessional practice, and to prepare engineers for service to industry, nation and empire. Indeed, they were often close to practical industry. But in several instances they blocked the appointment of professors in civil engineering *per se* as encroachments on their disciplinary territory. Rather, they chose themselves to extend, revive or inaugurate *academic* discussion of objects of national importance: of motive power, of naval architecture, of railways, and of electrical technologies re-housed in the laboratory of precision measurement.[35]

Take power as an example. In the late 1830s, the chemist James Finley Weir Johnston addressed gentlemanly engineers in Durham, at the heart of the coalfields supplying the factories of empire.[36] At King's College London, in the early 1840s, Henry Moseley refined Watt's indicator diagram, making the measure of power, in all its forms, the domain of the elite natural philosopher, and recasting the dynamics of the steam engine's beam, crank, fly-wheel and governor in mathematical analysis.[37] Sometimes such academic wares were hard to sell, even in the key centres of industry. When Lewis Gordon arrived at Glasgow College in 1841 he found his ambitions to train engineers in Watt's home town blocked by a conservative oligarchy opposed to the incursions of industry; still he emphasised the strategic importance of the exact measurement of power, through dynamometers.[38]

More successfully, from the 1840s Robert Willis gave Cambridge University's expertly coached mathematical wranglers a sanitised and abstract view of industry: by the 1850s, if not before, that view included 'the

Steam Engine'.[39] Above all, Rankine, in post as Regius Professor of Civil Engineering and Mechanics from 1855, charted the development of Watt's engine when 'science was to effect more in a few years than mere empirical progress had done in nineteen centuries'. Amongst the iconic images Rankine reproduced in the 'historical sketch' which began his university textbook on the steam engine were Bell's steam-powered vessel the *Comet* (1811–12), Stephenson's *Rocket* locomotive (1829) and, moreover, the famous Newcomen model, 'preserved by the University as the most precious of relics'.[40]

For these men, scientific training in the academy could indeed enhance engineering practice; and after the Great Exhibition, and especially from the 1860s, there were repeated re-evaluations of the best means of educating the 'industrial classes', the training of school science teachers, and the creation of engineers – especially electrical engineers. Engineering training could even be a lucrative export: Rankine's talented pupil, Henry Dyer, established a radical new programme of training as youthful Principal of the new Imperial College of Engineering in Tokyo.[41] The creation of a new species of expert occurred in an age which saw the rise of the expert professional at home and abroad – and not least the scientific naturalist, spearheaded by T.H. Huxley and pressure groups like the X-Club. Periodicals such as *Nature*, under the zealous stewardship of astronomer Norman Lockyer, pressurised government for additional resources and proactive intervention, often following European models.[42] Government commissions under Devonshire and Samuelson also grilled educationalists of all kinds. Most striking of all, perhaps, London's Imperial College received its charter in 1908.[43]

By that time the engineers had grouped themselves into powerful institutions. Founded in 1771 and a reflection of the Enlightenment love of clubs, the Society of Civil Engineers, renamed the 'Smeatonian' after the death of its prominent member John Smeaton, was one of the first of these in Britain. The 'Smeatonians' were the elite of civil engineers, men who had 'made it', but they met less to educate and reform than to dine and to direct – to fraternise, and to fulminate about the state of business and Britain.[44] The Institution of Civil Engineers, founded in 1818 by young engineers desperately seeking status (as the engineers continued to do through the century), soon had one of the old-guard, Thomas Telford, at the helm and a little later also boasted a charter which spoke grandly of harnessing the great forces of nature for the use and convenience of mankind. Although talk of college-educated engineers disquieted members of the Institution who insisted that the Institution was itself their school, the identity of the civil engineer remained closely linked to membership of the 'Civils'.[45]

Even by the 1840s, however, there were splits in institutional representation – notably with the formation of the Institution of Mechanical Engineers which catered particularly for railway practitioners, often based in the north of England and (allegedly) undervalued by the Civils and its power structures. By the time of the educational commissions of the 1860s and 1870s, engineering societies had proliferated – or, less charitably, the single profession had fragmented.[46] This pattern echoed the demise of the polymath of science exemplified by the Cambridge don William Whewell and mirrored a differentiation of expertise within science. The Society of Telegraph Engineers catered for increasing numbers of practitioners associated with that new form of communication; and at the end of the century, the growth of new electrical industries – like power and lighting – saw the Society transformed into the Institution of Electrical Engineers, modelled on the Civils (ch. 5).

With the rise of Clyde iron shipbuilding and marine engineering, especially from the 1850s (ch. 3), the establishment of the Institution of Engineers and Shipbuilders in Scotland (1857), founded by Rankine and his associates, marked the first such society in Britain. The North East Coast Institution of Engineers and Shipbuilders (1884) followed rather later. At a national level, the Institution of Naval Architects (1860) occupied pride of place. Together with the Admiralty, the Institution sponsored a new Royal School of Naval Architecture and Marine Engineering at South Kensington (1864) which merged into the Royal Naval College Greenwich nine years later. Eight students of naval architecture and eight marine engineering students enrolled annually from the Admiralty Dockyards.

The first intake included W.H. White who was later appointed as Admiralty Director of Naval Construction and achieved fame as warship designer for Sir William Armstrong whose Tyneside armaments empire added shipbuilding to its activities in the 1880s. Another early student was Francis Elgar who subsequently occupied the chair of naval architecture founded by John Elder's widow at the University of Glasgow in 1883. Elgar also headed Elder's old Glasgow (Govan) yard, now known as the Fairfield Shipbuilding & Engineering Company, as well as the famous Birkenhead yard of Lairds, now Cammell–Laird as a result of a merger with Sheffield steel and armourplate makers. Both yards focused strongly on Admiralty contracts as well as the high-class merchant ships of empire.[47]

Perhaps surprisingly, the Institution of Civil Engineers continued to fight shy of standard forms of training as a compulsory prerequisite for admission into the profession – although in the early part of the nineteenth century, this could be understood as a reflection of the great variety of routes taken into the profession by an older generation of members. Instead they opted to preserve the system of apprenticeship, or pupillage, despite the acid comments of men like Brunel, who took few pupils, unashamedly charged high premiums, and made no bones about the fact that he was not in the business of teaching the men under his care: they were to learn, by osmosis, from his example. But it should not be forgotten that engineers were readers as well as doers. Despite the caricature of the British engineers as empirically oriented but 'unschooled', one of the products of the long nineteenth century in Britain was a plethora of engineering in print.

Literary technologies

Cultural historians such as Elizabeth Eisenstein have long claimed print as an agent of change, linking changes in print technology and practice to the events of the Reformation and to the Scientific Revolution of 1500 to 1700.[48] Where Eisenstein emphasised print as a source of stability in knowledge, by facilitating mass reproduction of authoritative texts, Adrian Johns has disputed this simple account, focusing on the mess of practice in the making of natural knowledge, especially in England in the early modern period.[49] More specifically, Steven Shapin has exploited for the seventeenth century the term 'literary technology' as a tool of persuasion, and as a means of broadening audiences to warrant matters of natural fact.[50]

In our period, the trades in print were again transformed, by power presses, telegraphic messaging and steamship transport: railway distribution, station bookstalls, and journeys as opportunities for reading combined in a literary technology distinctive of its time.[51] Behind the myth of the heroic individual, there thus lies a culture of reading and writing engineering practitioners, keenly responsive to audience, to texts and to the possibilities of self-advertisement. Watt consulted his Desaguliers on steam; Wheatstone and Cooke issued pamphlets on electrical telegraphy; railways had their *Bradshaw* and a new 'battle of the books' over gauges; and Airy and Scoresby fought their own battles for magnetic authority in the pages of the metropolitan journal *The Athenaeum* (ch. 1).[52]

In our period, men of practice began to elaborate their own literary culture and canons. J.T. Desaguliers' monumental *A Course of Experimental Philosophy* (1734–44) became a classic in the field. This book, figured along Newtonian lines, was full of descriptions and images of the best practical machines of every district of eighteenth-century Britain.[53] There were also practical 'cyclopaedias', shadowing the technical parts of the famous French Enlightenment *Encyclopédie*, and indicating the ambition to schematise all useful knowledge between the covers of a dozen or fewer volumes.[54] Abraham Rees revised and updated the famous *Cyclopaedia* of Ephraim Chambers between 1779 and 1786; thereafter *Rees' Cyclopaedia*, as it was popularly known, would be much consulted as a resumé of the ever-changing state of British industry, with a new edition of 39 volumes appearing between 1819 and 1820. Trumping this, with no fewer than 133 volumes, was the *Cabinet Cyclopaedia* published by Longman under the editorship of Dionysius Lardner.[55] Such works vied with newcomers like that edited by Olinthus Gregory, evangelical, mathematical practitioner, and long-serving educator at the Royal Military Academy at Woolwich. Gregory's three-volume *Treatise of Mechanics* (1806) was internationally used; the *Ladies' Diary* he edited each year belied its title and provided a forum for mathematical problem solvers of all ages and both sexes; more ambitious but less successful was his vast survey of human genius, learning and industry, the *Pantologia* (12 vols, 1808–13).[56]

General scientific periodicals of the early nineteenth century, like the *Journal* of William Nicholson (*Nicholson's Journal*), Thomas Thomson's Glasgow-based *Annals of Philosophy* and Alexander Tilloch's *Philosophical Magazine*, joined with pamphlets and more substantial books to provide column inches for industry and the practical arts; newspapers reported large-scale public works, or the scandal surrounding them; there was often, indeed, a close if uneasy relationship between engineers and the press, one in which reporters were often happy to receive and adopt copy provided by the assistants of senior engineers explaining – and praising – their recent works. The great nineteenth-century review periodicals, amongst them the *Quarterly Review* and the *North British Review*, provided additional outlets for commentary on large-scale public works.[57]

The *Edinburgh Review* was the home, not least, for polemics issuing from the pen of the Scottish optical inventor and academic David Brewster. He wanted government support for engineering, and especially its training, on a par with the older professions of medicine, law and the church:

> If we demand from our lawyers a regular course of study in matters where our civil rights only are concerned, and claim from our medical advisers and religious instructors not only a long noviciate in their studies but a positive proficiency in their professional pursuits, shall we not, with equal reason, insist upon a thorough and profound knowledge of civil engineering in cases where property on the largest scale is at stake, where millions of lives are in peril, and where the highest national interests are involved?[58]

For Brewster, a life of scientific journalism provided a much-needed income stream but even with a career total surpassing 1000 assorted despatches (from the Loch Ness monster to Galileo as 'martyr of science') writing was a risky business. Brewster's *Edinburgh Encyclopaedia*, published with William Blackwood, was a dismal commercial failure.[59] Far more successful, Macvey Napier's *Encyclopaedia Britannica* (for which Brewster also wrote) digested contemporary knowledge, including practical knowledge; that category included an extended notice of engineering icon James Watt, extracted from James Watt junior.[60]

When changes in printing technology facilitated the production of cheaper books, this too created new consumables for reading engineers, and, in so doing, reformed the collective culture of British engineering. Historians have recently transformed our understanding of books and print culture and their relation to the sciences. As part of that programme, they have re-examined notions of 'popular science' during the mid-nineteenth century. But 'popular science' should not be seen as a diluted, or dumbed-down, form of 'elite science'; rather it can be understood as a category gradually defined during the middle third of the nineteenth century in opposition to simultaneously emerging categories of elite professional science.[61] The engineering knowledge being peddled in print at the same time had its own forums distinct from elite or popular science, responsive to them, but equally varied. These forms have received rather little attention; but we might speculate, in general terms, that 'popular' accounts of practical arts became similarly differentiated from exclusive accounts of engineering endeavour as engineering professionalisers withdrew into bodies and cliques avoiding direct engagement with a 'common culture' and increasingly autonomous.

From 1823 the London-based *Mechanics' Magazine*, edited by Thomas Hodgskin and the radical journalist Joseph Robertson, espoused an agenda celebrating the dignity and ingenuity of the artisan mechanic above the elite science of the universities and the 'Old Corruption' of metropolitan power centres. Costing only 3d, it was copiously illustrated.[62] The Institution of Civil Engineers (ICE) collaborated with publisher John Weale to bring out three volumes of *Transactions* (1836–42), reproducing its regulations and charter, rehearsing its history, celebrating exemplary lives (Smeaton, Rennie, Telford), and with extended technical papers substantiating its claim to be adding to the sum of engineering knowledge. From 1837 the *Minutes of Proceedings of the Institution of Civil Engineers* disseminated abstracts of papers presented at meetings and, emphasising the communal and critical nature of engineering endeavour, gave the gist of the 'conversations' that ensued. The other British engineering societies followed suit: buildings made institutions concrete; but paper 'proceedings' for Institutions of Mechanical Engineers, Naval Architects, and Electrical Engineers were frequent, portable, status-enhancing and attention-grabbing. Submitting 'papers' enhanced personal status; systems of gift exchange, of journals and reports, linked professional groups in cognate areas and to their international peers. As an independent foil to the ICE, Frederick William Laxton's monthly *Civil Engineer and Architects' Journal* (1837–68) promoted the professional status of the engineer through appropriate schooling complementing experience in the field; the *Builder*, likewise, formed paper bonds between architects and civil engineers. Later, two rival magazines – *The Engineer* in the 1850s and, shortly afterwards, *Engineering* – regularly served up state-of-the-art learning, housed debates and moderate controversy, developed confirmatory professional and practical histories and, thus, subtly bound together a sprawling community of busy engineers.[63]

But of course, periodicals, like technologies, might fail as well as succeed. The ease with which entrepreneurs could launch periodicals was rarely commensurate with sales. The second city of the Empire had its own weekly *Glasgow Mechanics' Magazine and Annals of Philosophy* from 1824. But although often reprinted, and despite being planted in what, on the face of it, was the most promising of industrial soils, it was dead before the end of the decade. The favoured outlet of Lewis Gompertz, visionary inventor avid to unburden the brute creation through mechanism, was the respectable *Repertory of Arts.* But another organ chosen for his work, the *Mechanics' Journal of Science and Art; and Magazine of Natural and Experimental Philosophy,* folded a year after its first issue in 1838. This was despite the promising title similar to the London *Mechanics' Magazine,* and the inventive fertility of its contributors: Gompertz of the leg-free bicycle or re-invented carriage wheel ('the scapers').[64]

Running beside debates about education in more successful journals were discussions about textbooks in engineering and the practical arts.[65] As engineers pushed to create uniformity in physical and engineering standards, engineering educators, if not state departments, called for uniformity in tuition – by providing canonical texts. In Cambridge, Willis claimed to have made more easily communicable to gentlemanly mathematicians the messy processes of mechanical invention with his *Principles of Mechanism* (1841). Very different was *Mechanics for Practical Men* (1833), a work whose issue was likely designed to coincide with a second edition of Gregory's *Mathematics for Practical Men* (1825; 2nd edition 1833), and written by the lowly mechanic James Hann. The attention it attracted helped to catapult Hann from the coal districts of Newcastle to King's College School in London, an institution which sat literally beneath King's College London and which provided the College with the numerous civil engineering students that kept it afloat financially. For Hann, as for many impoverished educators, publishing was a money-spinner; he had soon churned out books on the steam engine, spherical trigonometry, and much else besides, often in collaboration with the ICE's publisher, John Weale.[66]

A steady earner from textbook sales was Macquorn Rankine. Practically experienced in power, railway, telegraphic and, eventually, marine engineering, Rankine churned out long-lived if indigestible textbooks on all aspects of engineering once appointed professor in Glasgow. These works sought to standardise education and produce reliable engineers with a loyalty not simply to the firms in which they worked but to their alma mater, and to the literary and practical medium of 'engineering science'. Rankine wanted, and perhaps achieved, what his friend Gordon called 'the permanent Principia of Engineering' – an exhaustive statement of the principles of engineering science to stand beside Newton's foundational *Principia* for the physical sciences. He produced a monumental *Shipbuilding, Theoretical and Practical* which was truly a whale of a book. Combined with Rankine's other manuals, here was a veritable school of engineering texts.[67]

Engineering images

Engineers, like men and women of science, were adept in fashioning images of themselves. Cultural historians take a particular interest in such processes of self-presentation. For canonical examples in the history of science, we need only look back to the honourable Robert Boyle, working in the Royal

Society of London from the 1660s, finding the patterns of nature not in the books of the ancients but through the new, skill-intensive techniques of 'experiment', and looking to legitimate the knowledge produced by fashioning himself, like his companions, as a worldly Christian virtuoso.[68] Still more remarkable, perhaps, was the way in which Galileo promoted and developed the helio-centrism of Copernicus against the Aristotelian cosmology, as an adept courtier, attuned to the patterns of the Medici court and its subtle culture of patronage and gift exchange.[69] In the nineteenth century, Michael Faraday, too, laboured long and hard to represent himself as a 'philosopher' and sage, able through painstaking preparation of lectures, even to the extent of taking elocution lessons, to comment with apparent disinterest (and therefore objectively) on God's laws of nature revealed directly through him.[70]

We have seen something of William Scoresby's image as a trusted authority on ships' magnetism, an image moulded by his sea-going experience as captain and by his religious experience as a Christian (ch. 1). Following Scoresby's death, indeed, his friend Archibald Smith published Scoresby's posthumous *Journal of a Voyage to Australia and Round the World for Magnetical Research*. 'To the love of science and the devotion to the cause of humanity which led Dr. Scoresby to undertake this voyage at an advanced period of his life', Smith asserted in the introduction, 'he may be said to have fallen a martyr'. Not even the Astronomer Royal himself could compete with that image.[71]

In the context of engineering, there was again a spectrum of types simul-taneously social, practical and epistemic, inhabiting, as it were, a new geography of practical knowledge. Willis aped Faraday's childlike simplicity in his lecturing, eschewing the 'clap-trap' of dangerous enthusiasts, and standing, instead, as a true philosopher of mechanism, linking the mechanical with the mathematical order of Cambridge. Willis's friends in the higher echelons of the British Association, like Whewell, similarly recast themselves not as engineers, but as philosophers of engineering, evaluating its progress from above. Conversely, practical engineers, like Fairbairn, basked in the limelight of the BAAS, soaking up status. Intermediate figures, like Rankine in Glasgow, carved out new roles as scientific engineers, bridging the practice, and the need for immediate action, of the engineer and the theory, more cerebral and permanent, of the scholar – in harmony, and to their mutual benefit.[72] William Thomson, allied to Rankine and to William Ayrton, made a science of telegraphy, and a scientific type of the electrical engineer, pushing against the pressure of the self-taught Wildman Whitehouse, now a scapegoat for earlier telegraphic failures (ch. 5).

At odds with these appropriations from the academic centres of science, men like Brunel tapped norms of masculine daring and risk-taking to build images of uniqueness, originality and heroism. Status-conscious railway engineers like George Stephenson, and their acolytes, policed their own images – aware that attributions of moral degeneracy were bad for business.

Earlier, James Watt had shaken off his persona as general merchant haunting the thoroughfares of Glasgow, to present himself as philosopher steeped in principles both philosophical and moral, knowing that promulgating such an image of himself was good, not only for posterity, but for business.[73]

In this process of image-making, visual representations served to disseminate the patterns of the men and women of science, medicine and engineering, complementing the genre of aggrandising literary biography.[74] Historical actors were themselves active in making images, whilst strategically placed statuary and carefully posed photographs distilled, secured and enhanced reputations. Charles Darwin provides an example from science. He played a dominant role in manufacturing representations of himself – as man about town, as invalid, as bearded philosopher and eventually as sage – in portraits and in a series of signed *cartes de visite* taken by, or sent out to, admirers and celebrity hunters.[75]

Amongst our engineers, Watt famously commissioned a bust of himself, copied it with an apparatus he had designed, and saw it placed in centres of scientific and engineering power by that most zealous guardian of Watt's reputation, his son. Posthumously, a monument appeared, amid controversy, in Westminster Abbey, whilst larger-than-life statues dominated public squares in Glasgow and Leeds. Nineteenth-century artists then re-imagined Watt's moment of invention – whether inspired by the kettle, or by the troublesome Newcomen engine (ch. 2). Portraits of Marc Isambard and Isambard Kingdom Brunel helpfully identified them not with the tools of their trade but with their engineering products: for the father, the Thames Tunnel formed a fitting backdrop (above); for the son, conventional portraits in early years – coded to link the young Brunel with the surveys and plans of Bristol-based projects – were superseded by the famous photographs of Brunel, a little giant, dwarfed by the launching chains of the *Great Eastern*.[76] There, size mattered.

Not all engineers chose images of self by which to commemorate their works. With a powerful residue of austere puritanism, Alfred Holt opted for self-effacement. Not even the walls or windows of the new Ullet Road Unitarian Chapel, gentrified gothic successor to the plain architecture of Renshaw Street, displayed a memorial tablet to the engineer of long-distance ocean transport. Neither *Dictionary of National Biography* nor Edwardian life-and-letters immortalised this merchant prince of the Mersey. Neither title nor country house served to perpetuate the name to later generations. Instead, Holt, eschewing the rapid conversion of profits into the trappings of landed aristocracy, constructed an ocean fleet, the iconic Blue Funnel Line, with only its house-flag bearing the simple letters 'AH'.[77]

Images of the objects of practical engineering took many forms. Desaguliers reproduced extensive engravings of the machines he described; and the French *Encyclopédie* likewise depended on careful illustrations of crafts. Those images were attempts to capture the essential elements of – or to objectify – engineering forms and technical practices, rather in the manner of those illustrations of flora, fauna, or anatomy which showed no single specimen but an ideal, or 'mean' representative.[78] This reminds us, of course, that the mere existence of an image of a machine did not imply the existence of a material artefact (as many architectural prospects demonstrate); for the engineers, images were the visual tokens of possible futures, or even possible utopias. Most credible, although often not materialised, were the many audacious schemes given column inches – and giving work to the engravers – illustrated in the *Mechanics' Magazine* and other technical journals. In the age of exhibitions, what better way to disseminate the experience than through a catalogue – memento, keepsake and celebration in one? With the new railways, guidebooks routinely illustrated the best – most beautiful, most economically or politically significant – 'views' of the route and its environs, whilst showing also the stations and locomotives (ch. 4). Commissioned portraits of ships, like champion racehorses, graced the boardrooms of maritime magnates while builders' models, crafted to represent the finest details, lured passers-by to the office windows of imperial shipping lines in London's Cockspur Street or Haymarket.

On the one hand, popular broadsheets like the *Illustrated London News* and provocative journals like *Punch* enthusiastically celebrated or satirised the practical arts and public works of home and, increasingly, the exotica of orient and empire. On the other hand, mathematically schooled engineering educators (from the French geometer Gaspard Monge onwards) moved towards abstraction in professional engineering imagery. Martin Rudwick charts a similar dichotomy between the circumstantial and naturalistic in geological illustration and the abstract, decontextualised – and objectified. In the class room, and for the benefit of engineers seduced by the promises of engineering science, Rankine, for example, worked to reduce all the dirt, complexity and confusion of heat engines to (relatively) simple geometrical diagrams. He recognised not just the problem-solving value but also the persuasive power of the visual, when suitably managed.[79]

Prophecy, prayer and providence: what hath God wrought?

Studies of science, or natural knowledge, during the long nineteenth century show that the contexts of religious practice and of theology cannot be ignored. They show, moreover, that models of inherent 'conflict between science and religion', promoted by particular groups for specific purposes from the late nineteenth century, do not stand up to historical scrutiny.[80] If the case has been forcefully made for a rich interplay of science and religion, what then of engineering and religion?

At one level, the mission of empire included a religious mission. Macgregor Laird, bringer of commerce and Christianity to West Africa and brother of Birkenhead iron shipbuilder John Laird, wrote in 1837 that 'We have the power in our hands, moral, physical, and mechanical; the first, based on the Bible; the second, upon the wonderful adaptation of the Anglo-Saxon race to all climates...the third, bequeathed to us by the immortal Watt.'[81] Steamboats, able to penetrate the rivers of Africa and Asia, could indeed function as tools for the expansion of a simultaneously commercial and Christian empire, operating, as and when required, by force. Not surprisingly, perhaps, when Alfred Holt took over his first steamer, the second-hand *Dumbarton Youth*, he found her to contain a quantity of Bibles, arms, and – of more enduring value to the practical Unitarian ship-owner – some blue paint.[82] Western technologies, then, as much as or more than Western science, came to symbolise Western culture and its civilising mission, underpinned by evangelical Christian beliefs and values. Needless to say, Ottoman religious commentators, if not their pragmatic political masters, strongly resisted the crusading intervention of such 'moral, physical, and mechanical' powers.[83]

When Martin Luther called the printing press 'God's highest gift of grace' he was not the last to assess the power of a new means of communication in advancing the work of God.[84] Samuel Morse famously inaugurated an American age of electrical communication in 1844 with the portentous message: 'what hath god wrought'. Punctuated, according to the King James Version, as 'What hath God wrought!', Morse's exclamation expressed the blessing conferred by Providence on those freed from the bondage of an old land.[85] Twenty years later, *The Times*'s journalist W.H. Russell concluded his intimate study of the (failed) Atlantic telegraph cable of 1865 by claiming, like William Thomson in 1859, that success was now as inevitable as 'calculation' allowed; and indeed, for Russell, the telegraph would be an instrument in reuniting nations divided:

> next year will see the renewal of the enterprise of connecting the Old World with the New by an enduring link which, under God's blessing, may confer unnumbered blessings on the nations which the ocean has so long divided, and add to the greatness and the power which this empire has achieved by the energy, enterprise, and the perseverance of our countrymen, directed by Providence, to the promotion of the welfare and happiness of mankind. Remembering all that has occurred, how well-grounded hopes were deceived, just expectations frustrated, – there are still grounds for confidence, absolute as far as the nature of human affairs permits them in any calculation of future events to be, that the year 1866 will witness the consummation of the greatest work of civilised man, and the grandest exposition of the development of the faculties bestowed on him to overcome material difficulties.[86]

Not all readings of electrical telegraphy, however, straightforwardly confirmed religious verities: they might instead subtly question the providential action of God, the literal truth of His historic utterances, or the hopeless inferiority of man's intellect and power. Before the telegraph, God had convinced Job of his ignorance and manifold weaknesses, asking 'Canst thou send lightnings, that they may go, and say unto thee, Here we are?' But by 1860, as the *Atlantic Monthly* observed ambivalently, the fact of telegraphic communication meant that 'At the present day, every people in Christendom can respond in the affirmative.'[87]

That a nineteenth-century technology of communication might indeed fulfil biblical prophecy was made explicit in evangelical David Brewster's unfulfilled search for a single human originator of the railway.[88] Writing in the Free Church *North British Review*, Brewster found the vision of the ancients – whether poets or philosophers – had been too dim to descry it. Rather, if the railway had been anticipated at all, it was by the 'far-seeing eye of prophetical inspiration':

'Make straight in the desert', says Isaiah, 'a highway for our God. Every valley shall be exalted, and every mountain and hill shall be made low, and the crooked shall be made straight and the rough places plain, and the glory of the Lord shall be revealed'.[89]

Had Brewster's readers pondered the words immediately preceding this passage ('Prepare ye the way of the Lord'), the divine purpose of that 'highway' would have been yet further apparent. And indeed, to Brewster the millennial timetable of this long-anticipated 'system of railway' was clear when he cited a passage known to his readers in science to grace the immortal Bacon's *Great Instauration*: 'Daniel looks forward to the "time of the end, *when many shall run to and fro*, and knowledge shall be increased".'[90] Here was a compelling argument, surely, in Brewster's campaign to create a central role for British government in the 'disinterested supervision and able management' of a national railway network.[91]

Evangelicals, however, could stray from scriptural orthodoxy by invoking popular belief. Macgregor Laird, for example, told his wife Nell that the launch of a ship, the *Dayspring*, at his brother's yard had failed with the vessel becoming 'stuck on the way after running about 40 yds, through some fault of the foreman carpenter, and there she lies till tomorrow'. But the physical explanation masked something more sinister: 'a clergyman was on board the Dayspring when the attempt was made to launch her so we put it down to his being there, that she hung fire'. As though to underline the seriousness of his belief in such age-old seafaring lore as ill-fortune – or devil's work – wrought by the presence of a clergyman aboard ship, Laird also affirmed his unwillingness to see the new ship sail on a Friday.[92]

In contrast, the Cunard Line apparently had no worries about carrying clerical passengers. One such passenger in 1845, the Rev. Norman Macleod, was part of a three-minister deputation from the Church of Scotland General Assembly outward bound to visit congregations in British North America.

He recorded in his diary that also on board the *Acadia* was 'a missionary bronzed with the sun of India, Protestant clergy and Catholic [priests]'.[93] Such two-way traffic of clergymen, indeed, made for excellent business. But it needed special regulation. The founding partners in the Company there-fore laid down that Sundays at sea were to be marked only by the Church of England form of service, usually conducted by the captain or, if available, by a clergyman of an established church. This policy, ostensibly designed to prevent shipboard chaos from rival denominations staging simultaneous – and potentially disorderly – services throughout the vessel, alienated United States' clerics, excluded because there was no established church in their own land.[94]

Such regulation of religion at sea formed only one part of a strict set of management rules designed to discipline life, in the name of safety, aboard the Cunard ships. By the second half of the nineteenth century, the Line's reputation for safety had become legendary and formed one of its strongest selling points (ch. 3). George Burns's biographer noted in 1890 that from its foundation there had been 'singular exemption from misadventure'. This had, he continued, been attributed to 'a wonderful run of luck' or to 'a special interposition of Providence'. He further recorded that it had been rumoured that 'the sailing of every ship in the Cunard fleet was made the subject of special prayer, and that Mr Burns was wont to attribute his success to this source'. But he denied such appeals to direct divine interposition. Instead, he suggested that Burns had firmly believed in humankind's obligation to work within the natural order established and sustained by God: 'While trusting in Providence and believing implicitly in the power of prayer, ... [George Burns] was also a firm believer in doing work well, and in subordinating profit and speed to safety, comfort, and efficiency.'[95]

Burns believed that providential protection depended fundamentally upon God not intervening arbitrarily but, rather, acting through 'instruments', especially human instruments with the degree of 'personal fitness' requisite to carry out the divine will on the seas. His great mentor and founder of the Free Church, the Rev. Thomas Chalmers, had often preached on St Paul's account of his shipwreck: it was 'God's will that they should be saved by the exertions of the sailors – that they were the instruments.'[96] Samuel Cunard too shared these perspectives as he wrote to his other partner Charles MacIver in 1858: 'We have now been nearly twenty years connected in business, a large portion of our ordinary life, and we have been most providentially protected – see how many great houses have fallen around us during the last year, who were at the commencement of the year in affluence.'[97]

'Personal fitness' aboard ship centred upon the trustworthiness of the master. 'The trust of so many lives under the Captain's charge is a great trust', David and Charles MacIver stated in the seven-page 'Captain's Memoranda' (1848) which they issued to each master.[98] Implicit was the message that this was a sacred trust, not simply from the owners but from God. Perceptive passengers understood that this commandment endowed Cunard masters with the status of 'instruments', investing them with a moral leadership freighted with divine authority. As William Makepeace Thackeray, literary celebrity and investor in the first Atlantic telegraph, wrote of his recent experiences aboard a Cunard liner:

> And so through storm and darkness, through fog and midnight, the ship had pursued her steady way over the pathless ocean and roaring seas so surely that the officers who sailed her knew her place within a minute or two, and guided us with a wonderful providence safe on our way....We trust our lives to these seamen and how nobly they fulfil their trust! They are, under heaven, as a providence for us. Whilst we sleep, their untiring watchfulness keeps guard over us.[99]

From the mid-1850s a close informal network of natural philosophers, academic and practical engineers, and Presbyterian clergy existed in Glasgow.[100] The Rev. Macleod in particular forged close friendships with William Thomson, Macquorn Rankine, John Elder and John Burns (heir to the Cunard Company). Macleod himself expressed a strong liking for the heavy marine industry of the Clyde when he wrote of 'the waking up of the great city, the thundering of hammers from the boilers of great Pacific and Atlantic steamers – a music of humanity, of the giant march of civilisation; far grander to hear at morn than even the singing of larks'.[101] This veritable hymn to marine steam engines had its roots in a transatlantic voyage. 'You know my love of steam engines', he confided in his diary as he voyaged aboard the second of Cunard's original quartet of steamers, 'and certainly it has not been lessened by what I have seen in the *Acadia*.' The scene might have been a modern Dante's *Inferno*. But for Macleod, it was sublime: 'What a wonderful sight it is in a dark and stormy night to gaze down and see those great furnaces roaring and raging, and a band of black firemen laughing and joking opposite their red-hot throats! And then to see that majestic engine with its great shafts and polished rods moving so regularly night and day, and driving on this huge mass with irresistible force against the waves and storms of the Atlantic!' Not for MacLeod, then, messages of materialism and damnation, but a striking testimony in the engine-builder's skill to the skill of man's Maker: 'If the work glorifies the intellect of the human workman, what a work is man himself!'[102]

Macleod's love affair with steam was of course unsurprising in Presbyterian Scotland. Engineers routinely enacted this relationship. James Watt was motivated in his experiments on the Newcomen engine to save wasted steam. For him, saving waste was very much a moral duty in the context of stern Presbyterian values (ch. 2). Yet those austere values coincided with traditional Calvinist doctrines of original sin and innate depravity, productive of a deeply pessimistic attitude to this visible, material world. Resonating with

this culture was Watt's low opinion of himself and his worldly achievements: 'Today', he told William Small in January 1770, 'I entered into the
35th year of my life & I think have hardly done 35 pence worth of good in
the world but I cannot help it.'[103] Matthew Boulton, without such religious
allegiance, yet keenly aware of the close correlation between the positive
'reputation' of the steam engine and that of its inventor, advised Watt in
April 1781: 'I cannot help recommending it to you to pray morn[g] and even[g]
after the manner of your countrymen ["Oh, Lord, gie us a guid conceit
o'oorsels."] for you want nothing but a good opinion and confidence in
yourself and good health.'[104] Personal confidence, spiritual strength and
business success were mutually reinforcing.

Among Macleod's own generation, who venerated Watt as no other, Macquorn Rankine constructed the new role of the 'scientific engineer' as a
pursuit of perfection. Freed from the severest tenets of Scottish Calvinism,
Rankine worked, more optimistically, to set scientific standards of engineering
potential, especially but not only for heat engines, against which actual
engineering achievement could be measured. It was an agenda which paralleled similar pursuits in his (and his shipbuilder friend John Elder's) spiritual
life. That agenda was exemplified by the phrase 'the imitation of Christ'.[105]

South of the border, also, religious and engineering practices mutually
informed one another. The created natural world provided a bounty of
examples for the engineer – as human designer – to copy, however
imperfectly. He should strive to harness the great powers of nature without wasting God's gifts. The devout engineer must constantly recognise,
in order to optimise, God's providential arrangements. Cultivated in a
Christian framework, this was the man best fitted to serve state, empire – and
God. In London, Durham and Cambridge, engineers and educators struggled
to articulate these bonds between theology and practical art.

Teaching at Woolwich, Olinthus Gregory merged his evangelicalism with
practical mathematical training. Thus Gregory proposed in the 1830s:

> There is but one precise point from which we can take an undistorted view
> of a picture: all other points are too high or too low, too remote or too
> near. Perspective assigns that point in reference to the picture; but where
> shall we find the point from whence much that occurs in morals and in
> Providence, shall, in like manner, be observed free from distortion?
> Where, but in the regions of perfect purity, and light, and bliss.[106]

Once retired, Gregory penned a 'Note on "Genuine Christianity"' which
echoed the notion of Christian imitation but which might also have described
his career as teacher and guide to generations of military engineers. 'Genuine
Christianity is calculated to produce a twofold effect upon men; to bless them
and to make them blessings', Gregory asserted in 1839. Echoing Chalmers's
sermon on human beings as 'instruments' of God's will, he elaborated: 'It

first transforms men into the image of God; and then employs them, according to their respective spheres of influence, as his instruments, in illuminating and transforming others.'[107]

The Institution of Civil Engineers was the chief talking shop for elite practitioners, the venue in which sane professional engineering standards were fixed, and the platform from which professional virtues were disseminated. Amongst those virtues, according to its president James Walker speaking in January 1839, was a sincere faith. Thus he reflected, thinking of Edward Young,

> What the poet said of the undevout astronomer, that 'he is mad,' may apply with equal truth to the undevout Mechanic or Engineer; and it would be well if those delightful feelings were cultivated, and invariably associated with the study and practice of the Engineer, so that his mind might in every pursuit dwell upon the wondrous adaptations of nature to the wants and pleasures of the community, and both in its lowest and most improved state be led to the contemplation of the Power which formed, and the Goodness which so admirably fitted the whole for the use of His Creatures.[108]

When Walker spoke of the 'study' of the engineer, he was aware of a sudden growth in academic training for the profession, where each institution offered its own framework of religious – or secular – discipline. University College London (UCL), although secular in intent and favouring training for the new professions, failed to get a coherent engineering programme up and running. A nascent College for Civil Engineers in London attracted much attention but such a venture required support. The Conservative Duke of Buccleuch was a generous patron and eventually President of the new College; but when the Crown and the Whig Prime Minister Melbourne withheld the stamp of approval, *The Times* thundered that the sticking point had been the moderate allegiance of the College to the established church: 'The words "this College is based upon the principles of the established church," in the first regulation, was the choke-pear that the Liberal Premier could not swallow.'[109] In the years following Catholic Emancipation (1829) and during which the Committee of Council on Education was under fire for its refusal to impose religious orthodoxy on the new elementary schools, Melbourne was unwilling to support moves towards exclusion, however subtle, in the field of engineering pedagogy.

Clearly there were differences over the right way to form the spiritual natures of the nation's engineers. The young University of Durham, for example, embarked on its programme of training for 'academical engineers' in 1838. These new students were to be on an equal par with others studying for first degrees. They would enter the university armed with testimonials of character and having passed an exam covering, inter alia, the evidences and doctrines of Christianity. As in Oxbridge students must attend religious services and prayers. Subjected throughout to college discipline, they were trained to the Establishment in an institution whose major role was, after all, a religious seminary funded ultimately by the Anglican Church. The star student of the select class admitted to the rank of academical engineer in 1841 was one Samuel Smith: a prize winner in chemistry, geology and English, his subject in the last discipline was, naturally enough, the 'evidences of design in metalliferous deposits'. For Smith, engineering was in no way incompatible with theology, which he continued to study, becoming DD (Doctor of Divinity) in 1857.[110]

Durham was not alone. King's College London (KCL) had been set up, in righteous reaction to the alleged godlessness at UCL, as a 'seminary of sound learning and religious education, according to the doctrines of the United Church of England and Ireland'. Although, in a controversial 'half-open door policy', dissenters might attend individual courses at KCL without being asked to profess themselves members of the established national church, they could not be full 'members' of the College and, in any case, were subjected to the same religious discipline and procedure. All tutors, lecturers and professors were members of the 'United Church'. KCL soon had a thriving department of engineering, led by the Rev. Moseley accompanied by electrical telegrapher Charles Wheatstone, fashioning raw humanity into Christian engineering imperialists.[111] Olinthus Gregory's protégé James Hann achieved a religious trajectory from Catholicism to Church of England: once appointed to the King's College School, he admitted to the Bishop of London his Catholic upbringing and professed his subsequent disillusionment.[112]

In Cambridge, Willis concluded his *Principles of Mechanism* (1870) with an emphatic reaffirmation of the lessons to be drawn from natural theology. Man should be humble, since the best models for his mechanical ingenuity, however mathematically disciplined he might be, were to be found in divinely created nature. Thus even the 'universal flexure-joints and swivel joints' discussed in the final chapter of this grounding for the gentlemanly engineer were already manifest in the joints of crustacea and insects:

> Thus my series of mechanistic combinations has conducted me to an example from the numerous and marvellous constructions which characterise the machinery of the animated forms, with which the world has been peopled by its Beneficent, All-wise, and Merciful Creator, from the careful and reverent study of whose wondrous works, we derive all our practical science, under His Almighty guidance and protection, which is never withheld from those who humbly ask it, in the spirit of faith and truth.[113]

As the professional engineers and educators prepared the present and future servants of the state and empire for 'objects of national importance' (ch. 1), they responded to rival programmes, and factions, at home and abroad. If Carnot envied British economic and practical superiority, the British admitted, selectively, the benefits of education in the French style. The centralised system of *grandes écoles*, focused on the militaristic École Polytechnique in Paris, trained secular bureaucrats with a strong national identity dedicated to the pursuance of public works. British practical power had developed and had been sustained, however, within an ethos celebrating individualism in engineering as in science.[114] Yet individualistic empire-building might fracture national unity and erode imperial power. Cultivating Christian engineers through education, and promoting, through natural theology, a broad but valid consensus amongst engineering practitioners, could be seen as a response to these problems aligning individualist engineers more closely with imperial ambitions.

The younger Brunel, although a role model for aspiring engineers, shows how unfettered individualism might lead not only to a contempt for national standardisation in education and in practice (ch. 4) but also to religious waywardness. Although formally adherent to the Church of England, usually observing the Sabbath, attending church and (like Charles Darwin) supporting his local parish as *de facto* lord of the manor, Brunel did find a conflict, at least a practical one, between religion and engineering. Professional business easily reduced the opportunity for spiritual betterment of a contemplative kind. While very much in pursuit of 'greatness' and the 'sublime' (if not of perfection) in his engineering projects, Brunel laid himself open to the chastisement of friends, like John Horsley, when he failed, apparently, to link that pursuit to spiritual concerns. Horsley reprimanded Brunel for his slavish devotion to the works of mammon, disappointed to see in his life 'almost unparalleled devotion to your profession, to the exclusion, to far too great an extent, of that which was due to your God and even to your family'. Yet, Brunel had found solace in prayer when faced with difficulties. He impressed the advantage of that course upon his son, and derived comfort when his prayers '*appeared* to me to be granted'. The regular action of divine law, however, made it difficult for Brunel to argue that prayer was directly efficacious for *individuals*:

> I am not prepared to say that the prayers of individuals can be separately and individually granted, that would seem to be incompatible with the regular movements of the mechanisms of the Universe, and it would seem impossible to explain why prayer should now be granted, now refused.[115]

If the programmes of the engineering educators were attempts to discipline and unify disparate individuals according to national objectives, the growth in authority of engineering institutions later in the century, ironically, made possible professional fragmentation. In an age of clearly defined secular expertise, T.H. Huxley and his allies in the X-Club promoted scientific naturalism as an ideology of professional science; its adherents claimed the cultural authority to speak on matters of supreme importance previously the province of the established churches.[116] Social prophets familiar with Huxley and his programme nevertheless articulated fears over fragmentation along religious, professional and scientific lines alike. One such was Matthew Arnold who, in response to a partisan reporter in the *Nonconformist* who had quizzed Huxley on how he might ever, without religion, cure the chaos of Derby Day, replied: 'And how do you propose to cure it with such a religion as yours?'[117] Arnold refused to indict Huxley for a chaos to which others, especially competing religious denominations, had equally contributed.

Strongly echoing Carlyle, Matthew Arnold's famous critique of late nineteenth-century British culture, in all its fragmentation, complained of a world 'to a much greater degree than the civilization of Greece and Rome, mechanical and external'. Britain, above all nations, needed *culture* (in Arnold's sense of inward perfection) as a remedy for anarchy. For it was in Britain that the 'mechanical character, which civilization tends to take everywhere, is shown in the most eminent degree':

> Faith in machinery is, I said, our besetting danger; often in machinery most absurdly disproportioned to the end which this machinery, if it is to do any good at all, is to serve; but always in machinery, as if it had a value in and for itself. What is freedom but machinery? what is population but machinery? what is coal but machinery? what are railroads but machinery? what is wealth but machinery? what are, even, religious organisations but machinery?

Arnold made 'machinery' stand for all that was 'interested', partial, and dissipative in a world losing, or in danger of losing, its mission – including its spiritual mission – when faced with the clamour of the provincial, the materialistic, and the newly enfranchised, but inadequately educated, lower classes.[118]

Gentleman engineers and captains of industry

Throughout this study our key protagonists have been British engineers, many of whom came to play leadership roles in their respective industries. They thus became, in Thomas Carlyle's phrase, *captains* of industry, often trusted, like their maritime counterparts, on account of their perceived integration of social status *and* technological expertise. These men (and they were, indeed, mainly men) were notoriously sensitive about questions of status and respectability. Engineers, as members of a profession younger than medicine, the law, or the church, worked at achieving upward mobility on both personal and professional levels; to do so, they often assumed the trappings of 'gentlemen'.[119]

Yet the engineers, especially of the older generation, were keenly aware that a man born with gentlemanly status rather than one who had earned it through experience or perhaps education was unlikely to make the grade as an engineer. Watt and Boulton's sons were to be genteel and capable in company; but they were to have sufficient deliberate exposure to industry to have no 'gentlemanly' distaste for it. George Stephenson did not wish to 'make his son a gentleman', preferring to train him (and use him) on engineering projects, like the Stockton and Darlington Railway, over which he had control. Only grudgingly, it would seem, did he allow him to attend that most democratic and far from socially exclusive University of Edinburgh.[120]

I.K. Brunel, born into a family of wealth and status, as son of Marc Brunel, barely merited a mention from Smiles (in his *Self-Help*), who preferred those 'self-cultured' rather than those given status by birth; and yet Brunel himself, although adamant that his employees would be gentlemen and prospective apprentices would be of 'gentlemanly habits' *and* 'gentlemanly connections' thought, rather like Stephenson, that too much gentlemanly polish, independence and indeed education, might ill-equip a man to be an efficient and unquestioning servant.[121] Nevertheless, we can perhaps see, during the nineteenth century, a transition in which the engineers and industrialists worked to remake the notion of a 'gentleman' in their respective arenas: he was a man whose status was certainly a product of merit, be it proven achievement or through relevant education, rather than birth, even as new dynasties emerged in the late nineteenth century.

The engineers were of course not the only social group who reflected on their gentlemanly position. Men of science, too, underwent a change of status during the Victorian period and this change had its origins even earlier. Steven Shapin has addressed the apparent contradiction of being both 'scholar' and 'gentleman' in the early modern period: how might the man of science square the circle, engaging in cloistered unworldly pursuits yet living also the *vita activa* as a useful member of society?[122] Here, and in the future, it would be a useful ploy to emphasise the utilities of science. Eighteenth-century aristocratic patrons adeptly linked agricultural improvement, chemical science and enlightened patronage.[123] Early in the nineteenth century the man of science appeared in the guise of the 'gentlemanly specialist', someone with money for materials and equipment, with leisure for science, and with a reputation built on specialised knowledge – but not a paid professional (ch. 1).[124] This trustworthy, disinterested individual could easily claim to be engaged in the pursuit of knowledge 'for its own sake'; but this was not a position plausibly assumed by the engineers, however gentlemanly their posture.

Dominating the public presentation of science in the BAAS were the so-called 'gentlemen of science'. They set the agenda for the first fifteen years, seeing to it that science had its showcase throughout Britain and the

Empire (ch. 1). As Morrell and Thackray have shown, the BAAS was indeed run by a male coterie with women like Mary Somerville making occasional forays into the male demesne rather in the way marginal men such as the Quaker John Dalton – without Anglican, Oxbridge or Whig credentials – now and again proved the exceptions to the general trends. But 'dealing with the ladies' – whether or not to admit women to the meetings and sections – posed larger and more controversial problems for the gentlemen of science, although in the end women tended to be admitted because they enhanced the revenue generated from public attendance.[125]

The engineers too were not immediately admitted to the BAAS but by 1836 they had their Section G as a forum for Mechanical Science (ch. 1). High-profile engineers like William Fairbairn and John Scott Russell did well from the Association, receiving grants for extended experimental projects which loudly proclaimed that the BAAS was making science useful. But, despite this compromise between 'mechanical scientists' and the gentleman of science, there were increasing challenges. In the early part of the century, artisan botanists, fossil collectors, practical electricians and reforming phrenologists were only some of the groups with alternative agendas who questioned the hierarchy of sciences enshrined in the committees and sections of the BAAS.[126] In the later part of the century, the agenda was again partially wrestled from the hands of the gentleman of science by the scientific naturalists and the propagandists for social and scientific change led by Huxley and the astronomer Norman Lockyer.[127]

Questions of trust, credibility and authority ran through all of these debates about the position of engineers in wider social groupings such as the BAAS and in the life and economy of the nation and empire as a whole. The most ambitious of the early engineers valued gentlemanly status as a hallmark of trustworthiness. Most notably, Brunel established and maintained his authority in part by insisting upon (and constantly re-enacting) his gentlemanly status, especially when called into question in the wake of his not-infrequent failures. His attempts to manage projects overseas and by proxy dramatically illustrate the power – and pitfalls – of insisting upon 'gentlemanly habit' and 'conduct' in his assistants, without whom none of his spatially dispersed projects could have been completed.[128] That there were all-too frequent failures to live up to these ideals is well demonstrated in the case of Lardner, who found his gentlemanly credentials and authorial credibility seriously dented after he ran off with the wife of a military officer.[129]

From the late eighteenth century, successful engineers and captains of industry chose to represent their new-found status in the tangible form of country houses, often on a grand scale. The claim that this indicated a withdrawal from active entrepreneurship needs qualification.[130] Boulton's seat had been a working meeting place for the Lunar Society and a site in which to form connections so valuable to subsequent marketing ventures. Watt purchased a plot in 1790 and by the end of his life 'Heathfield', constructed by Albion Mill architect Samuel Wyatt, had 40 acres, garden, stables and additional lodges – and Watt's famous garret workshop (subsequently a shrine to the inventor).[131] Brunel projected an 'Italianate villa with a belvedere and colonnaded terrace' for his estate at Watcombe near Torquay, purchased in 1847; but although he landscaped the gardens he never completed the house.[132]

Famously Robert Napier rose from blacksmith beginnings to 'father of Clyde shipbuilding', an achievement marked by his move to a spectacular mansion at West Shandon, overlooking the Gareloch.[133] Even the self-effacing Alfred Holt moved to his new home, the classical-style 'Crofton' at Aigburth, outside Liverpool, once he had stabilised his Ocean Steamship Company.[134] The evangelical Burns family had fewer scruples about building a gothic mansion next to the historic Castle Wemyss on the south side of the Firth of Clyde. There, the newly created dynasty of Inverclyde could oversee their rivals, and entertain investors, admirals, cabinet ministers and royalty while extending their hospitality with cruises aboard one or other of their vessels.[135] Nothing, however, could quite compete with the home of former Belfast shipyard apprentice William John Pirrie who, as Viscount Pirrie of Belfast and autocratic chairman of the Harland & Wolff shipbuilding empire, purchased in 1909 the 2800-acre Witley Court estate, including extravagantly designed house and 500-acre park, in Surrey.[136]

Ceremonies and rituals not only accompanied life in these country residences, but also formed part of the engineering empires themselves. Launches of ships, especially the prestigious liners of empire, were cultural events widely reported in the local, national and technical presses. In the first half of the twentieth century it became commonplace for the largest ships to be named by royalty or the wives and daughters of imperial states-men. In some cases, the naming ceremony was performed 'at a distance' by radio telephone signal from the distant part of the Empire to which the new liner would trade. Thus in late 1934 – in the depths of the Depression – the Orient liner *Orion* was sent down the ways at Barrow-in-Furness after an electrical signal, relayed by land-line and radio, was sent 12,000 miles from Brisbane by the Duke of Gloucester. In 1947 a similar arrangement launched the Union Castle liner *Pretoria Castle* at Belfast, this time activated by the wife of Premier Jan Smuts in South Africa.[137]

In these cultures of increasing gentrification, women found a more formal, but perhaps – ironically – less influential, role. Ship-owning, especially among early steamship owners, like the Holts, was often a tightly knit family affair as in the days of sail. Women such as Emma Holt (née Durning), mother of four ship-owning brothers, continued throughout to exercise a powerful – and matriarchal – moral control over the family.[138] Similarly, George Burns married Jane Cleland, daughter of the famous Glasgow statisti-cian and political economist James Cleland (ch. 2), and throughout his business life not only confided in her his deepest spiritual and moral anxieties but received her guidance on all such matters.[139] In later years, however, women seemed more and more stereotyped by their theatrical roles on launch platforms and on other ceremonial occasions. And as if to emphasise the stricter demarcations and less fluid roles now in place, a memorandum left by Philip Henry Holt to be read at his death in 1913 made clear that 'Steamship property does not seem to me suitable for ladies.'[140]

By the early twentieth century, the highly gentrified culture of those railway and shipping empires which served the larger British Empire carried with it all the trappings of trustworthiness. Late Victorian and Edwardian company offices, such as the Cunard Building, the Liver Building, and the Mersey Docks Board Building which continue to render Liverpool's waterfront one of the finest in the world, made these enterprises seem as trustworthy, enduring and solid as the Bank of England. The inability of many such apparently trustworthy companies to weather the inter-war economic crises undermined these myths of solidity and durability. The most notorious casualty was the vast but unstable Royal Mail Group which included many of the most honoured names in British imperial shipping: White Star Line, Royal Mail Steam Packet Company, Union-Castle Line, Elder Dempster Lines, Glen & Shire Lines and Harland & Wolff (itself a large shipbuilding, ship-repairing and engineering empire with multiple sites in Belfast, Liverpool, Glasgow, Greenock, London and Southampton). Gentlemanly credentials alone, it seemed, were no guarantee of trustworthiness and the Group chairman, Lord Kylsant, was given a twelve-month jail sentence for issuing a misleading and deceptive prospectus.[141] In the fashion for gentrification, perhaps, Britain's captains of industry had forgotten that the credibility of their engineering empires, those large-scale technological systems explored in this book, had lain not simply with ceremony and grandeur but above all in cultures of technological expertise.

Notes

Introduction: technology, science and culture in the long nineteenth century

1. Bright 1911: xiii.
2. Wiener 1981; Edgerton 1996.
3. Singer 1954–58; and for a parallel in the history of science, Gunther 1923–67.
4. Ross 1962 ('scientist'); Cantor 1991 (Faraday).
5. Wilson 1855; Anderson 1992.
6. See esp. Schaffer 1991.
7. See esp. Brooke 1991: 16–51 ('science and religion').
8. Layton 1971; 1974.
9. Channell 1989 (for a bibliography of 'engineering science' following Layton); and for a critique see Marsden's article 'engineering science' in Heilbron *et al.* 2002.
10. Ashplant and Smyth 2001; for alternative 'varieties of cultural history' see Burke 1997; for the 'new cultural history', which adopts ideas from semiotics and deconstruction, see Hunt 1989; for a perceptive critique of the not always happy relationship between anthropological and cultural studies of science, and the new orthodoxy of social constructivist history of science, see Golinski 1998: 162–72.
11. Cooter and Pumfrey 1994.
12. Burke 2000.
13. See, for example, Briggs 1988; Lindqvist *et al.* 2000; Finn *et al.* 2000; Divall and Scott 2001; Macdonald 2002; Jardine *et al.* 1996.
14. For a recent histriographic manual open to both cultural history and history of science see Jordanova 2000b; for a recent attempt at 'big picture' history of science, technology and medicine – or STM – see Pickstone 2000.
15. Pacey 1999.
16. Bijker and Law 1992; Bijker *et al.* 1987; Staudenmaier 1989a.
17. Pinch and Bijker 1987: esp. 40–44.
18. Petroski 1993.
19. See Cowan 1983: 128–50.
20. Kirsch 2000.
21. See, for example, Gooday 1998.
22. Smiles 2002 [1866]: 3–4. Plutarch was the author of exemplary 'lives' of Greek heroes.
23. Winter 1994: 69–98.
24. Marsden 1998a.
25. Buchanan 2002a: 103–12 on Brunel's 'Disasters'.
26. Schaffer 1983; Morus 1992; 1993.
27. Shapin 1994: 16–22. On trust in economic action see Granovetter 1985.
28. Shapin 1994: 7–8. See also Smith *et al.* 2003b: 383–85 for further discussion.
29. Babbage 1846: 218–20. Middle-men is Babbage's term.
30. Bijker *et al.* 1987. Headrick 1981 does not explicitly address the issues raised by technological systems.

31. Gooday 2004; Wise 1995.
32. Headrick 1981 does not talk about the role of the sciences using history of science literature and historiography.

1 'Objects of national importance': exploration, mapping and measurement

1. Murchison 1838: xxxiii.
2. For example, Shapin 1996; Henry 2002a; Dear 2001 ('Scientific Revolution'); Hankins 1985; Outram 1995; Gascoigne 1994 ('Enlightenment'); Sorrenson 1996: 221–36 (Cook's scientific voyages); Schofield 1963 (Lunar Society of Birmingham); Thackray 1974: 672–709 (Manchester Lit. & Phil. Society).
3. For example, Shapin and Schaffer 1985 (on experimental culture surrounding the early Royal Society); Shapin 1994 (on trust and gentlemanly culture in seventeenth-century science); Howse 1980: 19–44 (on the early history of the Royal Observatory).
4. For example, Henry 2002b (on Bacon); Jacob 1976; Jacob and Jacob 1980: 251–67; Shapin 1981: 187–215 (on latitudinarian Anglicanism).
5. Hankins 1985: 2 (on Kant and 'Enlightenment'); Heilbron 1979 (on measurement in eighteenth-century electrical science); 1990: 207–42 (measurement and Enlightenment).
6. Foucault 1980: 69.
7. Beaune 1985; Revel 1991: 133.
8. Revel 1991: 139–40. Compare, on field operations, Livingstone 2003: 40–48.
9. Revel 1991: 141–43; Alder 1995: 39–71.
10. Revel 1991: 150–53. See especially Livingstone 2003: 153–63 ('Mapping territory'), 124–26 (maps and cartography). See also Harley 1988: 277–312 for a discussion of cartography and power.
11. Livingstone 2003: 124–25 shows how topographical maps in France, especially of Paris, produced the Cassini dynasty of astronomers.
12. Howse 1980: esp. 116–71; Morus 2000: 455–75.
13. Howse 1980: 141–42.
14. Conrad 1947 [1907]: 30–35.
15. Howse 1980: 21–27.
16. Quoted in Howse 1980: 27.
17. Howse 1980: 28–29.
18. Howse 1980: 45–56. Howse cites a range of early eighteenth-century British satirists (including Jonathan Swift) who regarded the quest for the longitude as akin to squaring the circle or the quest for perpetual motion.
19. Howse 1980: 45–79; Stewart 1992: esp. 183–211 ('The Longitudinarians'). See also Sobel 1996 (on Harrison). Howse 1980: 72 notes that Harrison received in total the huge sum of £22,550 under the 1714 Act.
20. Ginn 1991: 1–26 (eighteenth-century London instrument makers).
21. See Carter 1988; Sorrenson 1996 (on Cook and navigational calculations in relation to the charting of coasts).
22. *Nautical Magazine,* 28 October 1833 (quoted in Howse 1980: 80).
23. Howse 1980: 79–80 ('Time-balls').
24. Airy 1896: 40 (quoted in Ginn 1991: 208).
25. Schaffer 1988: 115–26. Airy's papers are included in the RGO Collection, Cambridge University Library. For wider cultural contexts on 'time discipline' and industrial

capitalism see Thompson 1967: 56–97. For the archiving mentality in imperial contexts see Richards 1993.

26. For example, Chapman 1998: 40–59.
27. Booth 1847: 4, 16 (quoted in Howse 1980: 87–88).
28. Anon 1851: 392–95; Howse 1980: 107.
29. Howse 1980: 94–105; Chapman 1998: 40–59; Morus 2000: 464–70.
30. Quoted in Howse 1980: 112, 99.
31. Howse 1980: 113–15, 138–51; Morus 2000: 469–70. Compare Bartsky 1989: 25–56 (standard time in the USA).
32. Although Thompson 1967 conveys a sense of an inexorable march of clock time in capitalist society, his article cites a rich range of alternative cultures of time, especially those – such as farming and fishing – which were 'task-orientated' (completion of a task like the harvest) rather than involving timed labour (paid according to hours worked). The former tended to regulate the work by natural cycles such as the length of day, the seasons or the tides while the latter was well suited to factory discipline and the mechanisation of society.
33. Esp. Ashworth 1994: 409–41.
34. Morrell and Thackray 1981: esp. 36–63; 1984: 299 (on the Marquis of Northampton's characterisation of the Royal Society and the BAAS as the 'two parliaments' of science. Northampton had been BAAS President in 1836 and was Royal Society President from 1838 until 1848, in which year he became BAAS President again).
35. Morrell and Thackray 1981: esp. 21–29, 63–94.
36. Morrell and Thackray 1984: 299.
37. Morrell and Thackray 1981: esp. 96–109, 396–411. The various seaports included: the commercial rivals of Liverpool (1837 and 1854) and Bristol (1836), the historic and naval town of Plymouth (1841), the industrially growing centres of Glasgow (1840 and 1855) and Newcastle (1839), and the more remote port city of Cork in southern Ireland (1843). These choices also reflected concern with ease of travel by rail and sea (chs 3, 4).
38. Morrell and Thackray 1981: esp. 267–75 (hierarchy of the sciences), 451–512 (the sections), 256–66 (Section G), 276–86 (the fringe sciences, excluded as unwelcome to the centralising ideology of the gentlemen of science with their programmes of mathematisation, measurement and standardisation).
39. *BAAS Report* 1 (1831): 11.
40. Morrell and Thackray 1981: esp. 372–447. On the geographical union promoted by the BAAS see Livingstone 2003: 108–09. On the origin of the term 'scientist' see Ross 1962: 65–85.
41. Ritchie 1995: 219–40 (*Beagle*); Reingold 1975: 51 (on Herschel's 'chartism').
42. On Enlightenment goals of quantification see Heilbron 1990: 207–42.
43. Herschel 1830: 122–23; Smith 1989: 28.
44. Ritchie 1967: esp. 208–18; Friendly 1977 (Beaufort); James 1997: 288–98; 1998–99: 53–60; 2000: 92–104 (Faraday and the lighthouses of Trinity House).
45. *BAAS Report* 3 (1833): xxxvi, 471, cited in Sedgwick 1833; Morrell and Thackray 1981: 515.
46. Morrell and Thackray 1981: 425–27, 515–17. See also Smith and Wise 1989: 583 (empire-wide tide gauges), 370–71 (William Thomson's tide predictor).
47. Forbes 1832: 199–201; Smith 1989: 30–31.
48. Morrell and Thackray 1981: 348–49, 517–23; Smith 1989: 31. See also Anderson 1999: 179–216; 2003: 301–33.
49. Cawood 1979: 493–518; Morrell and Thackray 1981: 353–70, 523–31.

50. For the earlier history of terrestrial magnetism and the problems of long-distance navigation see Pumfrey 2002.
51. Murchison 1838. On Murchison's Silurian fieldwork see esp. Secord 1981–82: 413–42; 1986a; Stafford 1989: 8–17.
52. Murchison 1838: xxxii.
53. Cawood 1979: 493–518; Morrell and Thackray 1981: 369, 523–31, esp. 524 (Sabine's 'magnetic empire').
54. Smith and Wise 1989: 755–63.
55. For example, Headrick 1981: 17–79.
56. Airy 1896: 134–36.
57. Thomas Jevons to John Laird, 1 December 1842, Laird Family Papers (microfilm), Merseyside Maritime Museum, Liverpool. Thomas was father of William Stanley Jevons, later famous as a leading economist and author of *The Coal Question* (ch. 3).
58. Winter 1994: 69–98, esp. 82.
59. Smith and Wise 1989: 765–66; Winter 1994: 82–83.
60. Harrowby 1854: lxxi. See Harrowby, *DNB*. A reform-mined aristocrat, Harrowby had strong geographical and statistical interests.
61. *Liverpool Mercury*, 29 September 1854; *The Times*, 30 September 1854.
62. Winter 1994: 83–87.
63. Henderson 1854.
64. Russell 1854; Smith *et al*. 2003b: esp. 381. In their faith, Unitarians denied the centrality of the doctrine of the trinity.
65. *Liverpool Mercury*, 29 September 1854. This advice did not appear in the *BAAS Report* which contained Scoresby's published versions as Scoresby 1854a,b,c.
66. *Liverpool Mercury*, 29 September 1854. Our italics.
67. *Liverpool Mercury*, 29 September 1854.
68. *Liverpool Mercury*, 29 September 1854.
69. James Joule to William Thomson, 30 January 1855 and 1 January 1858, J191 and J248, Kelvin Collection, University Library, Cambridge; Smith and Wise 1989: 766, 730.
70. Smith and Wise 1989: 733–40, 764, 769–70. For ships as 'laboratories' see Sorrenson 1996: 221–36; Livingstone 2003: 82. We also investigate the train as laboratory (ch. 4).
71. Smith and Wise 1989: 770–75 (on Thomson's system); Winter 1994: 75–76 (on Airy's contempt for captains).
72. Sir William Thomson to G.B. Airy, 3 March 1876 (quoted in Thomson 1910, 2: 708–10).
73. Smith and Wise 1989: 776–86.
74. Smith and Wise 1989: 786–98.
75. Herschel 1830: 287–88.
76. Porter 1977: 133–35.
77. Porter 1977: 135–38.
78. Porter 1977: 139–43; Rudwick 1985: 17–30. We should recall how the BAAS and Royal Society were later characterised as the Commons and the Lords (the two parliaments of science) respectively (above).
79. Babbage 1830: 45.
80. Rudwick 1985: 17–27. There are distinct parallels with the Institution of Civil Engineers (founded 1818) where individual reputations and common practices featured strongly in the society's activities.

81. Rudwick 1985: 37–41; Livingstone 2003: 40–48; Torrens 2002. Mary Anning provides a good example of a low-status, local fossil collector who derived income from her work in contrast to the metropolitan gentlemen of geology. See Torrens 1995: 257–84; Creese and Creese 1994: 23–54.
82. Rudwick 1985: 42–54.
83. Rudwick 1985: esp. 48, 54–60 (on correlations and sequences of rock strata); 1976: 148–95; 1992 (on the visual in geology); Secord 1986a (on geological disputes between gentlemen over classification of rock strata).
84. Sedgwick 1833: xxviii; Smith 1989: 33–34.
85. Morrell and Thackray 1981: 276–86.
86. William Hopkins to John Phillips, 19 October 1836, Phillips Correspondence, Oxford University Museum; Smith 1989: 34. We thank Jack Morrell for this reference. On Phillips see Morrell 2004 (forthcoming).
87. Secord 1986b: 223–41; Stafford 1989: 24–25.
88. Stafford 1989: 25
89. Secord 1981–82: 413–42; Stafford 1989: 8–19.
90. *DSB 9*: 583 (entry on Murchison by Martin Rudwick).
91. Stafford 1989: 13 (quoting Murchison), 19, 33–63 (Antipodean gold).
92. Stafford 1989: 25 (quoting Murchison), 30–31.
93. Stafford 1989: 202 (quoting von Buch).
94. Jevons 1866: 1–2.
95. Jevons 1866: v–vi.
96. Jevons 1866: 2–3; Armstrong 1863: li–lxiv.
97. Smith and Higginson 2001: 103–10.

2 Power and wealth: reputations and rivalries in steam culture

1. Carnot 1986 [1824]: 62.
2. Smith 1976 [1776], 1: 47, 10.
3. Wise with Smith 1989–90: 276–85 passim.
4. Carnot 1986 [1824]: 61. See Fox 1986: esp. 3 (French industrial contexts).
5. Smith 1976 [1776], 1: 17–21.
6. Smith 1976 [1776], 1: 20–21.
7. Smith 1976 [1776], 1: 21, 21n.
8. Ross 1995: esp. 146–47.
9. Carnot 1986 [1824]: 63.
10. Fox 1986: 2.
11. Carnot 1986 [1824]: 63; Kuhn 1961: 567–74; Fox 1986: 2–12.
12. Hills 2002a: esp. 29–38 (Greenock and Watt's immediate family), 40 (Presbyterianism), 50 (infrequent church attendance); 1999 (instrument making); Swinbank 1969 (Glasgow College workshop); Robinson 1969: 225 and Hills 1996: 67 (Delftfield Pottery); Bryden 1994 (merchant).
13. Hills 1996: 66.
14. Hills 2002a: 37 (Agnes Watt); Bryden 1994: 10; Hills 1996: 60; 2002a: 47.
15. Hills 2002a: 49 (Dick's role), 50–58 (London); Bryden 1970 (Jamaican observatories); Christie 1974 (Scottish universities in the Enlightenment); Jones 1969: 198 (Anderson); Hills 1996: 66 (Black).
16. Hills 1996: 63, 65.

17. Hills 2002a: 300 (Robison's steam designs).
18. Hills 2002a: 300; Cardwell 1965: 191–200.
19. Stewart 1992: 24–27 (Papin); Bennett *et al.* 2003; Iliffe 1995.
20. Tann 1979–80: 95 (Savery's patent).
21. Rolt and Allen 1977; Smith 1977–78.
22. Desaguliers 1734–44.
23. Lindqvist 1984 (Newcomen technologies on trial).
24. Quoted in Robison and Musson 1969: 2.
25. Hills 2002a: 301 (Papin's digester), 312–14 (Newcomen model), 297 (Shettleston engine). The various versions of the account are in Robison 1822, 2: 113–21 (Watt); Muirhead 1858: 88–90 (Black), 60–73 (Robison), 83–91 (Watt).
26. Watt's investigations are recorded in his 'Notebook of steam experiments', transcribed in Robinson and McKie 1970: 431–79.
27. Hills 1996–97; 2002a: 318–19.
28. Andrewes 1996 (Harrison); Hills 2002a (Smeaton and Harrison).
29. Quoted in Jones 1969: 203 (our emphasis).
30. Jones 1969: 203.
31. Law 1969 has investigated what remains of the original models.
32. For the definitive account of Watt's route to the separate condenser see Hills 1996–97; 1998a; 1998b and Hills 2002a: 294–378.
33. Quoted from 'Professor Robison's Narrative of Mr. Watt's Invention of the improved Engine versus Hornblower and Maberley 1796', reproduced in Robinson and Musson 1969: 23–38 (on 27–28).
34. Jones 1969: 197 (Black and Watt on latent heat of ice and steam).
35. Robinson 1969: 223.
36. Fleming 1952 discussed Watt's precise debt to Black.
37. Robison 1822.
38. Quoted from Watt's letter to Brewster in Robison 1822, 2: ix from Fleming 1952: 5.
39. For an important discussion of Watt's reputation see Miller 2000.
40. Robinson and Musson 1969: 42–43 (our emphasis); Hills 1996: 66, 72 (Watt's experience with atmospheric engines).
41. Hills 1996: 66 (Craig's death).
42. Hills 1996: 71; Campbell 1961; Watters 1998.
43. Robinson and Musson 1969: 2.
44. Hills 1996: 73, 76. 'Reputation' is Hills's term.
45. Hills 1996: 72 (quoting an undated note); Jones 1969: 201 (quoting Watt's note to Robison 1822, 2: 59).
46. The figure is quoted in Robinson 1969: 229, from Harris 1967.
47. Harvey 1973–74: 27.
48. Robinson 1969: 221–22; Robinson 1972; Davenport 1989.
49. Jones 1969: 201.
50. Hills 1996: 68 (quoting Watt to William Small, 9 September 1770); the evaluation is in Jones 1969: 207 (without further attribution).
51. Hills 1996: 67–68 (quoting Watt to William Small, 9 September 1770).
52. Schofield 1963; 1966: 145–46 (secrecy), 148 (Watt's introduction); Uglow 2002.
53. Thackray 1974.
54. Hills 2002a: 48.
55. Schofield 1966: 149 (without attribution).
56. Schofield 1966: 149 (quoting the author).

57. Schofield 1966: 152 (Grand Trunk Canal; pyrometer); Robinson 1969: 225 and Gittins 1996–97 (alkali experiments); Smith and Moilliet 1967 (Keir).
58. Dickinson 1937 (Boulton); Jones 1969: 207 (employment); Tann 1978: 363.
59. Schofield 1966: 148.
60. Boulton to Watt, 7 February 1769 (quoted in Robinson and Musson 1969: 62).
61. Hills 1996: 76. This is Hills's speculation.
62. Hills 1996: 69 (canal slump); 2002a: 29 (arrival in Birmingham); Jones 1969: 208 (Kinneil engine).
63. Jones 1969: 208 (boring).
64. Robinson 1964.
65. Schofield 1966: 151; Tann 1978 (marketing); 1979–80: 95 (additional patents).
66. Boulton to Watt, 7 February 1769 (quoted in Robinson and Musson 1969: 62).
67. Jones 1969: 208 (early engines); Dickinson 1936: 95 (Smeaton).
68. Marsden 2002: 137–38; Schofield 1966: 147, 149, 151–52 (factory tourism).
69. Dickinson 1937: 71 (quoting Keir).
70. Lardner 1836: 93.
71. Howard 2002.
72. Schofield 1966: 151.
73. Robinson and Musson 1969: 3.
74. Ferguson 1962.
75. Arago 1839c: 263.
76. Hills and Pacey 1972.
77. Schofield 1966: 156 (coin presses); Tann 1978: 380 (Soho corn mill). For the Albion Mill see Dickinson 1936: 146–50; Dickinson and Jenkins 1981 [1927]: 64–65, 162–67 and Dickinson 1937: 122–25.
78. Dickinson 1937: 208 (quoting Watt's memoir of Boulton); Dickinson and Jenkins 1981 [1927]: 167 (national object).
79. Gascoigne 1994: 207–13.
80. Dickinson and Jenkins 1981 [1927]: 165 (Banks, workmanship); Dickinson 1937: 123 (wonder, advertisement); Tann 1978: 380 (calling the Albion Mill 'a powerful advertisement'), 382 (foreign advice).
81. Rennie to Watt, 5 March 1791 (quoted in Dickinson and Jenkins 1981 [1927]: 167).
82. Schofield 1966: 157.
83. Dickinson 1936: 149 (quoting Watt to Boulton, 17 April 1786); see also Dickinson and Jenkins 1981: 64–65 (where the same letter is reproduced).
84. Dickinson 1936: 149–50 (quoting Watt to Boulton, 17 April 1786).
85. Tann 1974 (supply of engine parts).
86. Hills 1996: 77, 72, 76.
87. Boulton to Watt, 7 February 1769 (quoted in Robinson and Musson 1969: 62); also Hills 1996: 59.
88. Hills 1996: esp. 69–70 (quoting from Watt to Small, 9 September 1770, 7 and 24 November 1772), 77.
89. Robinson 1954: 301.
90. Robinson 1954: 303.
91. Robinson 1954: 309.
92. Jones 1999.
93. Tann 1979–80 (pirates).
94. Harvey 1973–74: 27.
95. Tann 1979–80.

96. Tann 1979–80: 95.
97. Dickinson 1936: 151 (quoting Watt to Boulton, 5 November 1785).
98. Dickinson 1936: 140 (quoting Watt to Boulton, 12 September 1786).
99. Tann 1979: 363.
100. Tann 1978: 367–69 (lobbying), 368 (Netherlands), 370 (proxies, quoting Boulton to Joseph Banks, 22 November 1790), 385 (King of Naples); 1979–80: 95 (lobbying), 364 (agents); van der Pols 1973–74 (Netherlands).
101. Harvey 1973–74: 27.
102. Tann 1978: 374 (high-profile erectors), 375–76 (mechanics abroad).
103. Tann 1978.
104. Redondi 1980.
105. For Erasmus Darwin see King-Hele 2003.
106. Schaffer 1995.
107. Marsden 1998a.
108. Cardwell 1965: 189–90 (Cagniard engine); Kuhn 1961 (on Carnot and the Cagniard engine).
109. Brunel senior patented a marine steam engine in the configuration of an inverted V – the type ultimately used on Isambard Kingdom Brunel's screw-propelled steamship *Great Britain*. See British patent no. 4683 (26 June 1822); Beamish 1862: 183; Clements 1970: 76, 257.
110. There are accounts of the 'gaz engine' in Beamish 1862: 185–87; Brunel 1870: 42–45 ('Experiments with Carbonic Acid Gas'); Rolt 1957: 41–42; Clements 1970: 77–78 and Buchanan 2002a: 20–22.
111. See Faraday 1823a,b. Faraday's result came whilst carrying out an experiment suggested by his mentor Davy; disputes over credit for the discovery culminated in Davy's opposition to Faraday's election to the Royal Society.
112. Clements 1970: 77 (quoting from Marc Brunel's diary, 30 May 1823).
113. Pugsley 1976: 2 (gaz engine); Beamish 1862: 188 (Brunel's hope for gaz).
114. See James 1991, letters *208, 218, 248, 250, 282, 617* and *639*. Henceforth James 1991: *N* signifies letter number *N* in James 1991.
115. James 1991: *208* (Marc Isambard Brunel to Michael Faraday, 15 September 1823).
116. James 1991: *218* (Isambard Kingdom Brunel to Michael Faraday, 4 February 1824).
117. Clements 1970: 77 (quoting from Marc Brunel's diary, 8 April 1824).
118. James 1991: *248* (Isambard Kingdom Brunel to Michael Faraday, 24 December 1824).
119. James 1991: *250* (Isambard Kingdom Brunel to Michael Faraday, 17 January 1825).
120. Clements 1970: 107 (quoting from Isambard Kingdom Brunel's diary, 9 March 1825).
121. Clements 1970: 77 (quoting from Marc Brunel's diary, 4 June 1825).
122. Vaughan 1991: 13 (expenses to July 1825).
123. 'Gas engines', British patent no. 5212, 16 July 1825; Clements 1970: 77 (quoting from the patent), 258.
124. Beamish 1862: 187.
125. James 1991: *282* (Marc Isambard Brunel to Michael Faraday, 8 February 1826), esp. James's note.
126. *Quarterly Journal of Science* 21 (1826): 131–32.
127. Rolt 1957: 42 (quoting from I.K. Brunel's diary, 4 April 1829). Vaughan 1991: 42 (transporting the engine).
128. James 1991: *617*n.
129. James 1991: *617* (Michael Faraday to John Barrow (Second Secretary to the Admiralty), 10 October 1832).
130. Beamish 1862: 187.

131. James 1991: *639* (Marc Isambard Brunel to Michael Faraday, 15 January 1833).
132. Rolt 1957: 42; Clements 1970: 202 and (partially) Vaughan 1991: 43 (quoting with minor discrepancies from I.K. Brunel's diary, 30 January 1833).
133. Beamish 1862: 187–88.
134. For a detailed historiographic discussion, and further references, see Marsden 1998a.
135. Hills 2002a: 40, 42–43 (Watt's reading); Robinson and Musson 1969: 42–43 ('Report by Messrs. Hart of Glasgow of conversations with Mr. Watt in 1817. Communicated by John Smith. 19 March 1845').
136. Kuhn 1961; Marsden 1998a.
137. On Ericsson, see Church 1890. On the rhetoric of power engineers, compare Robert Post's comment on the emergence *c.* 1850, of professionalisers in the United States with values different from empirical inventors: 'In matters of self-promotion, the professionalisers tended to operate more covertly than inventors, who as a rule had to brag in order to succeed' (Post 1983: 339).
138. Marsden 1998a: 380.
139. Marsden 1998a: 385.
140. Ferguson 1961.
141. Marsden 1998a: 387.
142. Marsden 1998a: 390.
143. Church 1890, 1: 233–303; 2: 1–181 ('monitors'), 260–301 (solar power).
144. Quoted in Anon 1859: 13.
145. Anon 1859: 3–4, 11, 15, 18, 20 (trade catalogue in Ben Marsden's collection).
146. Anon 1859: 12.
147. Dickinson 1837: 151.
148. Anon 1859: 4–10.
149. Rankine 1856; 1857; Channell 1982; Marsden 1992a,b.
150. Hutchison 1981.
151. Rankine to Napier, 7 February 1853 (quoted in Marsden 1992b: 145).
152. Morus 1998: 50–51 (Sturgeon's electro-magnet).
153. For an alternative account see Morus 1998: 184–91 ('Usurping the place of steam').
154. Jacobi 1836–37: 408.
155. Silliman 1838: 258 (Henry's beam engine).
156. This device reversed the direction of an electrical current, and thus the polarity of an electro-magnet, quickly and regularly – thereby, ultimately, allowing rotational power to be developed.
157. Taylor 1841: 4–5; Jacobi 1836–37: esp. 411–12, 444.
158. Sturgeon 1836–37: 78.
159. *Annals of Electricity* 1 (1836–37): 250.
160. Silliman 1838: 264; the patent is in Davenport 1838b: 347–49.
161. Davenport 1838b: 349 (emphasis ours).
162. Silliman 1838: 258 (emphasis ours).
163. *Annals of Electricity* 2 (1838): 158; Morus 1998: 108, 187 (Davenport's engines).
164. *Annals of Electricity* 2 (1838): 159.
165. Silliman 1838: 262–63.
166. Davenport 1838a: 286 (drill).
167. *Annals of Electricity* 2 (1838): 158–59.
168. Davenport 1838a: 286.
169. Davenport 1838b: 350.
170. [Editor of the *Journal of the Franklin Institute*], 'Speculations respecting electro magnetic propelling machinery', *Annals of Electricity* 3 (1838–39): 161–63.

171. Joule 1887, 1: 1–3 (quoting 3); for Joule on electro-magnetic engines, see Morus 1998: 187–90.
172. Joule 1887, 1: 14; see also Cardwell 1989: 30–31.
173. Taylor 1841: v–vi. The author may have been electrician W.H. Taylor: see Morus 1998: 191.
174. Taylor 1841: v–vi.
175. Taylor 1841: 17–18.
176. Callan 1836–37: 491–94.
177. Taylor 1841: 20–26, 35–36.
178. Anon 1841b: 252. See also Morus 1998: 190–91 (Davidson).
179. Anon 1841b: 252.
180. Anon 1841b: 252 (reproducing *Aberdeen Constitutional*, 6 November 1840); Jacobi 1840.
181. Post 1974 (Robertson's *Galvani*).
182. Jacobi 1840: 18–24.
183. Joule 1887, 1: 47–48; Cardwell 1989: 30–37 (Joule's electro-magnetic career); Smith 1998a: 53–76 (Joule, electro-magnetic engines and the mechanical value of heat).
184. Hills 1989: 101.
185. Hills 1989: 101, 105–08.
186. Fara 2002a: esp. 192–202.
187. See, for example, Robinson 1956: 261.
188. Dickinson and Jenkins 1981 [1927]: 81–89 (Watt portraiture).
189. Marsden 2002: 188–89.
190. Potts 1980 (Chantrey statue and bust); Arago 1839c: 284.
191. Torrens 1994: esp. 25–27.
192. Jones 1969: 203; Muirhead 1858: 83–91 (reproducing Watt's 1796 account).
193. Robison 1797.
194. Hills 2002a: 25.
195. Miller 2000: 4; Hills 2002a: 25; Farey 1827.
196. MacLeod 1998.
197. Arago 1839b: 191, 194; Robinson 1969: 225; compare Arago 1839c: 292–93.
198. Robert Peel's 'Speech on proposed monument to Watt', 1824, f. 164, Add 40366, British Library.
199. Quoted in Robinson 1969: 230; the draft of 1834 is f. 246, Add 47224, British Library.
200. Arago 1839c: 292–93 (quoting Liverpool, Huskisson and Mackinstosh).
201. Miller 2000: 1 (quoting Thomas Carlyle to John A. Carlyle, 10 August 1824).
202. Carlyle 1829: 456, 441–42.
203. Carlyle 1829: 449; Miller 2002: 1 (quoting Carlyle's 'Chartism').
204. Arago 1839c: 221, 251 (visits).
205. Arago 1839c: 221 (acquaintances and friends); Robinson 1956: 263; Hills 2002a: 25.
206. Arago 1839c: 225 (kettle anecdote); James Gibson to James Watt junior, 8 October 1834 (quoted in Robinson and Musson 1969: 22); Hills 2002a: 25 (Marion's anecdotes) and 41 (problem-solving); Robinson 1956 (Watt's use of kettle in experiments).
207. Arago 1839c: 256.
208. Arago 1839c: 278–79.
209. Arago 1839c: 250, 280.
210. Arago 1839c: 254.

211. Arago 1839c: 222, 247, 294.
212. Miller 2000: 1–24.
213. Schofield 1966: 156.
214. Robinson 1969: 225; and James Watt junior to Arago, 2 September 1834 (quoted in Hills 2002a: 29).
215. Arago 1839c: 228, 234, 250.
216. Arago 1839c: 292.
217. Arago 1839c: 287.
218. Arago 1839c: 294.
219. Arago 1839d: 298.
220. Arago 1839d: 298, 302.
221. Arago 1839c: 296.
222. Arago 1839c: 297.

3 Belief in steamers: making trustworthy the iron steamship

1. Carnot 1986 [1824]: 62.
2. Carnot 1986 [1824]: 62n; Bonsor 1975–80, 1: 204, 218–20.
3. Lindsay 1874–76, 3: 633.
4. Lindsay 1874–76, 4: 646.
5. Allen 1978: esp. 16–37.
6. For example, Munro 2003: 485–88 (on steamship technology).
7. Lindsay 1874–76, 4: 587–88.
8. *DNB* (William Schaw Lindsay).
9. Lindsay 1874–76, 1: xvii.
10. Lindsay 1874–76, 1: xx.
11. Lindsay 1874–76, 4: 582.
12. Lindsay 1874–76, 4: 188.
13. *The Illustrated London News*, 15 July 1843 (quoted in Armstrong 1975: 115). For later historical judgements see for example, Kirkaldy 1914: 65; Armstrong 1975: 115.
14. William to Francis Prodeaux, 28 July 1843 (quoted in Farr 1970: 20–21). Ewan Corlett claims that Prince Albert eventually succeeded with the bottle after Mrs Miles, wife of one of the directors of the Great Western Steamship Company and responsible for naming the *Great Western* exactly six years earlier, had failed twice. See Corlett 1990: 86. On the *Great Britain* see also Bonsor 1975–80, 1: 62–63; Fox 2003: 147–55.
15. Bonsor 1975–80, 1: 62–64 (*Great Britain*'s early career); Corlett 1990: 108–09; Winter 1994: 77 (controversy over causes of stranding); Kirkaldy 1914: 37–38 (proponents of iron ships).
16. Bonsor 1975–80, 1: 64–66 (the rescue, with Ewan Corlett as the driving force).
17. Thompson 1910, 1: 3–4; Smith and Wise 1989: 727–28.
18. Lindsay 1874–76, 4: 34–40; Armstrong 1975: 29–31. Relations of James Taylor, tutor to the family of Symington's original patron, Patrick Miller of Dalswinton, had claimed that Symington had patented the inventions of Mr Taylor. See Lindsay 1874–76, 4: 40.
19. Lindsay 1874–76, 4: 40–48.
20. Quoted in Lindsay 1874–76, 4: 53–54n.

21. Lindsay 1874–76, 4: 48–59, 587–90 (Fulton), 121–62 (on North American river-steamers); Armstrong 1975: 32–36 (on Fulton); Schivelbusch 1980: 89–112 (linking American river-steamer to railway culture).
22. Adams 1867: 476–77.
23. Shields 1949: 12–23; Riddell 1979: 1–51; Smith and Wise 1989: 21–22.
24. Tolstoy 1957 [1869]: 711.
25. Quoted in Lindsay 1874–76, 4: 63.
26. Lindsay 1874–76, 4: 60–67; Osborne 1995; Armstrong 1975: 37–40.
27. Quoted in Lindsay 1874–76, 4: 66n.
28. Lindsay 1874–76, 4: 67.
29. Lindsay 1874–76, 4: 73–77. Lindsay notes that these classes of watermen, 'semi-seafaring' men, provided a valuable reserve for the Royal Navy in time of national emergency.
30. Lindsay 1874–76, 4: 78–80; Armstrong 1975: 47.
31. Hodder 1890: 152–63; Napier 1904: 1–120; Armstrong 1975: 47–48.
32. Hodder 1890: 145–49.
33. Hodder 1890: 152–58.
34. Hodder 1890: 160–63.
35. Albion 1938 (on Black Ball Line sailing packets); Whipple 1980 (on American and British tea-clippers).
36. Shields 1949: 121–24; Pollard and Robertson 1979: 9–13; Whipple 1980: 102–23.
37. For example, Smith *et al.* 2003a,b.
38. See Morrell and Thackray 1981: 505–56.
39. Lindsay 1874–76, 4: 307–08. Only a handful of crew and passengers survived out of over 160 persons on board.
40. Bonsor 1975–80, 1: 41–44 (*Savannah*), 60–66 (*Great Western* and *Great Britain*); Buchanan 2002a: 58–60.
41. Bonsor 1975–80, 1: 54–59.
42. *The Times*, 10 December 1838 (quoted, with comment on the link to the *Great Britain*, in Bonsor 1975–80, 1: 57).
43. Booth 1844–46: 24–31. See esp. Smith *et al.* 2003b: 396–98.
44. Booth 1844–46: 25–26.
45. Booth 1844–46: 27–28.
46. Booth 1844–46: 30. Brunel's *Great Eastern* (below) also initially carried the name *Leviathan*, echoing references in the Book of Job (Job 41:1) to the greatest of sea monsters.
47. Booth 1844–46: 30–31. Booth's anxiety about wasted capital may also have been linked to anxieties about railway mania (ch. 4).
48. Morrell and Thackray 1981: 497 note that Woods played a role in the BAAS's investigations into the strength and properties of iron.
49. BAAS investigations of hull-designs (especially those of John Scott Russell) are examined in Morrell and Thackray 1981: 505–06.
50. *Liverpool Mercury*, 22 September 1854.
51. *Liverpool Mercury*, 26 September 1854.
52. Russell 1854; *Liverpool Mercury*, 26 September 1854.
53. Russell 1854. On Scott Russell's experimental investigations in naval architecture see Emmerson 1977: 14–19.
54. On Lardner's criticisms, notably before the BAAS at Bristol in 1836, of the *Great Western* see Morrell and Thackray 1981: 472–74. On the building of the *Great*

Eastern and the disputes between Scott Russell and Brunel see Emmerson 1977: 65–157; Buchanan 2002a: 113–33.

55. Quoted in Buchanan 2002a: 119 who notes briefly the relevance of viewing the construction – and operation – of the project as a 'system'.
56. Russell 1854.
57. *Liverpool Mercury,* 26 September 1854. On Fairbairn, Stephenson and the Britannia Bridge, see Pole 1970 [1877]: 195–213.
58. *Liverpool Mercury,* 26 September 1854.
59. Henderson 1854; *Liverpool Mercury,* 29 September 1854.
60. *Liverpool Mercury,* 29 September 1854. 'Nominal' horse-power tended to be calculated from engine dimensions, such as cylinder size, in contrast to 'indicated' horse-power which was measured by engine performance using indicator instruments and diagrams (ch. 2). Scientific engineers such as Rankine (below) saw the latter as a way of tightening up – with a more direct, even automatic method – the measurement of power.
61. Henderson 1854; *Liverpool Mercury,* 29 September 1854.
62. Buchanan 2002a: 117–18.
63. Banbury 1971: 242–52 (the shipyard); Emmerson 1977: 83 (gentlemanly appearance and manners).
64. Lindsay 1874–76, 4: 492–93. Lindsay also quotes from Scott Russell's letter to *The Times,* 20 April 1857, which details a similar division of credit. For a recent view on the subsequent arguments over credit, see Buchanan 2002a: 113–14. Compare Emmerson 1977: esp. 83.
65. Buchanan 2002a: 121 (quoting letters from Brunel to Scott Russell); Emmerson 1977: esp. 95, 100–02. For 'giving the lie' see Shapin 1994: 107–14.
66. Lindsay 1874–76, 4: 513n.
67. Lindsay 1874–76, 4: 496 (quoting Charles Atherton, chief engineer at the Royal Dockyard, Woolwich).
68. Buchanan 2002a: 122–33.
69. Bonsor 1975, 2: 579–85.
70. Banbury 1971: 171–74 (Fairbairn), 198–205 (Maudslay), 214–18 (Napier), 224–29 (Penn), 242–52 (Scott Russell); Pollard and Robertson 1979: 64. On Maudslay see esp. Cantrell 2002a: 18–38; Ince 2002: 166–84.
71. Hume and Moss 1975: 137 (steam yacht race). The Presbyterian theme is further explored in a forthcoming article by Crosbie Smith and Anne Scott.
72. Cain and Hopkins 2002: 278–84, esp. 283 (Bentinck reforms); Lindsay 1874–76, 4: 336–60 (East India Company and steam), esp. 359–60 (*Berenice*). By 1842 P&O had won the mail contract.
73. Napier 1904: 1–165. On Napier's yard as the source of a new generation of Clyde shipbuilders, see esp. Slaven 1975: 125–33.
74. Napier 1904: 135; Hyde 1975: 6. The dark warnings may have been fuelled by the problems with the *British Queen* noted in the previous section.
75. Grant 1967: esp. 17.
76. Hyde 1975: 1–5; Fox 2003: 84–93.
77. Hyde 1975: 5–6.
78. Samuel Cunard to Kidston & Company, 25 February 1839, Cunard Papers, Liverpool University Library. Quoted in Napier 1904: 124–25.
79. Hyde 1975: 6–8.
80. See Nenadic 1994: 122–56 (domestic cultures in Edinburgh and Glasgow). We thank Hugh Cunningham for this reference.

81. Hyde 1975: 8–11.
82. Hyde 1975: 12–15 outlines the division of labour but does not explore the cultural contexts. It is worth noting that the fourth ship, *Columbia*, cost almost £50,000 compared to the £32,000 per ship projected in the initial plan for three ships. See Hyde 1975: 338n.
83. Anon 1864: 483–523.
84. Anon 1864: 483–523. See esp. Smith, Higginson and Wolstenholme 2003a: 453–57 (Cunard *versus* Collins).
85. See esp. Marsden 1992b.
86. Lindsay 1874–76, 4: 69. His exception took note of the rising production in the north-east of England, especially on the Tyne.
87. Lindsay 1874–76, 4: 239. Italics his.
88. *New York Tribune*, 27 January 1873 (quoted in Babcock 1931: 146–47). See also Hodder 1890: 300–01; Fox 2003: 275.
89. George Holt, Diary 11 November 1844 to 31 December 1854, Papers of the Durning and Holt Families, 920 DUR/1/1–2, Liverpool Central Library: entries for 11 January and 25 April 1852; Lindsay 1874–76, 4: 305–10.
90. Marsden 1998a: 373–420, esp. 380–82, 385–90, 400–01. See also Ferguson 1961: 41–60; Bonsor 1975–80, 1: 334–35.
91. Prosser 1854; *Liverpool Mercury*, 29 September 1854.
92. Marsden 1998a: 390–400 argues that in fact the thermodynamicists used the controversy surrounding Ericsson as a way of promoting their science as a new science for new problems. On the sciences of thermodynamics and energy in this period see Smith 1998a: esp. 150–69.
93. *Liverpool Mercury*, 29 September 1854.
94. *Liverpool Mercury*, 29 September 1854.
95. *Liverpool Mercury*, 29 September 1854.
96. Marsden 1998a: esp. 390–400. See also Smith 1998a: 155.
97. *The Times*, 30 September 1854.
98. Rankine 1855: 19–20; Smith 1998a: 155–56.
99. Rankine 1855: 20–23; Smith 1998a: 156–57.
100. *The Times*, 30 September 1854.
101. Rankine 1854.
102. *The Times*, 30 September 1854.
103. Rankine 1854; Marsden 1992a: esp. 141–87; 1998a: 373–420, esp. 395–400 (Rankine-Napier air engine).
104. Gordon 1872–75: 296–306; Rankine 1871: 4–6. On Gordon's career see esp. Marsden 1998b.
105. Rankine 1871: 6, 31, 37–38; Bonsor 1975–80, 1: 276–77 (also quoting from the *North British Daily Mail*).
106. Rankine 1871: 38–39; Bonsor 1983: 144–65; Lingwood 1977: 97–114.
107. *The Edinburgh Academy Register* (Edinburgh: T. & A. Constable, 1914), 106–22. See [Scotts] 1911 for the privately published history of the firm. James Clerk Maxwell, Peter Guthrie Tait and William Thomson all held Scottish chairs of natural philosophy and together constructed a new science of energy which drew heavily on engineering concepts of 'work'. Rev. Lewis Campbell later co-authored Maxwell's biography. See Smith 1998a: 310–11. Edward Harland was co-founder of the world-famous Belfast shipbuilders, Harland & Wolff. See Hume and Moss 1986: 12–13.

108. W.J.M. Rankine to J.R. Napier, 22 November 1858, Glasgow University Archives; [Scotts] 1906: 33–36; Robb 1993, 2: 183.
109. On P&O's early compound engines from the Thames see Banbury 1971: 189–90.
110. Rankine 1871: 40–47. The *Mechanics' Magazine* 15 (1866): 1 reported that the *Constance* had arrived in Madeira first, but accused the Admiralty of tardiness in releasing the results of the trial.
111. J.P. Joule to William Thomson, January 1858, J248, Kelvin correspondence, Cambridge University Library.
112. See esp. Smith 1998a: esp. 1–14. The term 'energy sources' can be used after about 1850 without anachronism.
113. Jevons 1866: 122–37, esp. 129–34.
114. Smith *et al*. 2003a.
115. Holt 1911: 45–46; Lindsay 1874–76, 4: 468n.
116. Smith *et al*. 2003b: 398–405.
117. Alfred Holt, Circular regarding *Cleator*'s new engine, Papers of Alfred Holt, 920 HOL/2/10, Liverpool Central Library; Smith *et al*. 2003b: 409.
118. Smith *et al*. 2003b.
119. For accounts of the early Holt engine see Hyde 1956: 173; Le Fleming 1961: 8. See Marshall 1872: 453 for the later design. On Leslie's shipbuilding and its relation to Hawthorn's engine building see Clarke [1979].
120. Holt 1911: 46–48.
121. Hyde 1956: 19.
122. Alfred Holt, Diary, Book A, Papers of Alfred Holt, 920 HOL/2/52, Liverpool Central Library, 16–18 February 1866; Jevons 1866: x–xi.
123. Smith *et al*. 2003a: 464.
124. Holt 1911: 46–48; Alexander Kidd, 'Jottings from a sailor's life' (typescript), Ocean Archive, Merseyside Maritime Museum, Liverpool.
125. Alfred Holt, Circular letter dated 16 January 1866, Papers of Alfred Holt, 920 HOL/2, Liverpool Central Library; Smith *et al*. 2003b: 410–14.
126. Alfred Holt, Circular letter dated 16 January 1866, Papers of Alfred Holt, 920 HOL/2, Liverpool Central Library; Smith *et al*. 2003b: 414.
127. 'O.S.S. Co. General Book (1865–82)', Ocean Archive, Merseyside Maritime Museum, Liverpool, p. 1.
128. Alfred Holt, Diary, Book A, Papers of Alfred Holt, 920 HOL/2/52, Liverpool Central Library, 24–31 March 1866.
129. Alfred Holt, Diary, Book A, Papers of Alfred Holt, 920 HOL/2/52, Liverpool Central Library, 19 April 1866.
130. Alfred Holt, Diary, Book A, Papers of Alfred Holt, 920 HOL/2/52, Liverpool Central Library, 24 October 1866.
131. Lindsay 1874–76, 4: 434–37. On 'practical proof' and engineering see Marsden 1992b.
132. Alexander Kidd, 'Jottings from a sailor's life' (typescript), Ocean Archive, Merseyside Maritime Museum, Liverpool.
133. Le Fleming 1961: 38–55 (fleet list); Alfred Holt, Diary, Book A, Papers of Alfred Holt, 920 HOL/2/52, Liverpool Central Library; Smith *et al*. 2003b: 425–26 (n. 113 and 117).
134. Holt 1911: 46–48.
135. Holt 1877–78: 70–71 (our italics).

4   Building railway empires: promises in space and time

1. Babbage 1864: 319–20. From the 1820s and 1830s the words 'railroad' and 'railway' were used almost interchangeably in Britain and the United States. In what follows we use for simplicity 'railway'.
2. Williams 1904: 351–58 ('Adventures on the Line'), 352 (quoting Salisbury). Williams's book is a popular representation of imperial railways to Edwardian readers. The story of the Ugandan railway is also told in Burton 1994: 210–25; for the context of railways and imperialism in Central Africa, see Hanes 1991: esp. 48.
3. On technological choices see Evans 1981: 1–34. For a comprehensive survey of railways and their cultures see Simmons 1991. Guides to the vast literature on railway history include Ottley 1983 (and supplements); Simmons and Biddle 1997 and Freeman and Aldcroft 1985.
4. Babbage 1832: 306.
5. Dyos and Aldcroft 1969: 82–84 (traditional economic history).
6. Harris 1744: 76–78 (quoted in Smith and Wise 1989: 4).
7. Dyos and Aldcroft 1969: 66–84.
8. Especially Riddell 1979: 1–104. See also Dyos and Aldcroft 1969: 37–45 (river improvements), 45–65 (ports).
9. For example, Dyos and Aldcroft 1969: 85–116; Hadfield 1970.
10. Uglow 2002: 92–94, 107–121.
11. For example, Armytage 1976: 112.
12. Tann 1978.
13. Porter 1998: esp 71–104 (Gurney's steam carriages).
14. Schaffer 1996.
15. Porter 1998: 98.
16. Porter 1998: 105–32.
17. Cardwell 1994: 65, 101, 148, 210; Adas 1989: 135.
18. This forgotten aspect of railway construction would be remembered again when imperial railway builders and telegraph constructors attempted to establish tracks across little-charted and 'uncivilised' territory.
19. Trevithick 1872; Dickinson and Titley 1934; Cardwell 1994: 209–10; Armytage 1976: 112.
20. Dickinson and Titley 1934: 43–124.
21. Porter 1998: 95 (choice of Bath). Compare Morus 1996a: esp. 417–26 on machinery and 'exhibition culture'.
22. Cardwell 1994: 211; Armytage 1976: 113.
23. Kirby 1993: 2.
24. Cardwell 1994: 230; Marsden 1998b: 56–57; Wishaw 1842: 95–102 (Edinburgh, Leith and Dalkeith Railway).
25. Babbage 1864: 326.
26. See, for example, Burton 1980 (Rainhill trials).
27. Armytage 1976: 125–27; Cardwell 1994: 231 (origins of the L&MR). See also Carlson 1969; Thomas 1980; Ferneyhough 1980; Perkin 1971: 77–95.
28. Rastrick 1829.
29. In addition to Burton 1980, and comments in Cardwell 1994: 231, 233, 282, Vignoles 1889: 127–35 provides an extended account from one competitor.
30. Vignoles 1889: 128.
31. Vignoles 1889: 132 (conditions).

32. For the *Novelty* and Ericsson in England, see Bishop 1976–77: 44.
33. This rival account of the trials appears in Vignoles 1889: 127–35.
34. Vignoles 1889: 128.
35. See, for example, *Liverpool Mercury*, 6 October 1829.
36. Vignoles 1889: 129 (quoting *Liverpool Mercury*, 6 October 1829).
37. Vignoles 1889: 134.
38. Vignoles 1889: 130.
39. Tubular boilers enabled the heat of the furnace to produce steam from water surrounding a honeycomb of tubes rather than from water in a simple boiler. Their comparative efficiency was widely attributed to the greatly increased heating surface. See Pole 1970 [1877]: 389 (tubular boilers). The decision to use the tubular boiler in the steam locomotive has been the subject of dispute. For a recent discussion see Jarvis 1994: 35, 43n.
40. Vignoles 1889: 132.
41. Vignoles 1889: 133.
42. Vignoles 1889: 134.
43. Braithwate (quoting Ericsson) to Vignoles, 19 December 1829, in Vignoles 1889: 134–35.
44. The best recent study of Robert Stephenson is Bailey 2003.
45. Quoted, without further attribution, in Armytage 1976: 126.
46. For a recent study see Haworth 1994: 60–62 ('The obstacles posed by George Stephenson').
47. Cardwell 1994: 231 (our emphasis).
48. See Howse 1980: 87–88 (Booth and GMT); Booth 1980 (biography).
49. Booth 1830: 15; Jarvis 1994: 35–45 reassesses Booth's motivations for attacking landed aristocracy.
50. Booth 1830: 34–35.
51. Booth 1830: 50.
52. Booth 1830: 51.
53. Booth 1830: 57.
54. Booth 1830: 85–88.
55. Booth 1830: 88–92. Compare Carlyle 1829: 442.
56. Booth 1830: 93–94.
57. Babbage 1864: 313–16.
58. See Harrington 2001; 2003: 209–23.
59. Babbage 1864: 317–18; see also Freeman 1999: 57–89 ('March of intellect').
60. Babbage 1864: 319, 329.
61. Babbage 1864: 329–34.
62. Freeman 2001.
63. Cardwell 1994: 234 (enduring characteristics of L&MR).
64. Cardwell 1994: 233–34; see also Brewster 1849: 574 (quoting Scrivener on 'model').
65. Brewster 1849: 574.
66. *Railway Magazine; and Annals of Science*, n.s., 33 (1838): 369 (our emphasis).
67. Hills 2002b: 64. The working of the *Experiment* proved problematic – but clearly innovation had not ceased with the *Rocket*.
68. Barlow 1835.
69. De Pambour 1840.
70. Noble 1938: 139. On Brunel see Buchanan 2002a and, in addition, Rolt 1957 (which is uncritical) and Vaughan 1991 (which emphasises Brunel's wartier features).

The standard source on the Great Western Railway is MacDermot 1964. For Brunel's central role see Buchanan 2002a: 63–82; and Beckett 1980: esp. 33–87.

71. *[Prospectus for the] Great Western Railway, between Bristol and London* (London: Great Western Railway, 1833).
72. Marsden 2001.
73. Morrell and Thackray 1981: 472–74 (Lardner); Buchanan 2002a: 32, 73 and Rolt 1957: 184 (Buckland).
74. Wood 1825; on Wood's pupilage with and partiality in favour of George Stephenson, see Jarvis 1994: 35, 38; Porter 1998: 74.
75. Armytage 1976: 130.
76. On Brunel's liberal politics, see Buchanan 2002a: 173–90.
77. 'Electro-magnetic telegraphs', *The Railway Magazine; and Annals of Science*, n.s., 33 (1838), 365–66; 370.
78. Buchanan 2002a: 71 (telegraph).
79. Gilbert 1965; Buchanan 2002a: 1, 17.
80. See esp. Buchanan 1983a: 98–106 (*Great Eastern*).
81. Buchanan 2002a: 70–71 (quoting Brunel to Gooch, 2 January 1841), 244n.
82. Buchanan 2002a: 65.
83. Buchanan and Williams 1982; Buchanan 2002a: 43–62.
84. Buchanan 2002a: 59.
85. For example, Dyos and Aldcroft 1969: 132–39 (railway mania); Armytage 1976: 129. A scrip was a form of share certificate.
86. Anon 1845: 24.
87. I.K. Brunel quoted in Basalla 1988: 177; Hadfield 1967: 73.
88. Secord 2000: 9–40 ('A Great Sensation').
89. Buchanan 2002a: 70 (erratic locomotives).
90. Buchanan 2002a: 75 (locomotives), 161–62 (Gooch and Brunel).
91. Buchanan 2002a: 111 (quoting Gooch).
92. See Gordon 1849.
93. For the atmospheric railway see Hadfield 1967; Basalla 1988: 177–81 (as a passing technological 'fad'); Buchanan 2002a: 103–12 (as chief among Brunel's 'disasters'); and Atmore 2004: 245–79 ('Rope of air').
94. Williams 1904: 46–48 gives an account of the short-lived atmospheric working on the Exeter to Newton Abbott line in the late 1840s.
95. Brewster 1849: 609.
96. Booth 1830: 91.
97. For example, Dyos and Aldcroft 1969: 125–32 (first railway boom); Armytage 1976: 128.
98. See, for example, Dyos and Aldcroft 1969: 125–54, esp. 152–53 (map of national system 1852), 168–69 (map of national system 1914); Haworth 1994: 55 (Locke).
99. Walker 1969 (Brassey); Armytage 1976: 129.
100. Sekon 1895.
101. For example, Anon 1845: 4; Harding 1845: 4.
102. Babbage 1864: 326.
103. Anon 1845: 3.
104. Babbage 1864: 326 (our italics).
105. Babbage 1864: 334.
106. Babbage 1864: 326–28. The BAAS always sought to avoid unseemly controversies. See Morrell and Thackray 1981: 2–34.

107. Babbage 1864: 334–35. For a re-evaluation of Stephenson's role, see Jarvis 1994: 35–45.
108. Buchanan 2002a: 31; I.K. Brunel to Babbage [October/November 1838] and 17 November 1838, ff. 20, 33, Add. 37, 191, British Library.
109. Babbage 1864: 321; Williams 1904: 51–52 (end of the broad gauge in 1892).
110. Cole 1846: 3; Anon 1845: 9.
111. See, for example, Harding 1845: 1–3.
112. Anon 1845: 11; Harding 1845: 3.
113. Anon 1845: 4–7.
114. Anon 1845: 4.
115. See, for example, Rolt 1957: 59–61, 141; Buchanan 2002a: 59, 204–05.
116. Cole 1846: 3; Anon 1845: 7f.
117. Harding 1845: 5–6 (lack of foresight).
118. Cole 1846: 3 (atmospheric).
119. Anon 1846: title-page.
120. Cole 1846: title-page.
121. Cole 1846: 4. Don Quixote was the self-deluding hero of Cervantes' novel of that name.
122. See also Smith and Wise 1989: 684–98; Tunbridge 1992 for standards in electrical science.
123. Fairlie 1872: 1 (railway futures), 7–9 (rival US gauge; deficient Imperial railways).
124. See, for example, Bignell 1978.
125. Freeman 1999: 215–38.
126. Bury 1976 [1831]; Clayton 1970 [1831]; Shaw 1970 [1831].
127. See, for example, Walker 1830.
128. Schivelbusch 1980: 52–69 ('Panoramic travel'); Simmons 1991: 120–59 ('The artist's eye'); Tucker 1997. The beginning of the widespread use of the camera coincided with the second railway boom.
129. Carter 2001: 58.
130. Carter 2001: 61.
131. Carter 2001: 51–70.
132. Freeman 1999: 48 (Ruskin).
133. Carter 2001: 52.
134. Mogg 1841; Schivelbusch 1980: 70–88 ('The compartment'), esp. 73–77 ('The end of conversation while travelling').
135. Anon 1838: i–vii.
136. Wood 1825; for a general survey of technical literature which is, nevertheless, oddly silent on railway literature, see Emmerson 1973: 231–45 ('Engineering literature in the nineteenth century').
137. Winter 1998: 150–52.
138. For example, Simmons 1991: 345–47 ('Standard time'). For American parallels see Stephens 1989: 1–24.
139. Anon 1848: 15–19, on 18.
140. For example, Simmons 1991: 270–308 ('Leisure').
141. See esp. Simmons 1991: 309–44 ('Mobility'); 1993. See also Faith 1990: 286–304 ('Railway in town and city'); Freeman 1999: 121–47.
142. Armytage 1976: 131.
143. Simmons 1991: 35–36. Simmons notes that the architects of Central Station recognised the importance of curves to facilitate the easy flow of passengers to and from the trains.

144. Buchanan 1989: 88–124 (fragmentation of engineering societies); Armytage 1976: 130 (domination of canal engineers).
145. Weiss 1982: 141 (French engineering education).
146. Hilken 1967: 25; Guagnini 1993: 16–41 (training mechanical engineers); Whiting 1932: 104–05 (Durham); Vignoles 1889: 262–63; McDowell and Webb 1982: esp. 180–85 (Dublin engineering chair).
147. On Nasmyth see Armytage 1976: 126; Cantrell 2002b: 129–46, esp. 137 (Nasmyth quote). On Whitworth see Buchanan 2002b: 109–128.
148. Cantrell and Cookson 2002: 15 (Cookson's 'Introduction', quoting J. Nasmyth to Samuel Smiles, 31 March 1882).
149. Wide-ranging historical studies of railways and empire include Davis and Wilburn 1991; Burton 1994. Faith 1990: 144–82 ('Imperial railways') provides an introduction. On Indian railways see, for introduction, Headrick 1981: 180–91 and, for greater detail, Satow and Desmond 1980.
150. For example, Yenne 1986a: 247.
151. Brooks to Henry Adams, c. 1910, Adams Papers, Massachusetts Historical Society, Boston, MA. Published by permission of the Society.
152. On American cultures of invention see Hughes 1989.
153. For the transformation of North American railroads in the twentieth century see Yenne 1986.
154. Archibald Williams published fictional tales (e.g. 'The terrible submarine' [1901], after Jules Verne) and a series of books, primarily intended for boys, on the 'romance' of invention and engineering.
155. Williams 1904: 122–24. Schivelbusch 1980: 89–112 ('The American railroad') explores the thesis that American passenger trains followed the traditions for luxury and comfort set by river steamboats. He also notes important differences in engineering design compared to European railways, especially sharper curves (used as a fast, cheap alternative to slow, expensive tunnelling, for example) requiring a new type of rolling stock using a bogie at each end and enabling longer cars which, with enhanced internal space, could accommodate the various luxuries. On the introduction of electric lighting see Schivelbusch 1995.
156. Quoted in Secord 2003: 329. On Miller see Shortland 1996.
157. Zola 1890; for an alternative view see also Carter 2001: 117–42 ('Railway life: *La bête humaine*'); Baguley 1990 (Zola's naturalist fiction).
158. For example, Smith and Higginson 2001: 103–10.
159. For example, Johnson and Fowler 1986: 81–85 (Vanderbilts); Grodinsky 1957 (Gould); Geisst 2000: 11–34 ('The "Monopolist Menace"').
160. Adams 1867: 485–86. Kirkland 1965 provides a biography.
161. Explored more fully in Smith and Higginson 2004.
162. Adams 1870: 125, 134.
163. Adams 1870: 120. See Smith and Higginson 2004: 149–79.
164. Adams 1867: 480.
165. Adams 1867: 480–82.
166. Adams 1867: 484–85.
167. Adams 1867: 486–87. See, for example, Yenne 1986: 78–79 ('American railroads in the Civil War').
168. Adams 1867: 487–89.
169. Adams 1867: 489–90.

170. Adams 1867: 490.
171. Adams 1867: 491–92. The Adams brothers (esp. Henry and Brooks) frequently explored the ramifications of democracy in their later writings. See, for example, Henry's satirical novel *Democracy* and Adams 1920 (Brooks's pessimistic reflections on the 'degradation of the democratic dogma'). There were strong parallels between Charles Francis's railroad systems article and the contemporary discussions of culture collected in Matthew Arnold's *Culture and Anarchy*. Specifically, Arnold affected a viewpoint of disinterest, questioned the impact of democracy, and sought to preserve the gains of Enlightenment rationality whilst moderating its negative aspects. See Arnold 1993 [1861–78].
172. Adams 1867: 492–93.
173. Adams 1867: 493–94.
174. Adams 1867: 479.
175. Adams 1867: 497–98n. For the British parallels in the context of BAAS see Morrell and Thackray 1981: 291–96.
176. Adams 1869; 1871. See esp. Kirkland 1965: 34–64 ('Railroad reformer'); Geisst 2000: 18–21.
177. Quoted from Adams's diary in Kirkland 1965: 125–26.
178. 'The Pacific Railroads', typescript document, Adams Papers, Massachusetts Historical Society. We thank the MHS for permission to quote from this document.
179. Geisst 2000: 46; Adams 1867: 120.
180. Burton 1994: 131–86 (Asia), 226–48 (Australia); Davis and Wilburn 1991: 7–24 (British North America), 175–79 (Canadian Pacific Railway).
181. Burton 1994: 140 (Dalhousie), 227–28 (gauge).
182. Davis and Wilburn 1991: 179 (quoting Macdonald to Noirthcote, 1 May 1878).
183. Williams 1904: 66–67.
184. Williams 1904: 84 (quoting *The Railway Yearbook*); Musk 1989: 13 ('Canadian Pacific Spans the World'); Kohler 2004: 22 ('All Red Route'). Headrick 1981: 163 uses the 'All-Red Route' phrase in reference to the all-British character (represented by British territories coloured red in maps) of the telegraph cable across the Pacific and completed in 1902 (see ch. 5).
185. Williams 1904: 53–54; Skelton 1986: 133–34. In 1867 under the British North America Act the Confederation of Canadian States was established with Sir John A. Macdonald as Canada's first Prime Minister.
186. Williams 1904: 54.
187. Williams 1904: 55–56; Skelton 1986: 136–44.
188. Skelton 1986: 144.
189. Quoted in Williams 1904: 59.
190. Quoted in Williams 1904: 59–60.
191. Skelton 1986: 147–50.
192. *Montreal Gazette*, 28 June 1907 (quoted in Musk 1989: 30).
193. George Stephen to Sir John A. Macdonald, 29 January 1886, in Gibbon 1935: 303.
194. Stephen to Macdonald, 20 September 1886, in Gibbon 1935: 309–10.
195. Bonsor 1975–80, 3: 1286–87.
196. Stephen to Macdonald, 20 September 1886, in Gibbon 1935: 309–10.
197. Quoted in Gibbon 1935: 305.
198. Quoted in Musk 1989: 13–14.
199. Musk 1989.

5 'The most gigantic electrical experiment': the trials of telegraphy

1. *Glasgow Herald*, 21 January 1859. Reprinted in Thompson 1910: 389–90; Smith and Wise 1989: 652.
2. *Scientific American* 13 (1857–58): 285, cited in Hempstead 1989: 299.
3. According to the calculation of experimental natural philosopher Charles Wheatstone the electrical signal travelled at the extraordinary speed of 288,000 miles a second (the speed of light).
4. Hunt 1997. Stentor was the Greek herald in the Trojan war, with a voice as loud as fifty men.
5. Chappe first used the term tachygraphe to emphasise speed of writing rather than distance.
6. The history of the optical telegraph is treated fully in Field 1994: 315–47; see also Appleyard 1930: 263–98; McCloy 1952: 42–49; Daumas 1980: 376–78.
7. Appleyard 1930: 265 (forms of visual telegraph).
8. Appleyard 1930: 266.
9. McCloy 1952: 42–43.
10. McCloy 1952: 43, 46; Appleyard 1930: 266 and 276 (tachygraphe).
11. McCloy 1952: 43–44.
12. Daumas 1980: 377.
13. McCloy 1952: 44.
14. Derry and Williams 1960: 621; Shinn 1980.
15. Appleyard 1930: 267.
16. Appleyard 1930: 268.
17. Crosland 1992: 20–21.
18. McCloy 1952: 45 (L.M.N. Carnot, quoted without attribution); Appleyard 1930: 268 (enthusiasm).
19. Appleyard 1930: 270.
20. McCloy 1952: 52.
21. Appleyard 1930: 276.
22. McCloy 1952: 46 and Appleyard 1930: 272 (lottery funding).
23. Appleyard 1930: 272.
24. Appleyard 1930: 272–73.
25. Abraham suggested the use of flares to solve this, in Hughes's terms, 'reverse salient'. See Hughes 1983: 14; on weather and culture, see Janković 2000.
26. Derry and Williams 1960: 622.
27. Kirby *et al.* 1956: 337 (Louvre); Appleyard 1930: 267 (belfries).
28. Daumas 1980: 376.
29. Daumas 1980: 377 and Appleyard 1930: 265 (Delauney).
30. Appleyard 1930: 287.
31. McCloy 1952: 47–48 (codes); Daumas 1980: 376 (Breguet, telescopes).
32. Appleyard 1930: 274 (later Chappe lines).
33. Daumas 1980: 377–78 (message speeds). These were apparently conservative estimates: for even quicker times, see Appleyard 1930: 274.
34. McCloy 1952: 48–49.
35. McCloy 1952: 48; Kirby *et al.* 1956: 337.
36. Derry and Williams 1960: 622.
37. John Macdonald (quoted without attribution in Appleyard 1930: 281).

38. Cardwell 1994: 207; Derry and Williams 1960: 622–23; Appleyard 1930: 295 (final closure).
39. Kieve 1973: 13–28, esp. 16–17 (Ronalds).
40. Ronalds 1823: 2. In fact, electric telegraphs too remained susceptible to fog into the 1850s. See Wilson 1852: 52.
41. Ronalds 1823: 2, 4–6.
42. Morus 1998.
43. Schaffer 1983 (spectacle); Fara 2002b (Enlightenment electrical culture); Hankins 1985: 54–55 (Enlightenment sciences).
44. Derry and Williams 1960: 623–24.
45. Knight 1992; Golinski 1992: 188–235.
46. Derry and Williams 1960: 624; McNeil 1990: 714; Cardwell 1994: 223.
47. Snelders 1990: 228–40.
48. Morus 1998: 13–42 (Faraday), 43–69 (Sturgeon). On questions of display and natural philosophy see Morus 1991b: 20–40; 1992: 1–28; 1993: 50–69.
49. James 1991: *690* (Moll to Faraday, 15 November 1833).
50. 'Electro-magnetic telegraphs', *The Railway Magazine; and Annals of Science*, n.s., 33 (1838): 365–69.
51. Morus 1998: 194–230, esp. 198–220; 1996b: 339–78, esp. 349–67.
52. Bowers 2001 (on Wheatstone); Derry and Williams 1960: 625 (early telegraphic experiments); Smith and Wise 1989: 456, 677 and Morus 1991b: 25 (on Daniell). Electricians (including William Thomson) favoured the cell for telegraphic use in virtue of its steady current supply.
53. Morus 1998: 198; Bowers 2001: 122.
54. Morus 1998: 199. This last projection, in fact, seems incompatible with developing railway practice; a more plausible use, and that tried in practice, was to signal ahead (through tunnels, or up inclines) that the stationary engine should begin to 'pull'.
55. Morus 1998: 201.
56. Hubbard 1965: 36; Bowers 2001: 118–19 (who claims Wheatstone was not interested in commercial exploitation) and 124 (perpetuum mobile).
57. Bowers 2001: 124–25; see also Dawson 1978–79: 73–86 (early instrumentation). In the five-needle telegraph, pairs of needles pointed towards twenty letters of the alphabet arranged in the shape of a diamond on a vertical board.
58. Quoted in Bowers 2001: 120.
59. Morus 1998: 201–02 (preparations for demonstration).
60. Morus 1998: 202 (quoting Cooke); also quoted in Hubbard 1965: 52; for an earlier formulation of the difference between 'trying' and 'showing' an experiment, see Shapin 1988.
61. Hubbard 1965: 53.
62. Hubbard 1865: 54–55.
63. Morus 1998: 203–06 (Davy).
64. Hubbard 1965: 71.
65. Hubbard 1965: 72.
66. Hubbard 1965: 71–72.
67. Hubbard 1965: 75.
68. Morus 1998: 207.
69. These were the 'Difference Engine' (to generate mathematical tables) and the versatile 'Analytical Engine'. See Schaffer 1994 (on Babbage's 'intelligence'); Swade 2001. For Babbage's works see Campbell-Kelly 1989.

70. Brunel to Babbage, 14 October 1838, f. 11, Add 37, 191, British Library.
71. Hubbard 1965: 101.
72. Buchanan 2002a: 71; Hubbard 1965: 102.
73. See Brewster 1849: 570 for a positive contemporary critique of Sir Francis Bond Head's book, published in 1849.
74. Kieve 1973: 36–37.
75. Kieve 1973: 29–45, esp. 39 (Tawell's arrest); MacDermot 1964: 327, citing *Stokers and Pokers* as source.
76. McNeil 1990: 572 (horses).
77. MacDermot 1964: 328.
78. Morus 1998: 209 (quoting advertisement in *Railway Times* in 1842).
79. MacDermot 1964: 328.
80. Kieve 1973: 36–38.
81. Morus 1998: 221.
82. 'Block signalling', which Cooke had discussed with Stephenson in 1839, and which had been tried in 1841, was a system in which the telegraph was employed as a means of accepting the train into each section of line. It was eventually imposed by the Board of Trade. See Hubbard 1965: 114.
83. Kieve 1973: 52–53; Bartsky 1989; Morus 2000: 464–70.
84. Macleod 1988 (on patents); Cooke 1856–57; 1868; Morus 1998: 212–20.
85. Derry and Williams 1960: 626; Rolt 1970: 215.
86. For the development of the various telegraph companies, especially in relationship to railways see Kieve 1973: 46–72.
87. There were both male and female operators at the exchanges of the companies; telegraphy provided a respectable occupation for young females at the ETC. See Rolt 1970: 216; Briggs 1988: 377; Perkins 1998: 63–66; Bowers 2001: 152.
88. Rolt 1970: 217.
89. On Morse and the American context see Morus 1991b: 28–36; Bektas 2001: esp. 200–02. On Morse and machine-shop culture, see Israel 1989: esp. 58.
90. McNeil 1990: 714.
91. Morse and Vail gauged frequency of letters used by printers before determining the code: the most frequent letters had the simplest (shortest) code such as '...' for the letter 's'. For gender roles in telegraphy see Andrews 1990: 109–20; Jepsen 1995: 142–54; 1996: 72–80.
92. Bektas 2001: 201–02.
93. McNeil 1990: 714, 965; Briggs 1988: 376; Derry and Williams 1960: 626.
94. We discuss this further in the Conclusion.
95. Derry and Williams 1960: 626; McNeil 1990: 714.
96. McNeil 1990: 715, 965; Derry and Williams 1960: 626.
97. Kielbowicz 1994: esp. 95–96. Fleeming Jenkin, in 1866, also commented on the ambivalence of telegraphic news: 'Even the gain to individual merchants admits of doubt. By diminishing risks, telegraphy is sometimes thought to diminish profits. The mere convenience of sending a message quickly is outweighed in many minds by the annoyance of receiving at odd hours, scraps of news, often unintelligible from their conciseness.' See Jenkin 1866: 502 (quoted in Hempstead 1989: 298). For a discussion of telegraphy and diplomacy see Nickles 1999: 1–25.
98. Briggs 1988: 376.
99. Bowers 2001: 157–58.
100. *The Times*, 16 October 1840.
101. *The Times*, 26 November 1841.

102. Bowers 2001: 161–62.
103. Bright 1898: facing 6, plate III; Brett and Brett 1847.
104. *The Times*, 12 February 1849.
105. *The Times*, 7 February 1850.
106. Bright 1898: 248–331 (gutta percha), esp. 248–52 (its introduction as insulator); Finn 1973: 7; Hunt 1998: 87–88 (Siemens). Gutta percha, dubbed by Hunt 'an insulator for empire' was used for all manner of objects – including golf-balls and hats.
107. Hunt 1998: 88 (Faraday). Faraday's growing interest, and that of other British physicists, in electrical telegraphy found expression in a distinctive use of the language of 'fields' for electrical and magnetic phenomena. Gooding 1989: 183–223; Hunt 1991: 1–15. On Faraday's scientific language, and his discussions about it with William Whewell, see Schaffer 1991.
108. See *The Times*, 28, 29, 30 and 31 August 1850 (quotation).
109. Bright 1898: 9n; Russell 1865: 3 made a similar statement.
110. See, for example, *Annual Register*, 'Chronicle', October 1851: 164–66; see also Anon 1861.
111. Marsden 1998b: 92, 114; the cable is described and illustrated in *Bradshaw's General Shareholders' Guide, Manual, and Railway Directory, for 1853* (London: Adams and Manchester: Bradshaw and Blacklock, 1853), advertisements bound in.
112. Bright 1898: 10.
113. Anon 1887–88: 295–98.
114. *Annual Register*, 'Chronicle', October 1851: 166.
115. Read 1992: 6–7 (no primary source cited).
116. The image is reproduced in Barty-King 1979: 9.
117. Read 1992: 29–39; Blondheim 1994 ('news over the wires' in America).
118. Kielbowicz 1987: 26–41.
119. Briggs 1988: 377; Read 1992: 13, 15–16 (Reuters). For the developing relationship between telegraphy and business see Yates 1996: 149–93.
120. *Bradshaw's General Shareholders' Guide, Manual, and Railway Directory, for 1853* (London: Adams and Manchester: Bradshaw and Blacklock, 1853), advertisements bound in.
121. *Bradshaw's General Shareholders' Guide, Manual, and Railway Directory, for 1853* (London: Adams and Manchester: Bradshaw and Blacklock, 1853), advertisements bound in.
122. Russell 1865: 4; Bright 1898: 13–22. Wheatstone, who was fascinated by telegraphic 'ciphers' had recommended his 'Playfair Code' to Lord Palmerston for use during the Crimean War in 1854. See Bowers 2001: 169–72 (Wheatstone's codes and ciphers).
123. Wilson 1852: 49.
124. Smith and Wise 1989: 667.
125. Hunt 1991: 2–5.
126. Bright 1898: 25.
127. Anon 1861: 6–12.
128. Especially Bright 1898: 25–26.
129. Smith and Wise 1989: 446–53, 661–67, 675–78; Hunt 1996: 155–69.
130. On Charles Tilston Bright see Bright 1899.
131. Bright 1898: 29n (quoting Maury); McNeil 1990: 715 (on Field); Finn 1973: 9 (on capital).

132. Bright 1898: 26–28, 31.
133. Russell 1865: 11 (our italics).
134. Bright 1898: 31–33; Russell 1865: 11–12.
135. *New York Herald*, 20 April 1857 (quoted in Finn 1973: 43); Bright 1898: 23–24.
136. *The Engineer* 3 (1857): 82 (quoted in Smith and Wise 1989: 650–51).
137. Bright 1898: 36–41; Dibner 1964: 28–38; Finn 1973: 19.
138. *Scientific American* 13 (1857): 13 (quoted in Hempstead 1989: 299).
139. Dibner 1964: 37–38.
140. Especially Bright 1898: 44–47; Dibner 1964: 39–63; Finn 1973: 19–21.
141. *The Times*, 6 August 1858 (quoted in Bright 1898: 48).
142. Bright 1898: 49; Dibner 1964: 63–73 ('Celebrating the Cable Success').
143. Read 1992: 20–21 (on Reuters).
144. Bright 1898: 49–50.
145. Anon 1861: 11; Dibner 1964: 73–78 ('The Cable Fails').
146. Noakes 1999: 421–59.
147. Dibner 1964: 73–78. Reports of the number of messages varied: the Company Secretary George Saward reckoned 271; Bright's son gave 732 up to the final break.
148. Anon 1861: 7–12.
149. Gorman 1971; Headrick 1988: 119–22 (quoting Dalhousie on 120).
150. Joint Committee 1861: ix–x; Headrick 1981: 158–59 (Red Sea cable).
151. Joint Committee 1861: v–vi.
152. Headrick 1981: 159.
153. Dibner 1964: 79–81.
154. Brunel 1870: 487; see also Buchanan 2002a: 176.
155. Hunt 1996: 155–69 (reappraising Whitehouse as 'electrician') and see also De Cogan 1985: 1–15 (Whitehouse and the 1858 attempt).
156. Joint Committee 1861: x.
157. Joint Committee 1861: esp. xiii; Smith and Wise 1989: 678.
158. Joint Committee 1861: xxxvi; Hempstead 1989: 301.
159. Smith and Wise 1989: 678–80.
160. Rolt 1970: 94.
161. Smith and Wise 1989: 681; Rolt 1957: 394.
162. Russell 1865: 109. See also Dibner 1964: 93.
163. Bright 1898: 89, 97.
164. Dibner 1964: 117–20.
165. Bright 1898: 88–105; Dibner 1964: 84–149; Coates and Finn 1979 (1866 cable); Scowen 1976–77: 1–10 (oceanic cables).
166. Dibner 1964: 150; Hempstead 1995: S24.
167. Bright 1867: v.
168. Bright 1867: v.
169. Barty-King 1979.
170. In Thomson's instrument, a mirror reflected a beam of light from a lamp so that even a very small deflection of the coil on which the mirror was mounted gave a small perceptible motion to a spot of light.
171. Smith 1998b: 134–36 (Thomson); Bright 1867: 152–57 (Morse instruments).
172. Miller 1870: 118–19. Our thanks to Bruce Hunt for this reference.
173. Smith and Wise 1989: 698–712 (marketing instruments), 733–40 (yacht), 799–814 (peerage). A selection of the material culture of telegraphy can still be found preserved, notably, in the Smithsonian in Washington, DC, and in the National Museum of Science and Industry, London.

174. Marsden 1992a (engineering science).
175. Smith and Wise 1989: 670–78; Hunt 1996: 155–69 (theoretical versus practical approaches).
176. Smith 1998a: 277–78; Hunt 1994: 48–63 (standards). Thanks to Bruce Hunt for these insights.
177. Smith 1998a: 278–87. See also Lynch 1985. In this context of continuing concern with accurate measurement of electrical resistance Simon Schaffer has re-construed James Clerk Maxwell's Cavendish Laboratory in Cambridge as a 'manufactory of ohms' (paralleling the 'factory discipline' of the engineering laboratories): the culture of physics dovetailed with spaces and places designed as centres of calculation, where standards would be fixed and from which they could be distributed throughout the empire – such portable 'rulers' would make uniform and perhaps harmonise the smooth and rational running of the empire, formal and informal. See Schaffer 1992.
178. Rankine 1864; Tunbridge 1992.
179. In France the involvement of the state was characteristically greater than in Britain. Butrica 1987: 365–80.
180. Appleyard 1939; Reader 1987 (standard histories).
181. Reader 1991: 112 (gentlemanly ethos).
182. Pole 1888: 265 (Siemens at STE).
183. Buchanan 1989: 88–105 (on fragmentation).
184. Gooday 1991b: 73–111 (Ayrton's career); Brock 1981: 227–43; Gooday and Low 1998: 99–128 (Japanese connections).
185. Gooday 1991b.
186. Bright 1867: v.
187. Finn 1973: 8; for a detailed account Ahvenainen 1981.
188. Quoted in Thompson 1910: 869–70; Smith and Wise 1989: 805–06.
189. Smith and Wise 1989: 670.
190. Bright 1867: 125–26.
191. Finn 1973: 40.
192. Finn 1973: 41; Barty-King 1979: 113–40; Headrick 1981: 163 (on the 'all-red' line).
193. For example, Musk 1989: 1–20 (CPR steamships); Falkus 1990: 27, 52 (Holts' services).
194. Quoted in Moyal 1987: 39.
195. Moyal 1987: 35–54, esp. 35–44.
196. Forrest 1875: 258–62.

Conclusion: cultures of technological expertise

1 Beamish 1862; Lampe 1963; Chrimes *et al.* n.d.; Howie and Chrimes 1987.
2. Rolt 1957: 40–61; Buchanan 2002a: 25–26; Ashworth 1998: 63–79.
3. See, for example, Buchanan 2002a: 122–26.
4. *The Porcupine* 2 (1861): 123. This short-lived Liverpool magazine was subtitled *A Journal of Current Events – Social, Political, and Satirical.*
5. The main biographical source is Forbes 1844–50: 68–78; John Robison to unknown correspondent, 14 January 1840, MS 19989, ff. 123 National Library of Scotland; Robison 1839.
6. Farish 1821; Hilken 1967: 38–44 (on Farish); Willis to Thomas Coates, 4 November 1834, SDUK Papers, UCL (tourism); Willis 1851 (system of apparatus); Marsden 2004b.

7. Babbage 1832.
8. Ure 1835; Pacey 1992: 201–06 (Ure on automation).
9. Marsden 1998b: 92 (on Gordon); Hunt 1998 (on gutta percha); Marsden 2002: 137 (Birmingham Chamber of Manufacturers was formed in part to deal with the problems of foreign espionage in the 1780s).
10. Pole 1888.
11. Marsden 1998a: 383–84; Pole 1888: 44–50, 67; Siemens 1845: 324–29.
12. Gooday 2004: 82–127.
13. For the early years and the politics between the first visits, see Morrell and Thackray 1981. The later years of the Association have been less well served, but see Howarth 1931. The BAAS visited Montreal (1884), Toronto (1897), South Africa (1905), Winnipeg (1909) and Australia (1914). See *BAAS Reports*.
14. Adas 1989: 221–36 ('The machine as civilizer').
15. Ruskin 1903–08, 28: 105.
16. Quoted in Adas 1989: 233; Ruskin 1903–08, 28: 105. Sixty pounds per square inch was the pressure used on the Liverpool & Manchester Railway and by Alfred Holt in his early compound-engined steamers to China (ch. 3).
17. Morus *et al.* 1992. For a brief literature review see Brain 2000: 241–42.
18. Smith 1998b (Glasgow natural philosophy); Marsden 1998b (Glasgow mechanics and engineering); Becher 1986 (Cambridge 'voluntary science'); Morrell 1969a (Edinburgh practical chemistry); Anderson 1992 (Edinburgh 'technology').
19. See, for example, Morus 1996a: 403–34; 1996b: 342–49; 1998: 70–98.
20. Anon 1840 (Glasgow Models Exhibition); Morrell and Thackray 1981: 202–22 (Glasgow meeting), esp. 212–13; Smith and Wise 1989: 52–55.
21. Inkster and Morrell 1983: 11–54 (metropolitan and provincial science).
22. For the Great Exhibition see Purbrick 2001; Auerbach 1999; Briggs 1988: esp. 52–102 ('The great Victorian collection'); Gibbs-Smith 1981.
23. See, for example, Lubar and Hindle 1986: 248–68 (American exhibit at Great Exhibition); Dalzell 1960; and especially Post 1983; for the oration Chapin 1854. We thank Yakup Bektas for essential advice.
24. Greeley 1853; Carstensen and Glidemeister; Richards 1853.
25. Whitworth 1854; Rosenberg 1969; Post 1983: 349 (quoting Lyell) and 353; Slotton 1994 (Dallas Bache).
26. Edgerton 1996.
27. Morrell and Thackray 1981: esp. 47–52 (on 'the decline of science' debates).
28. See, on servicemen and practical science, Miller 1986.
29. Babbage 1851; 1864: 149.
30. Cardwell 1972: 111–55, esp. 111.
31. Robison to Watt, 3 May 1797 (quoted in Robinson and McKie 1970: 273).
32. Smiles 2002 [1866]: 261.
33. Marsden 1998b.
34. Walker 1841: 25–26; compare Buchanan 1989: 165.
35. Gooday 1991b.
36. Durham University 1840; Preece 1982.
37. Moseley 1843: 363–94.
38. Marsden 1998b.
39. Hilken 1967: 50–57 (Willis), esp. 54 (steam engine); Warwick 2003: 49–113 (new coaching system).
40. Rankine 1876: xx–xxvii.
41. Brock 1981; Gooday and Low 1998 (Japanese connection); De Maio 2003.

42. Barton 1976; 1990; 1998 (X-Club); Gooday 1991a (*Nature*).
43. Cardwell 1972: esp. 197–98.
44. Watson 1989.
45. Buchanan 1989: 50–68.
46. Buchanan 1989: 69–124.
47. Buchanan 1989: 215–20 (institutions); Pollard and Robertson 1979: 142–45 (education). Earlier schools of naval architecture (1811–32) and mathematics and naval construction (1848–53), both run by Cambridge mathematicians, were comparatively short-lived, though the former produced a future Admiralty Chief Constructor, Isaac Watts, and the latter another Chief Constructor, E.J. Reed, and a future head of Fairfield, William Pearce.
48. Eisenstein 1979; 1983.
49. Johns 1998.
50. Shapin 1984: 481–520.
51. See, for example, Secord 2000: 24–34, 138, 192.
52. *The Athenaeum*, 28 October, 9 and 16 December 1854, 3 February 1855.
53. Stewart 1992: 213–54.
54. Yeo 2001.
55. Secord 2000: 50–51; Hays 1981.
56. Anon 1823 (Gregory) and *DNB*.
57. Morrell 1969b: 246–47.
58. Brewster 1839: 4–5.
59. Brock 1984.
60. Yeo 2001: esp. 260–64 (Macvey Napier).
61. See esp. Secord 2000: esp. 449–70 (popular science); Frasca-Spada and Jardine 2000; Fyfe 2000; Topham 2000 (books and the sciences in history); Cooter and Pumfrey 1994 (popularisation and science).
62. See Morus 1996a: esp. 411.
63. Ferguson 1989.
64. For Gompertz see *DNB*; on the 'scapers' see Gompertz 1824; for additional inventions, see Gompertz 1850.
65. For a parallel case in chemistry, see Lundgren and Bensaude-Vincent 2000; for a preliminary study of engineering education in print, see Emmerson 1973: 231–45.
66. Marsden 2004a; Willis 1841; Hann 1833; Hann in *DNB*. Jon Topham provided useful insights regarding Weale.
67. Marsden 1992a,b; Rankine in *DNB*; Gordon 1872–75: 305 (Principia); Watts *et al.* 1866; for an extension of 'engineering science' in Oxford as literary, practical and research discipline, see Morrell 1997: 30, 94–106.
68. Shapin 1991: 279–327.
69. Biagioli 1993.
70. See esp. Cantor 1991; Morus 1992.
71. See Scoresby 1859: vii (Introduction by Smith). Airy responded in *The Athenaeum*, 12 November 1859.
72. Marsden 2004b (on Willis); Morrell and Thackray 1981: esp. 408–10, 497–98 (on Fairbairn); Marsden 1992a: 319–46 (on Rankine).
73. Miller 2000 (on Watt's reputation); Marsden 2002: 183–210 (images of Watt).
74. Jordanova 2000a; on aggrandisement through biography, see especially Nasmyth 1897; Siemens 1966.
75. Browne 1998.

76. Buchanan 2002a: esp. plates 1, 2, 18, 19.
77. Smith *et al.* 2003a: esp. 447, 469.
78. Daston and Galison 1992.
79. Rudwick 1976; for the visual in palaeontology see Rudwick 1992; Marsden 1992b: 160–63; for the visual and engineering design see Ferguson 1992.
80. For an exemplary survey, see Brooke 1991. See Moore 1979 for the historical construction of the conflict thesis.
81. Quoted in Headrick 1981: 17.
82. Smith *et al.* 2003b: 400.
83. Adas 1989: esp. 221–36; Bektas 1995 (technological crusade to the Ottoman Empire).
84. Burke and Briggs 2002: 28.
85. Numbers 23: 23.
86. Russell 1865: 103–04. Genesis 10: 5 states: 'By these were the isles of the Gentiles divided in their lands; every one after his tongue, after their families, in their nations.'
87. Job 38: 35; Anon 1860: 290 (quoted in Hempstead 1995: S19).
88. Baxter 1984.
89. Isaiah 40: 3, 4; Brewster 1849: 571. In the 1830s New Englanders had similarly invoked Isaiah when they enthused over the dominion of nature brought about by canals. See Hughes 2004: 17–43, esp. 30–31.
90. Daniel 12: 4.
91. Brewster 1849: 581.
92. Macgregor to Nell Laird, 26 April 1857, DX/258/1/27, transcripts of Macgregor Laird letters to his wife Nell, Merseyside Maritime Museum Archives, Liverpool.
93. Macleod 1876, 1: 234, 238.
94. Crosbie Smith and Anne Scott (paper presented at the Liverpool meeting of the British Society for the History of Science, June 2004).
95. Hodder 1890: 298–99.
96. Acts 27: 22, 31; Chalmers 1836–42, 9: 154–61 (St Paul's shipwreck). On Chalmers see esp. Smith 1998a: 15–22; Hilton 1991: esp. 15–17, 31–33.
97. Samuel Cunard to Charles MacIver, 1 January 1858, Cunard papers, Liverpool University Library.
98. D.&C. MacIver, 'Captain's Memoranda', D.128/2/4, Cunard papers, Liverpool University Library.
99. Quoted in Babcock 1931: 84.
100. Smith 1998a: esp. 23–30, 150–52 (on Macleod and the 'North British' engineers and natural philosophers).
101. Macleod 1876, 2: 20; Smith 1998a: 30.
102. Macleod 1876, 1: 238.
103. Watt to Small, 31 January 1770 (quoted in Dickinson 1936: 70).
104. Boulton to Watt, 16 April 1781 (quoted in Dickinson and Jenkin 1981: 54); on Boulton's lack of explicit religious activity, see Robinson 1954.
105. Marsden 1992a,b; Smith 1998a: 151–52.
106. Olinthus Gregory, 'Note on "Precise Point"', 18 February 1833, Autograph Letter Series, Wellcome Library.
107. Olinthus Gregory, 'Note on "Genuine Christianity"', April 1839, Autograph Letter Series, Wellcome Library.
108. Walker 1839: 17–18. The quotation is from Edward Young's (1683–1765) *Night Thoughts*, Night ix, line 771. The line was also cited by William Herschel.

109. *The Times*, 7 November 1839.
110. Preece 1982; Whiting 1932: 82–83.
111. Hearnshaw 1929: 25–45 (reaction to the 'University of London' (as University College was originally known)), 46–69 (foundation of King's College), 110 (Wheatstone and Moseley), 146–50 (engineering department), quoting 69.
112. David T. Hann, '[James Hann] The Self Taught Mathematical Genius', unpublished typescript, copy at King's College London archives.
113. Marsden 2004 (forthcoming); Willis 1870: 458–63 (quoting 463).
114. Morrell 1971.
115. Marsden 2001; Buchanan 1989: 187–88; 2002: 213–17 (quoting Brunel on 214, his emphasis); Rolt 1957: 362–63 (Horsell's letter); and see also Rolt 1957: 419–20 (an alternative view on Brunel and prayer). Horsell's letter was in fact never sent.
116. Smith 1998a (for 'scientists of energy'); Desmond 1998 (scientific naturalists); Barton 1998 (X-Club).
117. Arnold 1993: 70.
118. Arnold 1993: 63.
119. See esp. Buchanan 1983; 1989: 192–95.
120. Haworth 1994: 56; Davie 1961.
121. See, for example, letters from Brunel and his secretary quoted in Buchanan 2002a: 89 (character, confidence) and 165 (habit, connection).
122. Shapin 1991.
123. Golinski 1992.
124. Rudwick 1985: 17–30.
125. Morrell and Thackray 1981: esp. 148–57 ('Dealing with the ladies'); Neeley 2001 (Somerville).
126. Secord 1994, 1996; Morus 1993.
127. Gooday 1991a
128. Buchanan 2002a: 103–12 (Brunel's 'disasters').
129. Hays 1981.
130. Wiener 1981.
131. Dickinson 1936: 160–61, 187–91.
132. Buchanan 2002a: 197–98.
133. Napier 1904.
134. Alfred Holt, Diary, Book A, Papers of Alfred Holt, 920 HOL/2/52, Liverpool Central Library.
135. Inverclyde Collection, Mitchell Library, Glasgow.
136. Moss and Hume 1986: 141, 227, 247–48.
137. *The Shipbuilder & Marine Engine Builder* 42 (1935): 28–30 (*Orion*); Moss and Hume 1986: 360–62 (*Pretoria Castle*).
138. See Smith *et al.* 2003a: 448n; 2003b: 385.
139. Hodder 1890: 60, 94–98, 175–76, 283.
140. 'Memorandum left by Mr Philip H. Holt', 920 DUR 14/4, Durning family papers, Liverpool Central Library.
141. Green and Moss 1982; Moss and Hume 1986: 245–82.

Bibliography

Manuscripts

Full details of manuscript sources are given in the notes.

The following collections have been consulted:
Airy Papers, Royal Greenwich Observatory Collection, Cambridge University Library.
Autograph Letter Series (Olinthus Gregory), Wellcome Library.
Babbage Papers, British Library.
Cunard Papers, Liverpool University Library.
Durning and Holt Family Papers, Liverpool Central Library.
Inverclyde Collection, Mitchell Library, Glasgow.
Kelvin Correspondence, Cambridge University Library.
Laird Family Papers, Merseyside Maritime Museum, Liverpool.
Macgregor Laird Letters, Merseyside Maritime Museum Archives, Liverpool.
Napier Family Papers, Glasgow University Archives.
Ocean Archive, Merseyside Maritime Museum, Liverpool.
Peel Papers, British Library.
Phillips Correspondence, Oxford University Museum.
John Robison Papers, National Library of Scotland.
SDUK Papers, University College London.

Periodicals

Annals of Electricity, Magnetism, & Chemistry; and Guardian of Experiment Science
Athenaeum
Bradshaw's Railway Almanack, Directory, Shareholders' Guide, and Manual
Cambridge University Reporter
Chambers's Edinburgh Journal
Civil Engineer and Architects' Journal
Edinburgh New Philosophical Journal
Edinburgh Review
Engineer
Engineering
Glasgow Herald
Illustrated London News
Liverpool Mercury
Mechanics' Magazine
Minutes of Proceedings of the Institution of Civil Engineers
New York Herald
New York Tribune
North American Review
North British Daily Mail
North British Review
Philosophical Transactions of the Royal Society of London
The Porcupine: A Journal of Current Events – Social, Political, and Satirical

Proceedings of the Institution of Mechanical Engineers
Proceedings of the Royal Institution
Proceedings of the Royal Society of Edinburgh
Punch
Quarterly Journal of Science and the Arts
Quarterly Review
Railway Magazine; and Annals of Science
Report of the British Association for the Advancement of Science
Scientific American
The Times
Transactions of the Liverpool Polytechnic Society
Vanity Fair
The Witness

Journals

Ambix
American Historical Review
American Journal of Sociology
Annals of Science
British Journal for the History of Science
Bulletin of the Scientific Instrument Society
Bulletin of the United States National Museum
Critical Inquiry
Engineering Science and Education Journal
Glasgow University Gazette
Historia Scientiarum
Historical Journal
History
History and Technology
History of Science
History of Technology
History Today
IEEE Power Engineering Review
Institution of Electrical Engineers Proceedings A
Interdisciplinary Science Reviews
Isis
Journal of American Studies
Journal of Economic History
Newcomen Society Transactions
Notes and Records of the Royal Society of London
Osiris
Past & Present
Representations
Science in Context
Sea Breezes
Ships Monthly
Social History
Social History of Medicine
Social Studies of Science

Studies in History and Philosophy of Science
Technology and Culture
Transactions of the Newcomen Society
Transactions of the Architectural and Archaeological Society of Durham and Northumberland
Victorian Studies

Books

[Adams, Charles F.]. 1867. 'The railroad system', *North American Review* 104: 476–511.
Adams, Charles F. 1869. 'A chapter of Erie', *North American Review* 109: 30–106.
——. 1870. 'Railway problems in 1869', *North American Review* 110: 116–50.
——. 1871. 'An Erie raid', *North American Review* 112: 240–91.
Adams, Henry. 1920. *The Degradation of the Democratic Dogma*. Introduction by Brooks Adams. New York: Macmillan.
Adas, Michael. 1989. *Machines as the Measure of Men. Science, Technology, and Ideologies of Western Dominance*. Ithaca and London: Cornell University Press.
Agar, Jon. 2003. *Constant Touch. A Global History of the Mobile Phone*. Cambridge: Icon.
Ahvenainen, Jorma. 1981. *The Far Eastern Telegraphs. The History of Telegraphic Communications between the Far East, Europe and America before the First World War*. Helsinki: Suomalainen Tiedeakatemia.
Airy, George B. 1896. *Autobiography of Sir George Biddell Airy*. Ed. Wilfred Airy. Cambridge: Cambridge University Press.
Albion, Robert Greenhalgh. 1938. *Square-Riggers on Schedule. The New York Sailing Packets to England, France, and the Cotton Ports*. Princeton: Princeton University Press.
Alder, Ken. 1995. 'A revolution to measure: the political economy of the metric system in France'. In M. Norton Wise, ed., *The Values of Precision*, pp. 39–71. Princeton: Princeton University Press.
Allen, Oliver E. 1978. *The Windjammers*. Amsterdam: Time-Life Books.
Anderson, John M. 1998. 'Morse and the telegraph: another view of history', *IEEE Power Engineering Review* 18 (7): 28–29.
Anderson, Katharine. 1999. 'The weather prophets: science and reputation in Victorian meteorology', *History of Science* 37: 179–216.
——. 2003. 'Looking at the sky: the visual context of Victorian meteorology', *British Journal for the History of Science* 36: 301–32.
Anderson, R.G.W. 1992. ' "What is technology?": education through museums in the mid-nineteenth century', *British Journal for the History of Science* 25: 169–84.
Andrewes, William J.H. (ed.). 1996. *The Quest for Longitude*. Cambridge, MA: Collection of Scientific Instruments, Harvard University.
Andrews, Melodie. 1990. ' "What the girls can do": the debate over the employment of women in the early American telegraph industry', *Essays in Economic and Business History* 8: 109–20.
Anon. 1823. 'Memoir of Olinthus Gregory', *Imperial Magazine* 5: 777–92.
——. 1838. *Railroadiana. A New History of England, or Picturesque, Biographical, Legendary and Antiquarian Sketches. Descriptive of the Vicinity of the Railways*. London: Simpkin, Marshall.
——. 1840. *Catalogue of the Exhibition of Models and Manufacturings, &c. at the Tenth Meeting of the British Association for the Advancement of Science*. Glasgow: Robert Weir.
——. 1841a. 'Originality of discovery', *Chambers's Edinburgh Journal* 10: 1–2.

——. 1841b. 'Electro-magnetic power', *Chambers's Edinburgh Journal* 10: 252.

——. 1845. *The Narrow and Wide Gauges Considered; also, Effects of Competition and Government Supervision.* London: Effingham Wilson.

——. 1846. *£. s. d. The Broad Gauge the Bane of the Great Western Railway Company, with an Account of the Present and Prospective Liabilities Saddled on the Proprietors by the Promoters of that Peculiar Crotchet.* London: Ollivier.

——. 1848. 'The electric telegraph, and uniformity of time', *Bradshaw's Railway Almanack, Directory, Shareholders' Guide, and Manual for 1848:* 15–19.

——. 1851. 'Railway-time aggression', *Chambers's Edinburgh Journal* 15: 392–95.

——. 1858. 'A derivation and illustration', *Punch* 34: 89.

——. 1859. *Ericsson's Caloric Engine Manufactured by the Massachusetts Caloric Engine Company, South Groton, MS.* Boston: W. and E. Howe.

——. 1860. 'The progress of the electric telegraph', *Atlantic Monthly* 5: 290–97.

——. 1861. *The North Atlantic Telegraph; via the Faröe Isles, Iceland, and Greenland.* London: Stanford.

——. 1864. 'Ocean steam navigation', *North American Review* 99: 483–523.

——. 1887–88. '[Obituary of Thomas Russell Crampton]', *Minutes of Proceedings of the Institution of Civil Engineers* 94: 295–98.

Appleyard, Rollo. 1930. *Pioneers of Electrical Communication.* London: Macmillan.

——. 1939. *The History of the Institution of Electrical Engineers (1871–1931).* London: Institution of Electrical Engineers.

Arago, François. 1839a. *Life of James Watt, with Memoir on Machinery Considered in Relation to the Prosperity of the Working Classes. With contributions from Henry Brougham and Francis Jeffrey.* Edinburgh: Black.

——. 1839b. *Historical Eloge of James Watt.* Trans. J.P. Muirhead. London: John Murray.

——. 1839c. 'Biographical memoir of James Watt, one of the eight Associates of the Academy of Sciences', *Edinburgh New Philosophical Journal* 27: 221–97.

——. 1839d. 'On machinery considered in relation to the prosperity of the working classes', *Edinburgh New Philosophical Journal* 27: 297–310.

Armstrong, Richard. 1975. *Powered Ships. The Beginnings.* London and Tonbridge: Ernest Benn.

Armstrong, William G. 1863. 'Presidential address', *Report of the British Association for the Advancement of Science* 33: li–lxiv.

Armytage, W.H.G. 1976. *A Social History of Engineering.* London: Faber.

Arnold, Matthew. 1993 [1861–78]. *Culture and Anarchy and Other Writings.* Ed. Stefan Collini. Cambridge: Cambridge University Press.

Ashplant, T.G. and Gerry Smyth (eds). 2001. *Explorations in Cultural History.* London and Sterling, VA: Pluto Press.

Ashworth, William J. 1994. 'The calculating eye: Baily, Herschel, Babbage and the business of astronomy', *British Journal for the History of Science* 27: 409–41.

——. 1998. ' "System of terror": Samuel Bentham, accountability and dockyard reform during the Napoleonic wars', *Social History* 23: 63–79.

——. 2001. ' "Between the trader and the public": British alcohol standards and the proof of good governance', *Technology and Culture* 42: 27–50.

Atmore, Henry. 2004. 'Railway interests and the "rope of air"', *British Journal for the History of Science* 37: 245–79.

Auerbach, Jeffrey A. 1999. *The Great Exhibition of 1851. A Nation on Display.* New Haven: Yale University Press.

Babbage, Charles. 1830. *Reflections on the Decline of Science in England, and on Some of its Causes.* London: Fellowes. Reprinted New York: Kelley, 1970.

——. 1832. *On the Economy of Machinery and Manufactures*. London: Charles Knight.

——. 1846. *On the Economy of Machinery and Manufactures*. Fourth edition. London: John Murray.

——. 1851. *The Exposition of 1851 or, Views of the Industry, Science, and the Government of England*. Second edition. London: John Murray.

——. 1864. *Passages from the Life of a Philosopher*. London: Longman.

Babcock, F. Lawrence. 1931. *Spanning the Atlantic*. New York: Knopf.

Baguley, David. 1990. *Naturalist Fiction. The Entropic Vision*. Cambridge: Cambridge University Press.

Bailey, Michael R. (ed.). 2003. *Robert Stephenson. The Eminent Engineer*. Aldershot: Ashgate.

Banbury, Philip. 1971. *Shipbuilders of the Thames and Medway*. Newton Abbot: David & Charles.

Barlow, Peter. 1835. *Second Report Addressed to the Directors and Proprietors of the London and Birmingham Railway Company, Founded on an Inspection of, and Experiments made on, the Liverpool and Manchester Railway*. London: Fellowes.

Barton, Ruth. 1976. 'The X Club: science, religion, and social change in Victorian England'. PhD diss., University of Pennsylvania.

——. 1990. ' "An influential set of chaps": the X-Club and Royal Society politics 1864–85', *British Journal for the History of Science* 23: 53–81.

——. 1998. ' "Huxley, Lubbock, and half a dozen others": professionals and gentlemen in the formation of the X Club, 1851–1864', *Isis* 89: 410–44.

Bartsky, I.R. 1989. 'The adoption of standard time', *Technology and Culture* 30: 25–56.

Barty-King, Hugh. 1979. *Girdle Round The Earth. The Story of Cable and Wireless and its Predecessors to Mark the Group's Jubilee, 1929–1979*. London: Heinemann.

Basalla, George. 1988. *The Evolution of Technology*. Cambridge: Cambridge University Press, Cambridge.

Baxter, Paul. 1984. 'Brewster, evangelism and the disruption of the Church of Scotland'. In A.D. Morrison-Low and J.R.R. Christie, eds, *'Martyr of Science'. Sir David Brewster 1781–1868*, pp. 45–50. Edinburgh: Royal Scottish Museum.

Beamish, Richard. 1862. *Memoir of the Life of Sir Marc Isambard Brunel*. London: Longman.

Beaune, C. 1985. *Naissance de la nation France*. Paris: Gallimard.

Becher, Harvey. 1986. 'Voluntary science in nineteenth-century Cambridge University to the 1850s', *British Journal for the History of Science* 19: 57–87.

Beckett, Derrick. 1980. *Brunel's Britain*. Newton Abbot: David & Charles.

Bektas, Yakup. 1995. 'The British technological crusade to post-Crimean Turkey: electric telegraphy, railways, naval shipbuilding and armament technologies'. PhD diss., University of Kent at Canterbury.

——. 2000. 'The Sultan's messenger: cultural constructions of Ottoman telegraphy, 1847–1880', *Technology and Culture* 41: 669–96.

——. 2001. 'Displaying the American genius: the electromagnetic telegraph in the wider world', *British Journal for the History of Science* 34: 199–232.

Bennett, Jim, Michael Cooper, Michael Hunter and Lisa Jardine. 2003. *London's Leonardo. The Life and Work of Robert Hooke*. Oxford: Oxford University Press.

Berg, Maxine. 1980. *The Machinery Question and the Making of Political Economy, 1815–1848*. Cambridge: Cambridge University Press.

Biagioli, Mario. 1993. *Galileo, Courtier. The Practice of Science in the Culture of Absolutism*. Chicago and London: University of Chicago Press.

Bignell, Philippa. 1978. *Taking the Train. Railway Travel in Victorian Times*. London: HMSO for the National Railway Museum.

Bijker, Wiebe E. and John Law (eds). 1992. *Shaping Technology/Building Society.* Cambridge, MA and London: MIT Press.

——, Thomas P. Hughes and Trevor Pinch (eds). 1987. *The Social Construction of Technological Systems. New Directions in the Sociology and History of Technology.* Cambridge, MA and London: MIT Press.

Bishop, P.W. 1976–77. 'John Ericsson (1803–89) in England', *Transactions of the Newcomen Society* 48: 41–52.

Blondheim, Menahem. 1994. *News over the Wires. The Telegraph and the Flow of Public Information in America, 1844–1897.* Cambridge, MA: Harvard University Press.

Bonsor, N.R.P. 1975–80. *North Atlantic Seaway.* 5 vols. Newton Abbot and Jersey Channel Islands: David & Charles and Brookside Publications.

——. 1983. *South Atlantic Seaway.* Jersey Channel Islands: Brookside Publications.

Booth, Henry. 1830. *An Account of the Liverpool and Manchester Railway, Comprising a History of the Parliamentary Proceedings, Preparatory to the Passing of the Act, a Description of the Railway, in an Excursion from Liverpool to Manchester, and a Popular Illustration of the Mechanical Principles Applicable to Railways. Also an Abstract of the Expenditure from the Commencement of the Undertaking, with Observations on the Same.* Liverpool: Wales and Baines. Reprinted London: Frank Cass, 1969.

——. 1844–46. 'On the prospects of steam navigation &c', *Transactions of the Liverpool Polytechnic Society* 2: 24–31.

——. 1847. *Uniformity of Time, Considered Especially with Reference to Railway Transit and the Operations of the Electric Telegraph.* London and Liverpool: J. Weale.

Booth, Henry. 1980. *Henry Booth: Inventor – Partner in the Rocket and Father of Railway Management.* Ilfracombe: Stockwell.

Bowers, Brian. 1975. *Sir Charles Wheatstone FRS, 1802–1875.* London: HMSO.

——. 1982. *A History of Electric Light and Power.* Stevenage and New York: Peter Peregrinus in association with the Science Museum.

——. 2001. *Sir Charles Wheatstone FRS, 1802–1875.* Second edition. London: Institution of Electrical Engineers/Science Museum.

Brain, Robert. 2000. 'Exhibitions'. In Arne Hessenbruch, ed., *Reader's Guide to the History of Science*, pp. 241–42. London and Chicago: Fitzroy Dearborn.

Brett, Jacob and John Watkins Brett. 1847. *Copy of a Letter submitted to the Government in July 1845. Printed by Brett's Electric Telegraph.* London: Printed for the Authors.

[Brewster, David]. 1839. 'Review of "Life and Works of Thomas Telford"', *Edinburgh Review* 70: 1–47

[Brewster, David]. 1849. 'The railway systems of Great Britain', *North British Review* 11: 569–617.

Briggs, Asa. 1988. *Victorian Things.* Harmondsworth: Penguin.

Bright, Charles. 1898. *Submarine Telegraphs. Their History, Construction, and Working.* London: Crosby Lockwood.

——. 1899. *The Life Story of Sir Charles Tilston Bright, Civil Engineer. With Which is Incorporated the Story of the Atlantic Cable and the First Telegraph to India and the Colonies.* London: Constable.

——. 1911. *Imperial Telegraphic Communication.* London: P.S. King.

Bright, Edward B. 1867. *The Electric Telegraph. By Dr Lardner.* London: James Walton.

Broadie, Alexander (ed.). 1997. *The Scottish Enlightenment. An Anthology.* Edinburgh: Canongate.

Brock, W.H. 1981. 'The Japanese connexion: engineering in Tokyo, London, and Glasgow at the end of the nineteenth century', *British Journal for the History of Science* 14: 227–43.

——. 1984. 'Brewster as a scientific journalist'. In A.D. Morrison-Low and J.R.R. Christie, eds, *'Martyr of Science'. Sir David Brewster 1781–1868*, pp. 37–42. Edinburgh: Royal Scottish Museum.

Brooke, John Hedley. 1991. *Science and Religion. Some Historical Perspectives.* Cambridge: Cambridge University Press.

Browne, Janet. 1998. 'I could have retched all night: Charles Darwin and his body'. In Christopher Lawrence and Steven Shapin, eds, *Science Incarnate. Historical Embodiments of Natural Knowledge*, pp. 240–87. Chicago and London: University of Chicago Press.

Brunel, Isambard. 1870. *The Life of Isambard Kingdom Brunel, Civil Engineer.* Longman: London.

Bryden, D.J. 1970. 'The Jamaican observatories of Colin Campbell FRS and Alexander MacFarlane FRS', *Notes and Records of the Royal Society* 24: 265–68.

——. 1994. 'James Watt, merchant: the Glasgow years, 1754–1774'. In Denis Smith, ed., *Perceptions of Great Engineers. Fact and Fantasy*, pp. 9–21. London: Science Museum.

Buchanan, R. Angus. 1972. *Industrial Archaeology in Britain.* Harmondsworth: Penguin.

——. 1983a. 'The *Great Eastern* controversy: a comment', *Technology and Culture* 24: 98–106.

——. 1983b. 'Gentleman engineers: the making of a profession', *Victorian Studies* 26: 407–29.

——. 1989. *The Engineers. A History of the Engineering Profession in Britain, 1750–1914.* London: Jessica Kingsley.

——. 1992. 'The atmospheric railway of I.K. Brunel', *Social Studies of Science* 22: 231–43.

——. 2002a. *Brunel. The Life and Times of Isambard Kingdom Brunel.* London and New York: Hambledon and London.

——. 2002b. 'Joseph Whitworth'. In John Cantrell and Gillian Cookson, eds, *Henry Maudslay & the Pioneers of the Machine Age*, pp. 109–28. Stroud and Charleston: Tempus.

—— and Michael Williams. 1982. *Brunel's Bristol.* Bristol: Redcliffe.

Burke, Peter. 1997. *Varieties of Cultural History.* Cambridge: Polity Press.

——. 2000. *A Social History of Knowledge. From Gutenberg to Diderot.* Cambridge: Polity.

—— and Asa Briggs. 2002. *A Social History of the Media. From Gutenberg to the Internet.* Cambridge: Polity.

Burton, Anthony. 1980. *The Rainhill Story. The Great Locomotive Trial.* London: BBC.

——. 1994. *The Railway Empire.* London: John Murray.

Bury, T.T. 1976 [1831]. *Coloured Views of the Liverpool and Manchester Railway.* Oldham: Broadbent.

Butrica, Andrew J. 1987. 'Telegraphy and the genesis of electrical engineering institutions in France, 1845–1895', *History and Technology* 3: 365–80.

Butterfield, Herbert. 1931. *The Whig Interpretation of History.* London: Bell.

Cahan, David (ed.). 1993. *Hermann von Helmholtz and the Foundations of Nineteenth-Century Science.* Berkeley: University of California Press.

Cain, P.J. and A.G. Hopkins. 2002. *British Imperialism, 1688–2000.* Harlow: Pearson Education. First edition 1993.

Callan, N.J. 1836–37. 'On a method of connecting electro-magnets so as to combine their electric powers, &c; and on the application of electro-magnetism to the working of machines', *Annals of Electricity* 1: 491–94.

Campbell, R.H. 1961. *Carron Company.* Edinburgh: Oliver and Boyd.

Campbell-Kelly, Martin (ed.). 1989. *Works of Charles Babbage*. 11 vols. London: Pickering.

Cannon, Susan F. 1978. *Science in Culture. The Early Victorian Period*. Folkestone and New York: Dawson and Science History Publications.

Cantor, G.N. 1991. *Michael Faraday: Sandemanian and Scientist. A Study of Science and Religion in the Nineteenth Century*. Macmillan: Basingstoke.

Cantrell, John. 2002a. 'Henry Maudslay'. In John Cantrell and Gillian Cookson, eds, *Henry Maudslay & the Pioneers of the Machine Age*, pp. 18–38. Stroud and Charleston: Tempus.

——. 2002b. 'James Nasmyth'. In John Cantrell and Gillian Cookson, eds, *Henry Maudslay & the Pioneers of the Machine Age*, pp. 129–46. Stroud and Charleston: Tempus.

—— and Gillian Cookson (eds). 2002. *Henry Maudslay & the Pioneers of the Machine Age*. Stroud and Charleston: Tempus.

Cardwell, D.S.L. 1965. 'Power technologies and the advance of science, 1700–1825', *Technology and Culture* 6: 188–207.

——. 1972. *The Organisation of Science in England*. Revised edition. London: Heinemann Educational Books.

——. 1989. *James Joule. A Biography*. Manchester: Manchester University Press.

——. 1994. *The Fontana History of Technology*. Hammersmith: Fontana.

Carey, James W. 1983. 'Technology and ideology: the case of the telegraph', *Prospects: An Annual of American Cultural Studies* 8: 303–25.

Carlson, Robert E. 1969. *The Liverpool & Manchester Railway Project 1821–1831*. Newton Abbot: David & Charles.

Carlyle, Thomas. 1829. 'Signs of the times', *Edinburgh Review* 49: 439–59.

Carnot, Sadi. 1986 [1824]. *Reflexions on the Motive Power of Fire. A Critical Edition with the Surviving Scientific Manuscripts*. Trans. and ed. R. Fox. Manchester and New York: Manchester University Press and Lilian Barber Press.

Carstensen, Georg, Johan Bernhard and Charles Gildemeister. 1854. *New York Crystal Palace. Illustrated Description of the Building*. New York: Riker, Thorne.

Carter, Ian. 2001. *Railways and Culture in Britain. The Epitome of Modernity*. Manchester: Manchester University Press.

Carter, Paul. 1988. *The Road to Botany Bay. An Exploration of Landscape and History*. New York: Knopf.

Cawood, John. 1979. 'The magnetic crusade: science and politics in early Victorian Britain', *Isis* 70: 493–518.

Chalmers, Thomas. 1836–42. *The Works of Thomas Chalmers*. 25 vols. Glasgow: Collins.

Chambers, Ephraim and Abraham Rees. 1779–86. *Cyclopaedia. An Universal Dictionary of Arts and Sciences*. 5 vols. London: Rivington.

Chambers, Robert. 1994. *Vestiges of the Natural History of Creation and Other Evolutionary Writings*. Ed. J.A. Secord. Chicago: University of Chicago Press.

Channell, David F. 1982. 'The harmony of theory and practice: the engineering science of W.J.M. Rankine', *Technology and Culture* 23: 39–52.

——. 1989. *The History of Engineering Science. An Annotated Bibliography*. New York and London: Garland.

Chapin, E.H. 1854. *The American Idea, and What Grows Out of It. An Oration, Delivered in the New York Crystal Palace, July 4, 1854*. Boston: A. Tompkins.

Chapman, Allan. 1998. 'Standard time for all: the electric telegraph, Airy, and the Greenwich time service'. In F.A.J.L. James, ed., *Semaphores to Short Waves*, pp. 40–59. RSA: London.

Chrimes, Michael, Julia Elton, John May and Timothy Millett. n.d. *The Triumphant Bore. A Celebration of Marc Brunel's Thames Tunnel*. London: Institution of Civil Engineers.

Christie, J.R.R. 1974. 'The origins and development of the Scottish scientific community, 1680–1760', *History of Science* 12: 122–41.

Church, William Conant. 1890. *The Life of John Ericsson*. 2 vols. London: Sampson Low, Marston, Searle and Rivington.

Clark, William, Jan Golinski and Simon Schaffer (eds). 1999. *The Sciences in Enlightened Europe*. Chicago and London: University of Chicago Press.

Clarke, J.F. [1979]. *Power on Land and Sea: 160 Years of Industrial Experience on Tyneside – A History of Hawthorn Leslie and Company Ltd., Engineers and Shipbuilders*. Newcastle: Smith Print Group.

Clayton, Alfred B. 1970 [1831]. *Views on the Liverpool and Manchester Railway, Taken on the Spot*. Newcastle: Graham.

Clements, Paul. 1970. *Marc Isambard Brunel*. London: Longmans.

Coates, Vary T. and Bernard Finn (eds). 1979. *A Retrospective Technology Assessment. Submarine Telegraphy – The Transatlantic Cable of 1866*. San Francisco: San Francisco Press.

Cole, Henry. 1846. *Railway Eccentrics. Inconsistencies of Men of Genius Exemplified in the Practice and Precept of Isambard Kingdom Brunel, and the Theoretical Opinions of Charles Alexander Saunders*. London: Ollivier.

Collini, Stefan. 1999. *English Pasts. Essays in History and Culture*. Oxford: Oxford University Press.

Conrad, Joseph. 1947 [1907]. *The Secret Agent. A Simple Tale*. Collected edition of Conrad's works. London: J.M. Dent.

Constable, Thomas. 1877. *Memoir of Lewis D.B. Gordon, F.R.S.E.* Edinburgh: Constable.

Cooke, W.F. 1856–57. *The Electric Telegraph. Was it Invented by Professor Wheatstone?* 2 vols. London: Printed for the author.

——. 1868. *Authorship of the Practical Electric Telegraph of Great Britain*. Bath: Peach.

Cookson, Gillian and Colin A. Hempstead. 2000. *A Victorian Scientist and Engineer. Fleeming Jenkin and the Birth of Electrical Engineering*. Aldershot: Ashgate.

Cooter, Roger and Stephen Pumfrey. 1994. 'Separate spheres and public places: reflections on the history of science popularization and science in popular culture', *History of Science* 32: 237–67.

Corlett, Ewan. 1990. *The Iron Ship. The Story of Brunel's SS 'Great Britain'*. London: Conway Maritime Press. First published 1975.

Cossons, Neils (ed.). 2000. *Perspectives on Industrial Archaeology*. London: Sciences Museum.

Cowan, Ruth Schwartz. 1983. *More Work for Mother. The Ironies of Household Technology from the Open Hearth to the Microwave*. New York: Basic Books.

Creese, Mary R.S. and Thomas M. Creese. 1994. 'British women who contributed to research in the geological sciences in the nineteenth century', *British Journal for the History of Science* 27: 23–54.

Crosland, Maurice. 1992. *Science under Control. The French Academy of Sciences 1795–1914*. Cambridge: Cambridge University Press.

Daiches, David, Peter Jones and Jean Jones (eds). 1996. *The Scottish Enlightenment 1730–1790. A Hotbed of Genius*. Edinburgh: Saltire Society.

Dalzell, Robert F. jun. 1960. *American Participation in the Great Exhibition of 1851*. Amherst, MA: Amherst College Press.

Daston, Lorraine and Peter Galison. 1992. 'The image of objectivity', *Representations* 40: 81–128.

Daumas, Maurice (ed.). 1980. *A History of Technology and Invention Progress Through the Ages. Volume III. The Expansion of Mechanization 1725–1860*. Trans. Eileen B. Hennessy. London: John Murray.

Davenport, A.N. 1989. *James Watt and the Patent System*. London: British Library.

Davenport, Thomas. 1838a. 'Davenport's recent experiments in electro-magnetic machinery', *Annals of Electricity* 2: 284–86.

——. 1838b. 'Specification of a Patent for the application of electro-magnetism to the propelling of machinery; granted to Thomas Davenport, of Brandon, Rutland County, Vermont, February, 1837', *Annals of Electricity* 2: 347–50.

Davie, G.E. 1961. *The Democratic Intellect. Scotland and her Universities in the Nineteenth Century*. Edinburgh: Edinburgh University Press.

Davis, Clarence B. and Kenneth J. Wilburn (eds). 1991. *Railway Imperialism*. New York, Westport and London: Greenwood Press.

Dawson, Keith. 1978–79. 'Early electro-magnetic telegraph instruments', *Newcomen Society Transactions* 50: 73–86.

De Cogan, D. 1985. 'Dr E.O.W. Whitehouse and the 1858 trans-Atlantic cable', *History of Technology* 10: 1–15.

De Maio, Silvana. 2003. 'The development of an educational system at the beginning of the Meiji Era: reference models from Western countries', *Historia Scientiarum* 12: 183–99.

De Pambour, François Marie Guyonneau. 1840. *A Practical Treatise on Locomotive Engines, Founded of a Great Many New Experiments on the Liverpool and Manchester, and Other Railways*. London: Weale.

Dear, Peter. 2001. *Revolutionizing the Sciences. European Knowledge and its Ambitions, 1500–1700*. Basingstoke: Palgrave.

Derry, T.K. and Trevor I. Williams. 1960. *A Short History of Technology from the Earliest Times to A.D. 1900*. Oxford: Clarendon Press.

Desaguliers, J.T. 1734–44. *A Course of Experimental Philosophy*. 2 vols. London: John Senex.

Desmond, Adrian. 1989. *The Politics of Evolution. Morphology, Medicine, and Reform in Radical London*. Chicago and London: University of Chicago Press.

——. 1998. *Huxley. From Devil's Disciple to Evolution's High Priest*. London: Penguin.

Dibner, Bern. 1964. *The Atlantic Cable*. New York, Toronto and London: Blaisdell Publishing. First published 1959.

Dickinson, H.W. 1936. *James Watt. Craftsman and Engineer*. Cambridge: Cambridge University Press.

——. 1937. *Matthew Boulton*. Cambridge: Cambridge University Press.

—— and Arthur Titley. 1934. *Richard Trevithick. The Engineer and the Man*. Cambridge: Cambridge University Press.

—— and Rhys Jenkins. 1981. *James Watt and the Steam Engine*. Ashbourne: Moorland Publishing. First published 1927.

Divall, Colin and Andrew Scott. 2001. *Making Histories in Transport Museums*. London and New York: Leicester University Press.

[Durham University]. 1840. *The Durham University Calendar for 1841*. Durham: Francis Humble.

Dyos, H.J. and Aldcroft, D.H. 1969. *British Transport. An Economic Survey from the Seventeenth Century to the Twentieth*. Leicester: Leicester University Press.

Edgerton, David. 1996. *Science, Technology and the British Industrial 'Decline', 1870–1970*. Cambridge: Cambridge University Press.

Eisenstein, Elizabeth L. 1979. *The Printing Press as an Agent of Change. Communications and Cultural Transformation in Early Modern Europe.* 2 vols. Cambridge: Cambridge University Press.

——. 1983. *The Printing Revolution in Early Modern Europe.* Cambridge: Cambridge University Press.

Emmerson, George S. 1973. *Engineering Education. A Social History.* Newton Abbot: David & Charles.

——. 1977. *John Scott Russell. A Great Victorian Engineer and Naval Architect.* London: John Murray.

Evans, Francis T. 1981. 'Roads, railways, and canals: technical choices in 19th-century Britain', *Technology and Culture* 22: 1–34.

Fairlie, Robert R. 1872. *Railways or No Railways. Narrow Gauge, Economy with Efficiency v. Broad Gauge, Costliness with Extravagance.* London: Effingham Wilson.

Faith, Nicholas. 1994. *The World the Railways Made.* London: Random House. First published 1990.

Falkus, Malcolm. 1990. *The Blue Funnel Legend. A History of the Ocean Steamship Company, 1865–1973.* Basingstoke: Macmillan.

Fara, Patricia. 2002a. *Newton. The Making of Genius.* Basingstoke and Oxford: Macmillan.

——. 2002b. *An Entertainment for Angels. Electricity in the Enlightenment.* Cambridge: Icon.

Faraday, Michael. 1823a. 'On fluid chlorine', *Philosophical Transactions* 113: 160–64.

——. 1823b. 'On the condensation of several gases into liquids', *Philosophical Transactions* 113: 189–98.

Farey, John, jun. 1827. *A Treatise on the Steam Engine. Historical, Practical and Descriptive.* London: Longman.

Farish, William. 1821. *A Plan of a Course of Lectures on Arts and Manufactures, More Particularly Such as Relate to Chemistry.* Cambridge: J. Smith.

Farr, Grahame. 1970. *The Steamship Great Britain.* Bristol: Bristol Branch of the Historical Association.

Ferguson, Eugene S. 1961. 'John Ericsson and the age of caloric', *Bulletin of the United States National Museum* 228: 41–60.

——. 1962. 'Kinematics of mechanisms from the time of Watt', *Bulletin of the United States National Museum* 228: 185–230.

——. 1989. 'Technical journals and the history of technology'. In Stephen H. Cutcliffe and Robert C. Post, eds, *In Context. History and the History of Technology. Essays in Honor of Melvin Kranzberg*, pp. 53–70. Bethlehem, London and Toronto: Lehigh University Press.

——. 1992. *Engineering and the Mind's Eye.* Cambridge, MA and London: MIT Press.

Ferneyhough, Frank. 1980. *Liverpool & Manchester Railway 1830–1980.* London: Hale.

Field, Alexander J. 1994. 'French optical telegraphy, 1793–1855: hardware, software, administration', *Technology and Culture* 35: 315–47.

Finn, Bernard S. 1973. *Submarine Telegraphy. The Grand Victorian Technology.* London: Science Museum.

——, Robert Bud and Helmuth Trischler (eds). 2000. *Exposing Electronics.* Amsteldijk: Harwood Academic Publishers.

Fisch, Menachem and Simon Schaffer (eds). 1991. *William Whewell. A Composite Portrait.* Oxford and New York: Clarendon Press.

Fleming, Donald. 1952. 'Latent heat and the invention of the Watt engine', *Isis* 43: 3–5.

Forbes, James David. 1832. 'Report on meteorology', *Report of the British Association for the Advancement of Science* 2: 196–258.

——. 1844–50. 'Biographical notice of the late Sir John Robison, K.H., Sec. R. S. Ed.', *Proceedings of the Royal Society of Edinburgh* 2: 68–78.

Forrest, John. 1875. *Explorations in Australia*. London: Sampson Low.

Foucault, Michel. 1977. *Discipline and Punish. The Birth of the Prison*. Trans. Alan Sheridan. London: Allen Lane.

——. 1980. *Power/Knowledge: Selected Interviews and Other Writings 1972–1977*. Ed. C. Gordon. Brighton: Harvester.

Fox, Robert. 1986. 'Introduction'. In Robert Fox, ed., *Sadi Carnot. Reflexions on the Motive Power of Fire. A Critical Edition with the Surviving Scientific Manuscripts*, pp. 1–57. Manchester: Manchester University Press.

Fox, Stephen. 2003. *The Ocean Railway. Isambard Kingdom Brunel, Samuel Cunard, and the Revolutionary World of the Great Atlantic Steamships*. London: Harper Collins.

Frasca-Spada, Marina and Nick Jardine (eds). 2000. *Books and the Sciences in History*. Cambridge: Cambridge University Press.

Freeman, Michael. 1999. *Railways and the Victorian Imagination*. New Haven and London: Yale University Press.

——. 2001. 'Tracks to a new world: railway excavation and the extension of geological knowledge in mid-nineteenth-century Britain', *British Journal for the History of Science* 34: 51–65.

—— and Derek Aldcroft. 1985. *Atlas of British Railway History*. London: Croom Helm.

Friedel, Robert and Paul Israel. 1986. *Edison's Electric Light. Biography of an Invention*. New Brunswick: Rutgers University Press.

Friendly, Alfred. 1977. *Beaufort of the Admiralty. The Life of Sir Francis Beaufort 1774–1857*. London: Hutchinson.

Fyfe, Aileen. 2000. 'Reading children's books in eighteenth-century dissenting families', *Historical Journal* 43: 453–74.

Garnham, S.A. 1934. *The Submarine Cable. The Story of the Submarine Telegraph Cable, from its Invention down to Modern Times*. London: Sampson Low.

Gascoigne, John. 1994. *Joseph Banks and the English Enlightenment. Useful knowledge and Polite Culture*. Cambridge: Cambridge University Press.

Geisst, Charles R. 2000. *Monopolies in America. Empire Builders and their Enemies from Jay Gould to Bill Gates*. Oxford: Oxford University Press.

Gibbon, John M. 1935. *Steel of Empire. The Romantic History of the Canadian Pacific, the North West Passage of Today*. London: Rich and Cowan.

Gibbs-Smith, C.H. 1981. *The Great Exhibition of 1851*. Second edition. London: HMSO. First published 1950.

Gilbert, K.R. 1965. *The Portsmouth Block-Making Machinery. A Pioneering Enterprise in Mass Production*. HMSO: London.

Gillispie, C.C. (ed.). 1970–80. *Dictionary of Scientific Biography*. 16 vols. New York: Scribner.

Ginn, W.T. 1991. 'Philosophers and artisans: the changing relationship between men of science and instrument makers 1820–1860'. PhD diss., University of Kent at Canterbury.

Gittins, L. 1996–97. 'The alkali experiments of James Watt and James Keir, 1765–1780', *Transactions of the Newcomen Society* 68: 217–29.

Golinski, Jan. 1992. *Science as Public Culture. Chemistry and Enlightenment in Britain, 1760–1820*. Cambridge: Cambridge University Press.

——. 1998. *Making Natural Knowledge. Constructivism and the History of Science*. Cambridge: Cambridge University Press.

Gompertz, Lewis. 1824. *Moral Inquiries on the Situation of Man and of Brutes. On the Crime of Committing Cruelty on Brutes, and of Sacrificing them to the Purposes of Man; with Further Reflections. Observations on Mr. [Richard] Martin's Act, on the Vagrant Act, and on Tread Mills; to which are Added Some Improvements in Scapers, or Substitutes for Carriage Wheels; a New Plan of the Same, and Some Other Mechanical Subjects.* London: The author.

——. 1850. *Mechanical Inventions and Suggestions on Land and Water Locomotion, Tooth Machinery.* Second edition. London: W. Horsell.

Gooday, Graeme. 1991a. 'Nature in the laboratory: domestication and discipline with the microscope in Victorian life science', *British Journal for the History of Science* 24: 307–41.

——. 1991b. 'Teaching telegraphy and electrotechnics in the physics laboratory: William Ayrton and the creation of an academic space for electrical engineering, 1873–84', *History of Technology* 13: 73–111.

——. 1998. 'Re-writing the "book of blots": critical reflections on histories of technological "failure"', *History and Technology* 19: 265–91.

——. 2004. *The Morals of Measurement. Accuracy, Irony and Trust in Late Victorian Electrical Practice.* Cambridge: Cambridge University Press.

—— and Morris F. Low. 1998. 'Technology transfer and cultural exchange: western scientists and engineers encounter late Tukugawa and Meiji Japan', *Osiris* 13 (second series): 99–128.

Gooding, David. 1989. '"Magnetic curves" and the magnetic field: experimentation and representation in the history of a theory'. In David Gooding, Trevor Pinch and Simon Schaffer, eds, *The Uses of Experiment. Studies in the Natural Sciences*, pp. 183–223. Cambridge: Cambridge University Press.

Gordon, Lewis. 1849. *Railway Economy. An Exposition of the Advantages of Locomotion by Locomotive Carriages instead of the Present Expensive System of Steam Tugs.* Edinburgh: Sutherland and Knox and London: Simpkin, Marshall.

——. 1872–75. 'Obituary notice of Professor Rankine', *Proceedings of the Royal Society of Edinburgh* 8: 296–306.

Gorman, Mel. 1971. 'Sir William O'Shaughnessy, Lord Dalhousie, and the establishment of the telegraph system in India', *Technology and Culture* 12: 581–601.

Granovetter, Mark. 1985. 'Economic action and social structure: the problem of embeddedness', *American Journal of Sociology* 91: 481–510.

Grant, Kay. 1967. *Samuel Cunard. Pioneer of the Atlantic Steamship.* London, New York and Toronto: Abelard-Schuman.

Greeley, Horace. 1853. *Art and Industry as Represented in the Exhibition at the Crystal Palace, New York, 1853–4.* New York: Redfield.

Green, Edwin and Michael Moss. 1982. *A Business of National Importance. The Royal Mail Shipping Group 1902–1937.* London and New York: Methuen.

Grodinsky, Julius. 1957. *Jay Gould. His Business Career, 1867–1892.* Philadelphia: University of Pennsylvania Press.

Guagnini, Anna. 1993. 'Worlds apart: academic instruction and professional qualifications in the training of mechanical engineers'. In Robert Fox and Anna Guagnini, eds, *Education, Technology and Industrial Performance in Europe, 1850–1939*, pp. 16–41. Cambridge: Cambridge University Press and Editions de la Maison des Sciences de l'Homme.

Gunther, R.T. 1923–67. *Early Science in Oxford.* Oxford: Oxford Historical Society.

Hadfield, Charles. 1967. *Atmospheric Railways. A Victorian Venture in Silent Speed.* Newton Abbot: David & Charles.

Hadfield, Ellis Charles Raymond. 1970. *British Canals. An Illustrated History*. Fourth edition. Newton Abbot: David & Charles.

Hanes, W. Travis III. 1991. 'Railway politics and imperialism in Central Africa, 1889–1953'. In Clarence B. Davis and Kenneth J. Wilburn, eds, *Railway Imperialism*, pp. 41–69. New York, Westport and London: Greenwood Press.

Hankins, Thomas L. 1985. *Science and the Enlightenment*. Cambridge: Cambridge University Press.

Hann, James with Isaac Dodds. 1833. *Mechanics for Practical Men, Containing Explanations of the Principles of Mechanics; the Steam Engine ... Strength and Stress of Materials ... Hydrostatics, and Hydraulics. With a Short Dissertation on Rail-Roads*. Newcastle upon Tyne: Printed for the authors by MacKenzie and Dent.

Harding, Wyndham. 1845. *Railways. The Gauge Question. Evils of a Diversity of Gauge, and a Remedy*. London: John Weale.

Harley, J.B. 1988. 'Maps, knowledge, and power'. In Denis Cosgrove and Stephen Daniels, eds, *The Iconography of Landscape. Essays on the Symbolic Representation, Design and Use of Past Environments*, pp. 277–312. Cambridge: Cambridge University Press.

Harrington, Ralph. 2001. 'The railway accident: trains, trauma and technological crises in nineteenth-century Britain'. In Mark S. Micale and Paul Lerner, eds, *Traumatic Pasts. History, Psychiatry and Trauma in the Modern Age 1870–1930*, pp. 31–56. Cambridge: Cambridge University Press.

——. 2003. 'On the tracks of trauma: railway spine reconsidered', *Social History of Medicine* 16: 209–23.

Harris, J.R. 1967. 'Employment of steam power in the eighteenth century', *History* 52: 133–48.

Harris, Walter. 1744. *The Antient and Present State of the County of Down*. Dublin: Alexander Reilly.

Harrowby, Earl of. 1854. 'Presidential address', *Report of the British Association for the Advancement of Science* 24: lv–lxxi.

Harvey, W.W. 1973–74. 'Mr Symington's improved atmospheric engine', *Transactions of the Newcomen Society* 46: 27–32.

Haworth, Victoria. 1994. 'Inspiration and instigation: four great railway engineers'. In Denis Smith, ed., *Perceptions of Great Engineers. Fact and Fantasy*, pp. 55–83. London: Science Museum.

Hays, J.N. 1981. 'The rise and fall of Dionysius Lardner', *Annals of Science* 38: 527–42.

Headrick, Daniel R. 1981. *The Tools of Empire. Technology and European Imperialism in the Nineteenth Century*. Oxford: Oxford University Press.

——. 1988. *The Tentacles of Progress. Technology Transfer in the Age of Imperialism, 1850–1940*. Oxford: Oxford University Press.

——. 2000. *When Information Came of Age. Technologies of Knowledge in the Age of Reason and Revolution, 1700–1850*. New York: Oxford University Press.

Hearnshaw, F.J.C. 1929. *The Centenary History of King's College London 1828–1928*. London: Harrap.

Heilbron, J.L. 1979. *Electricity in the 17th and 18th Centuries. A Study of Early Modern Physics*. Berkeley and London: University of California Press.

——. 1990. 'The measure of Enlightenment'. In T. Frangsmyr, J.L. Heilbron and R.E. Rider, eds, *The Quantifying Spirit in the Eighteenth Century*, pp. 207–42. Berkeley, Los Angeles and Oxford: University of California Press.

—— James Bartholomew, J.A. Bennett, F.L. Holmes, Rachel Laudan and Giuliano Pancaldi (eds). 2002. *Oxford Companion to the History of Modern Science*. New York: Oxford University Press.

Helmholtz, Hermann von. 1873–81. *Popular Lectures on Scientific Subjects*. London: Longman.

Hempstead, Colin A. 1989. 'The early years of oceanic telegraphy: technology, science and politics', *Institution of Electrical Engineers Proceedings A* 136: 297–305.

——. 1991. 'An appraisal of Fleeming Jenkin (1833–1885)', *History of Technology* 13: 119–44.

——. 1995. 'Representations of transatlantic telegraphy', *Engineering Science and Education Journal* 4: S17–S25.

Henderson, Andrew. 1854. 'On ocean steamers and clipper ships', *Report of the British Association for the Advancement of Science* 24: 152–56.

Henry, John. 2002a. *The Scientific Revolution and the Origins of Modern Science*. Second edition. Basingstoke: Palgrave.

——. 2002b. *Knowledge is Power. How Magic, the Government and an Apocalyptic Vision Inspired Francis Bacon to Create Modern Science*. Cambridge: Icon Books.

Herschel, John F.W. 1830. *Preliminary Discourse on the Study of Natural Philosophy*. London: Longman, Rees, Orme, Brown & Green. Reprinted New York and London: Johnson Reprint Corporation, 1966.

Hessenbruch, Arne (ed.). 2000. *Reader's Guide to the History of Science*. London and Chicago: Fitzroy Dearborn.

Hilken, T.J.N. 1967. *Engineering at Cambridge University 1783–1965*. Cambridge: Cambridge University Press.

Hills, Richard L. 1989. *Power From Steam. A History of the Stationary Steam Engine*. Cambridge: Cambridge University Press.

——. 1996. 'James Watt, mechanical engineer', *History of Technology* 18: 59–79.

——. 1996–97. 'The origins of James Watt's perfect engine', *Transactions of the Newcomen Society* 68: 85–107.

——. 1998a. 'How James Watt invented the separate condenser (part I)', *Bulletin of the Scientific Instrument Society* 57: 26–29.

——. 1998b. 'How James Watt invented the separate condenser (part II)', *Bulletin of the Scientific Instrument Society* 58: 6–10.

——. 1999. 'James Watt's barometers', *Bulletin of the Scientific Instrument Society* 60: 5–10.

——. 2002a. *James Watt. Volume 1: His Time in Scotland, 1736–1774*. Ashbourne: Landmark Publishing.

——. 2002b. 'Richard Roberts'. In John Cantrell and Gillian Cookson, eds, *Henry Maudslay & the Pioneers of the Machine Age*, pp. 54–73. Stroud and Charleston: Tempus.

—— and A.J. Pacey. 1972. 'The measurement of power in early steam-driven textile mills', *Technology and Culture* 13: 25–43.

Hilton, Boyd. 1991. *The Age of Atonement. The Influence of Evangelicalism on Social and Economic Thought 1785–1865*. Oxford: Clarendon Press. First published 1988.

Hincks, John. 1832. *Sermons and Occasional Services, Selected from the Papers of the Late Rev. John Hincks, with a Memoir of the Author, by John H. Thom*. Ed. J.H. Thom. London: Longmans.

Hobsbawm, Eric. 1962. *The Age of Revolution. Europe, 1789–1848*. London: Weidenfeld and Nicolson.

——. 1969. *Industry and Empire*. Harmondsworth: Penguin.

Hodder, Edwin. 1890. *Sir George Burns, Bart. His Time and Friends*. London: Hodder & Stoughton.

Holt, Alfred. 1877–78. 'Review of the progress of steam shipping during the last quarter of a century [and discussion]', *Minutes of Proceedings of the Institution of Civil Engineers* 51: 2–135.

——. 1911. *Fragmentary Autobiography of Alfred Holt*. Privately printed.

Houghton, Walter. 1957. *The Victorian Frame of Mind, 1830–1870*. New Haven and London: Yale University Press.

Howard, Bridget. 2002. *Mr Lean and the Engine Reporters*. Redruth: Trevithick Society.

Howarth, O.J.R. 1931. *The British Association for the Advancement of Science. A Retrospect, 1831–1931*. London: BAAS.

Howie, Will and Mike Chrimes (eds). 1987. *Thames Tunnel to Channel Tunnel: 150 Years of Civil Engineering. Selected Papers from the Journal of the Institution of Civil Engineers Published to Celebrate its 150th Anniversary*. London: Thomas Telford.

Howse, D. 1980. *Greenwich Time and the Discovery of the Longitude*. Oxford and New York: Oxford University Press.

Hubbard, Geoffrey. 1965. *Cooke and Wheatstone and the Invention of the Electric Telegraph*. London: Routledge & Kegan Paul.

Hughes, Thomas P. 1983. *Networks of Power. Electrification in Western Society, 1880–1930*. Baltimore: Johns Hopkins University Press.

——. 1987. 'The evolution of large technological systems'. In Wiebe E. Bijker, Thomas P. Hughes and Trevor Pinch, eds, *The Social Construction of Technological Systems. New Directions in the Sociology and History of Technology*, pp. 51–82. Cambridge, MA and London: MIT Press.

——. 1989. *American Genesis. A Century of Invention and Technological Enthusiasm, 1870–1970*. New York: Viking.

——. 2004. *Human-Built World. How to Think about Technology and Culture*. Chicago and London: University of Chicago Press.

Hume, John and Michael Moss. 1975. *Clyde Shipbuilding from Old Photographs*. London and Sydney: B. T. Batsford.

Hunt, Bruce. 1991. 'Michael Faraday, cable telegraphy and the rise of field theory', *History of Technology* 13: 1–19.

——. 1994. 'The ohm is where the art is: British telegraph engineers and the development of electrical standards', *Osiris* 9: 48–63.

——. 1996. 'Scientists, engineers and Wildman Whitehouse: measurement and credibility in early cable telegraphy', *British Journal for the History of Science* 29: 155–69.

——. 1997. 'Doing science in a global empire: cable telegraphy and electrical physics in Victorian Britain'. In Bernard Lightman, ed., *Victorian Science in Context*, pp. 312–33. Chicago: University of Chicago Press.

——. 1998. 'Insulation for an empire: gutta-percha and the development of electrical measurement in Victorian Britain'. In F.A.J.L. James, ed., *Semaphores to Short Waves*, pp. 85–104. RSA: London.

Hunt, Lynn (ed.). 1989. *The New Cultural History*. Berkeley: University of California Press.

Hutchison, Keith. 1981. 'W.J.M. Rankine and the rise of thermodynamics', *British Journal for the History of Science* 14: 1–26.

Hyde, Francis E. 1956. *Blue Funnel. A History of Alfred Holt and Company of Liverpool from 1865 to 1914*. Liverpool: Liverpool University Press.

——. 1975. *Cunard and the North Atlantic 1840–1973. A History of Shipping and Financial Management*. London and Basingstoke: Macmillan.

Iliffe, Rob. 1995. 'Material doubts: Hooke, artisan culture and the exchange of information in 1670s' London', *British Journal for the History of Science* 28: 285–318.

Ince, Laurence. 2002. 'Maudslay, Sons & Field, 1831–1904'. In John Cantrell and Gillian Cookson, eds, *Henry Maudslay & the Pioneers of the Machine Age*, pp. 166–84. Stroud and Charleston: Tempus.

Inkster, Ian and Jack Morrell. 1983. *Metropolis and Province. Science in British Culture 1780–1850*. London: Hutchinson.

Israel, Paul. 1989. *From the Machine Shop to the Industrial Laboratory. Telegraphy and the Changing Context of American Invention, 1830–1920*. New Brunswick: Rutgers University Press.

———. 1998. *Edison. A Life of Invention*. New York: John Wiley.

——— *et al.* (eds). ongoing. *The Papers of Thomas A. Edison*. Baltimore and London: Johns Hopkins University Press.

Jacob, James R. and Margaret C. Jacob. 1980. 'The Anglican origins of modern science: the metaphysical foundations of the Whig constitution', *Isis* 71: 251–67.

Jacob, Margaret C. 1976. *The Newtonians and the English Revolution, 1689–1720*. Ithaca: Cornell University Press.

Jacobi, M.H. 1836–37. 'On the application of electro-magnetism to the moving of Machines', *Annals of Electricity* 1: 408–15, 419–44.

———. 1840. 'On the principles of electro-magnetic machines', *Report of the British Association for the Advancement of Science* 10: 18–24.

James, F.A.J.L. (ed.). 1991. *The Correspondence of Michael Faraday*. London: Institution of Electrical Engineers.

———. 1997. 'Faraday in the pits, Faraday at sea: the role of the Royal Institution in changing the practice of science and technology in nineteenth-century Britain', *Proceedings of the Royal Institution* 68: 277–301.

———. 1998. *Semaphores to Short Waves*. RSA: London.

———. 1998–99. '"The civil engineer's talent": Michael Faraday, science, engineering and the English lighthouse service, 1836–1865', *Transactions of the Newcomen Society* 70: 153–60.

———. 2000. 'Michael Faraday and lighthouses'. In I. Inkster, ed., *The Golden Age. Essays in British Social and Economic History, 1850–1870*, pp. 92–104. Aldershot: Ashgate.

Janković, Vladimir. 2000. *Reading the Skies. A Cultural History of English Weather, 1650–1820*. Manchester: Manchester University Press.

Jardine, Nick, J.A. Secord and E.C. Spary (eds). 1996. *Cultures of Natural History*. Cambridge: Cambridge University Press.

Jarvis, Adrian. 1994. 'The story of the story of Robert Stephenson'. In Denis Smith, ed., *Perceptions of Great Engineers. Fact and Fantasy*, pp. 35–45. London: Science Museum.

———. 1997. *Samuel Smiles and the Construction of Victorian Values*. Thrupp: Sutton.

Jenkin, H.C. Fleeming. 1866. 'Submarine telegraphy', *North British Review* 45: 459–505.

Jepsen, Thomas C. 1995. 'Women telegraphers in the railroad depot', *Railroad History* 173: 142–54.

———. 1996. 'Women telegraph operators on the Western Frontier', *Journal of the West* 35: 72–80.

Jevons, W. Stanley. 1866 [1865]. *The Coal Question. An Inquiry Concerning the Progress of the Nation, and the Probable Exhaustion of Our Coal Mines*. Second edition. London: Macmillan. First edition 1865.

Johns, Adrian. 1998. *The Nature of the Book. Print and Knowledge in the Making*. Chicago and London: University of Chicago Press.

Johnson, Fletcher and Allen Fowler. 1986. 'The Vanderbilts: the first family of Eastern rail tycoons'. In B. Yenne, ed., *The History of North American Railroads*, pp. 81–85. London: Bison Books.

Joint Committee. 1861. *Report of the Joint Committee appointed by the Lords of the Committee of Privy Council for Trade and the Atlantic Telegraph Company to Inquire into the*

Construction of Submarine Telegraph Cables. Together with the Minutes of Evidence and Appendix. Presented to Both Houses of Parliament by Command of Her Majesty. London: Eyre and Spottiswoode.

Jones, Jean, Hugh S. Torrens and Eric Robinson. 1994–95. 'The correspondence between James Hutton (1726–1797) and James Watt (1736–1819), with two letters from Hutton to George Clerk-Maxwell (1715–1784)', *Annals of Science* 51: 637–53 and 52: 357–82.

Jones, Peter M. 1999. 'Living the Enlightenment and the French Revolution: James Watt, Matthew Boulton, and their sons', *Historical Journal* 42: 157–82.

Jones, R.V. 1969. 'The 'plain story' of James Watt: the Wilkins Lecture 1969', *Notes and Records of the Royal Society of London* 24: 194–220.

Jordanova, Ludmilla. 2000a. *Defining Features. Scientific and Medical Portraits 1660–2000.* London: Reaktion Books in association with the National Portrait Gallery.

——. 2000b. *History in Practice.* London: Arnold.

Joule, J.P. 1887. *The Scientific Papers of James Prescott Joule.* 2 vols. London: The Physical Society of London.

Kielbowicz, Richard B. 1987. 'News gathering by mail in the age of the telegraph: adopting a new technology', *Technology and Culture* 28: 26–41.

——. 1994. 'The telegraph, censorship, and politics at the outset of the Civil War', *Civil War History* 40: 95–118.

Kieve, Jeffrey L. 1973. *The Electric Telegraph. A Social and Economic History.* Newton Abbot: David & Charles.

King-Hele, Desmond (ed.). 2003. *Charles Darwin's The life of Erasmus Darwin.* Cambridge: Cambridge University Press.

Kingery, W. David (ed.). 1996. *Learning from Things. Method and Theory of Material Culture Studies.* Washington: Smithsonian Institution Press.

Kingsley, Charles. 1851. *Yeast. A Problem.* London: King.

Kirby, Richard Shelton, Sidney Withington, Arthur Burr Darling and Frederick Gridley Kilgour. 1956. *Engineering in History.* New York, Toronto and London: McGraw Hill.

Kirby, Maurice W. 1993. *The Origins of Railway Enterprise. The Stockton and Darlington Railway, 1821–1863.* Cambridge: Cambridge University Press.

Kirkaldy, Adam W. 1914. *British Shipping. Its History, Organisation and Importance.* London: Kegan Paul, Trench, Trubner. Reprinted Newton Abbot: David & Charles, 1970.

Kirkland, Edward C. 1965. *Charles Francis Adams, Jr. 1835–1915. The Patrician at Bay.* Cambridge, MA: Harvard University Press.

Kirsch, David A. 2000. *The Electric Vehicle and the Burden of History.* New Brunswick and London: Rutgers.

Knight, David. 1986. *The Age of Science. The Scientific World-View in the Nineteenth Century.* Oxford: Blackwell.

——. 1992. *Humphry Davy. Science and Power.* Oxford: Blackwell.

Kohler, Peter C. 2004. 'Empresses of the sea', *Ships Monthly* 39: 22–24.

Kuhn, Thomas S. 1961. 'Sadi Carnot and the Cagnard engine', *Isis* 52: 567–74.

Lampe, David. 1963. *The Tunnel. The Story of the World's First Tunnel Under a Navigable River Dug Beneath the Thames 1824–42.* London: Harrap.

Landes, David S. 1983. *Revolution in Time. Clocks and the Making of the Modern World.* Cambridge, MA: Harvard University Press.

Lardner, Dionysius. 1836. *The Steam Engine Familiarly Explained and Illustrated, with its Applications to Navigation and Railways.* Sixth edition. London: Taylor and Walton.

Law, R.J. 1969. *James Watt and the Separate Condenser.* London: HMSO.

Lawrence, Christopher and Steven Shapin (eds). 1998. *Science Incarnate. Historical Embodiments of Natural Knowledge.* Chicago: University of Chicago Press.

Layton, Edwin T., jun. 1971. 'Mirror-image twins: the communities of science and technology in 19th-century America', *Technology and Culture* 12: 562–80.

——. 1974. 'Technology as knowledge', *Technology and Culture* 15: 31–41.

Le Fleming, H.M. 1961. *Ships of the Blue Funnel Line.* Southampton: Adlard Coles.

Lefebvre, Henri. 1991. *The Production of Space.* Trans. D. Nicholson-Smith. Oxford: Blackwell. First published 1974.

Lightman, Bernard (ed.). 1997. *Victorian Science in Context.* Chicago: University of Chicago Press.

Lindqvist, Svante. 1984. *Technology on Trial. The Introduction of Steam Power Technology into Sweden, 1715–1736.* Stockholm: Almqvist.

——, Marika Hedin and Ulf Larsson (eds). 2000. *Museums of Modern Science.* Canton: Science History Publications.

Lindsay, W.S. 1874–76. *History of Merchant Shipping and Ancient Commerce.* 4 vols. London: Sampson Low, Marston, Low and Searle. Reprinted New York: AMS Press, 1965.

Lingwood, J.E. 1977. 'The steam conquistadores: a history of the Pacific Steam Navigation Company', *Sea Breezes* 51: 97–115.

Livingstone, David N. 2003. *Putting Science in its Place. Geographies of Scientific Knowledge.* Chicago: University of Chicago Press.

Lubar, Steven and Brooke Hindle. 1986. *Engines of Change. The American Industrial Revolution, 1790–1860.* Washington, DC and London: Smithsonian Institution.

—— and W. David Kingery (eds). 1995. *History from Things. Essays on Material Culture.* Washington: Smithsonian Books.

Lundgren, Anders and Bernadette Bensaude-Vincent (eds). 2000. *Communicating Chemistry. Textbooks and their Audiences, 1789–1939.* Canton, MA: Science History Publications.

Lyall, Heather. 1991. *Vanishing Glasgow. Through the Lens of George Washington Wilson, T & R Annan and Sons, William Graham, Oscar Marzaroli and Others.* Aberdeen: AUL Publishing.

Lynch, A.C. 1985. 'History of the electrical units and early standards', *Institution of Electrical Engineers Proceedings A* 132: 564–73.

MacDermot, E.T. 1964. *History of the Great Western Railway.* Revised C.R. Clinker. London: Ian Allan. First published London: Great Western Railway, 1927–31.

Macdonald, Sharon. 2002. *Behind the Scenes at the Science Museum.* Oxford and New York: Berg.

MacKenzie, Donald and Judy Wajcman (eds). 1999. *The Social Shaping of Technology.* Second edition. Buckingham: Open University Press. First edition 1985.

Macleod, Donald. 1876. *Memoir of Norman Macleod, D. D.* 2 vols. London: Daldy, Isbister.

MacLeod, Christine. 1988. *Inventing the Industrial Revolution. The English Patent System, 1660–1800.* Cambridge: Cambridge University Press.

——. 1998. 'James Watt, heroic invention and the idea of the industrial revolution'. In Maxine Berg and Kristine Bruland, eds, *Technological Revolutions in Europe. Historical Perspectives*, pp. 96–115. Cheltenham: Edward Elgar.

Marsden, Ben. 1992a. 'Engineering science in Glasgow: economy, efficiency and measurement as prime movers in the differentiation of an academic discipline', *British Journal for the History of Science* 25: 319–46.

——. 1992b. 'Engineering science in Glasgow: W.J.M. Rankine and the motive power of air'. PhD diss., University of Kent at Canterbury.

——. 1996. 'Fighting cruelty: Lewis Gompertz and the Animals' Friend Society for the Prevention of Cruelty to Animals' (paper delivered at University of Manchester).

——. 1998a. 'Blowing hot and cold: reports and retorts on the status of the air-engine as success or failure, 1830–1855', *History of Science* 36: 373–420.

——. 1998b. ' "A most important trespass": Lewis Gordon and the Glasgow chair of civil engineering and mechanics 1840–1855'. In Crosbie Smith and Jon Agar, eds, *Making Space for Science. Territorial Themes in the Shaping of Knowledge*, pp. 87–117. Basingstoke: Macmillan.

——. 2000. 'The professional and the professorial: engineering under cover in the early Victorian universities' (paper delivered at St Louis 'Three Societies' conference).

——. 2001. 'Re-reading Isambard Kingdom Brunel: a case study in cultures of reading and writing in nineteenth-century British engineering' (paper delivered at Brunel University).

——. 2002. *Watt's Perfect Engine. Steam and the Age of Invention*. Cambridge: Icon.

——. 2004a. 'James Hann', *Dictionary of National Biography*. Oxford: Oxford University Press (forthcoming).

——. 2004b. ' "The progeny of these two Fellows": Robert Willis, William Whewell and the sciences of mechanism, mechanics and machinery in early Victorian Britain', *British Journal for the History of Science* (forthcoming).

Marshall, F.C. 1872. 'On the progress and development of the marine engine', *Proceedings. Institution of Mechanical Engineers [1872]*: 449–509.

McCloy, Shelby T. 1952. *French Inventions of the Eighteenth Century*. Lexington: University of Kentucky Press.

McDowell, R.B. and Webb, D.A. 1982. *Trinity College Dublin, 1592–1952. An Academic History*. Cambridge: Cambridge University Press.

McNeil, Ian (ed.). 1990. *An Encyclopedia of the History of Technology*. London and New York: Routledge.

Miller, David P. 1986. 'The revival of the physical sciences in Britain', *Osiris* (new series) 2: 107–34.

——. 2000. ' "Puffing Jamie": the commercial and ideological importance of being a "philosopher" in the case of the reputation of James Watt', *History of Science* 38: 1–24.

Miller, Hugh. 1842. 'The two conflicts', *The Witness*, 25 May 1842.

Miller, R. Kalley. 1870. 'The proposed chair of natural philosophy', *Cambridge University Reporter* 1: 118–19.

Mogg, Edward. 1841. *Mogg's Great Western Railway and Windsor, Bath, and Bristol Guide*. London: Edward Mogg.

Moore, James R. 1979. *The Post-Darwinian Controversies. A Study of the Protestant Struggle to Come to Terms with Darwin in Great Britain and America, 1870–1900*. Cambridge and New York: Cambridge University Press.

Morrell, Jack. 1969a. 'Practical chemistry in the University of Edinburgh, 1799–1843', *Ambix* 16: 66–80.

——. 1969b. 'Thomas Thomson: professor of chemistry and university reformer', *British Journal for the History of Science* 4: 245–65.

——. 1971. 'Individualism and the structure of British science in 1830', *Historical Studies in the Physical Sciences* 3: 183–204.

——. 1997. *Science at Oxford 1914–1939. Transforming an Arts University*. Oxford: Clarendon Press.

——. 2004. *John Phillips and the Business of Victorian Science*. Aldershot: Ashgate.
—— and Arnold Thackray. 1981. *Gentlemen of Science. Early Years of the British Association for the Advancement of Science*. Oxford: Clarendon Press.
—— (eds). 1984. *Gentlemen of Science. Early Correspondence of the British Association for the Advancement of Science*. London: Royal Historical Society.
Morus, Iwan Rhys. 1991a. 'Correlation and control: William Robert Grove and the construction of a new philosophy of scientific reform', *Studies in History and Philosophy of Science* 22: 589–621.
——. 1991b. 'Telegraphy and the technology of display: the electricians and Samuel Morse', *History of Technology* 13: 20–40.
——. 1992. 'Different experimental lives: Michael Faraday and William Sturgeon', *History of Science* 30: 1–28.
——. 1993. 'Currents from the underworld: electricity and the technology of display in early Victorian England', *Isis* 84: 50–69.
——. 1996a. 'Manufacturing nature: science, technology and Victorian consumer culture', *British Journal for the History of Science* 29: 403–34.
——. 1996b. 'The electric Ariel: telegraphy and commercial culture in early Victorian England'. *Victorian Studies* 39: 339–78.
——. 1998. *Frankenstein's Children. Electricity, Exhibition, and Experiment in Early Nineteenth-Century London*. Princeton: Princeton University Press.
——. 2000. ' "The nervous system of Britain": space, time and the electric telegraph in the Victorian age', *British Journal for the History of Science* 33: 455–75.
——, Simon Schaffer and J.A. Secord. 1992. 'Scientific London'. In Celina Fox, ed., *London. World City, 1800–1840*, pp. 129–42. New Haven: Yale University Press.
Moseley, Henry. 1843. *The Mechanical Principles of Engineering and Architecture*. London: Longman.
Moss, Michael and John R. Hume. 1986. *Shipbuilders to the World. 125 Years of Harland and Wolff, Belfast 1861–1986*. Belfast: Blackstaff Press.
Moyal, Ann. 1987. 'The history of telecommunication in Australia: aspects of the technological experience, 1854–1930'. In Nathan Reingold and Marc Rothenberg, eds, *Scientific Colonialism. A Cross-Cultural Comparison*, pp. 35–54. Washington and London: Smithsonian Institution Press.
Muirhead, J.P. 1846. *Correspondence of the Late James Watt on his Discovery of the Theory of the Composition of Water*. London: John Murray.
——. 1854. *The Origins and Progress of the Mechanical Inventions of James Watt*. 3 vols. London: John Murray.
——. 1858. *Life of James Watt*. London: John Murray.
Munro, J. Forbes. 2003. *Maritime Enterprise and Empire. Sir William Mackinnon and his Business Network, 1823–1893*. Woodbridge: Boydell Press.
Murchison, Roderick. 1838. 'Presidential address', *Report of the British Association for the Advancement of Science* 8: xxxi–xliv.
—— and Edward Sabine. 1840. 'Address', *Report of the British Association for the Advancement of Science* 10: xxxv–xlviii.
Musk, George. 1989. *Canadian Pacific. The Story of the Famous Shipping Line*. Newton Abbot: David & Charles. First published 1981.
Napier, James. 1904. *Life of Robert Napier of West Shandon*. Edinburgh and London: Blackwood.
Nasmyth, James. 1897. *James Nasmyth, Engineer. An Autobiography*. Ed. Samuel Smiles. London: Murray.
Neeley, Kathryn A. 2001. *Mary Somerville. Science, Illumination, and the Female Mind*. Cambridge: Cambridge University Press.

Nenadic, Stana. 1994. 'Middle-rank consumers and domestic culture in Edinburgh and Glasgow 1720–1840', *Past & Present* 145: 122–56.

Nickles, David Paul. 1999. 'Telegraph diplomats: the United States' relations with France in 1848 and 1870', *Technology and Culture* 40: 1–25.

Noakes, Richard J. 1999. 'Telegraphy is an occult art: Cromwell Fleetwood Varley and the diffusion of electricity to the other world', *British Journal for the History of Science* 32: 421–59.

Noble, Celia. 1938. *The Brunels. Father and Son*. London: Cobden-Sanderson.

Ord-Hume, Arthur W.J.G. 1977. *Perpetual Motion. The History of an Obsession*. London: Allen and Unwin.

Osborne, Brian D. 1995. *The Ingenious Mr Bell. A Life of Henry Bell (1767–1830). Pioneer of Steam Navigation*. Glendaruel: Argyll Publishing.

Ottley, George. 1983. *A Bibliography of British Railway History*. Second edition. London: HMSO.

Outram, Dorinda. 1995. *The Enlightenment*. Cambridge and New York: Cambridge University Press.

Pacey, Arnold. 1992. *The Maze of Ingenuity. Ideas and Idealism in the Development of Technology*. Second edition. Cambridge, MA and London: MIT Press.

——. 1999. *Meaning in Technology*. Cambridge, MA and London: MIT Press.

Perkin, Harold. 1971. *The Age of the Railway*. Newton Abbot: David & Charles.

Perkins, Veronica Davis. 1998. 'Whose line is it anyway? Women, opportunity and change, 1830–1920'. In F.A.J.L. James, ed., *Semaphores to Short Waves*, pp. 60–70. RSA: London.

Petroski, Henry. 1993. *The Evolution of Useful Things*. London: Pavilion.

Pickstone, John. 2000. *Ways of Knowing. A New History of Science, Technology and Medicine*. Manchester: Manchester University Press.

Pinch, Trevor J. and Wiebe E. Bijker. 1987. 'The social construction of facts and artifacts: or how the sociology of science and the sociology of technology might benefit each other'. In Wiebe E. Bijker, Thomas P. Hughes and Trevor Pinch, eds, *The Social Construction of Technological Systems. New Directions in the Sociology and History of Technology*, pp. 17–50. Cambridge, MA and London: MIT Press.

Pole, William (ed.). 1877. *The Life of Sir William Fairbairn, Bart. Partly Written by Himself*. London: Longmans, Green. Reprinted ed. A.E. Musson. Newton Abbot: David & Charles, 1970.

——. 1888. *The Life of Sir William Siemens*. London: John Murray.

Pollard, Sidney and Paul Robertson. 1979. *The British Shipbuilding Industry, 1870–1914*. Cambridge, MA and London: Harvard University Press.

Pool, Ithiel de Sola (ed.). 1977. *The Social Impact of the Telephone*. Cambridge, MA: MIT Press.

——. 1983. *Forecasting the Telephone. A Retrospective Technology Assessment*. Norward: Ablex Publishing.

Porter, Dale H. 1998. *The Life and Times of Sir Goldsworthy Gurney, Gentleman Scientist and Inventor, 1793–1875*. London: Associated University Presses.

Porter, Roy S. 1977. *The Making of Geology. Earth Science in Britain, 1660–1815*. Cambridge: Cambridge University Press.

Post, Robert C. 1974. 'Electro-magnetism as a motive power: Robert Davidson's *Galvani* of 1842', *Railroad History* 130: 5–22.

——. 1983. 'Reflections of American Science and Technology at the New York Crystal Palace Exhibition of 1853', *Journal of American Studies* 17: 337–56.

Potts, Alex. 1980. *Sir Francis Chantrey 1781–1841. Sculptor of the Great.* London: National Portrait Gallery.

Preece, Clive. 1982. 'The Durham engineer students of 1838', *Transactions of the Architectural and Archaeological Society of Durham and Northumberland* 6: 71–74.

Prosser, Thomas. 1854. 'On unchanged steam', *Report of the British Association for the Advancement of Science* 24: 159.

Pugsley, Sir Alfred (ed.). 1976. *The Works of Isambard Kingdom Brunel. An Engineering Appreciation.* London and Bristol: Institution of Civil Engineers and University of Bristol.

Pumfrey, Stephen. 1991. 'Ideas above his station: a social study of Hooke's curatorship of experiments', *History of Science* 19: 1–44.

——. 2002. *Latitude and the Magnetic Earth. The True Story of Queen Elizabeth's Most Distinguished Man of Science.* Cambridge: Icon.

Purbrick, Louise (ed.). 2001. *The Great Exhibition of 1851. New Interdisciplinary Essays.* Manchester: Manchester University Press.

Rankine, W.J. Macquorn. 1854. 'On the means of realizing the advantages of the air-engine', *Report of the British Association for the Advancement of Science* 24: 159–60.

——. 1855. 'On the means of realizing the advantages of the air-engine', *Edinburgh New Philosophical Journal* 1 (new series): 1–32.

——. 1856. *Introductory Lecture on the Harmony of Theory and Practice in Mechanics, Delivered to the Class of Civil Engineering and Mechanics in the University of Glasgow.* London and Glasgow: Richard Griffin.

——. 1857. *Introductory Lecture on the Science of the Engineer, Delivered to the Class of Civil Engineering and Mechanics in the University of Glasgow.* London and Glasgow: Richard Griffin.

——. 1864. 'On units of measure', *Report of the British Association for the Advancement of Science* 34: 188.

——. 1871. *A Memoir of John Elder. Engineer and Shipbuilder.* Edinburgh and London: Blackwood.

——. 1876. *A Manual of the Steam Engine and Other Prime Movers.* Eighth edition. London: Charles Griffin.

Rastrick, John. 1829. *Liverpool and Manchester Railway. Report to the Directors on the Comparative Merits of Loco-Motive and Fixed Engines, as a Moving Power.* Second edition. Birmingham: Wrightson.

Read, Donald. 1992. *The Power of News. The History of Reuters, 1849–1989.* Oxford: Oxford University Press.

Reader, W.J. 1987. *A History of the Institution of Electrical Engineers 1871–1971.* London: Peter Peregrinus.

——. 1991. ' "The engineer must be a scientific man": the origins of the Society of Telegraph Engineers', *History of Technology* 13: 112–18.

Redondi, Pietro. 1980. *L'accueil des idées de Sadi Carnot et la technologie française de 1820 à 1860.* Paris: Vrin.

Reingold, Nathan. 1975. 'Edward Sabine'. In C.C. Gillispie, ed., *Dictionary of Scientific Biography*, pp. 49–53. 16 vols. New York: Charles Scribner.

Reingold, Nathan and Marc Rothenberg (eds). 1987. *Scientific Colonialism. A Cross-Cultural Comparison.* Washington and London: Smithsonian Institution Press.

Revel, Jacques. 1991. 'Knowledge of the territory', *Science in Context* 4: 133–61.

Richards, Thomas. 1993. *The Imperial Archive. Knowledge and the Fantasy of Empire.* London: Verso.

Richards, William Carey. 1853. *A Day in the New York Crystal Palace, and How to Make the Most of It*. New York: G.P. Putnam.

Riddell, John F. 1979. *Clyde Navigation. A History of the Development and Deepening of the River Clyde*. Edinburgh: John Donald.

Ritchie, G.S. 1995. *The Admiralty Chart. British Naval Hydrography in the Nineteenth Century*. Durham: 1995. First edition. London: Hollis & Carter, 1967.

Ritvo, Harriet. 1987. *The Animal Estate. The English and Other Creatures in the Victorian Age*. Cambridge, MA and London: Harvard University Press, 1987.

Robb, Johstone Fraser. 1993. 'Scotts of Greenock. Shipbuilders and engineers 1820–1920. A family enterprise'. PhD diss., University of Glasgow.

Robinson, Eric. 1954. 'Training captains of industry: the education of Matthew Robinson Boulton (1770–1842) and the younger James Watt (1769–1848)', *Annals of Science* 10: 301–13.

——. 1956. 'James Watt and the tea kettle: a myth justified', *History Today* 6: 261–65.

——. 1964. 'Matthew Boulton and the art of parliamentary lobbying', *Historical Journal* 7: 209–29.

——. 1969. 'James Watt, engineer and man of science'. *Notes and Records of the Royal Society of London* 24: 221–32.

——. 1972. 'James Watt and the law of patents', *Technology and Culture* 13: 115–39.

—— and A.E. Musson. 1969. *James Watt and the Steam Revolution. A Documentary History*. London: Adams and Dart.

—— and Douglas McKie (eds). 1970. *Partners in Science. Letters of James Watt and Joseph Black*. London: Constable.

Robison, John. 1797. 'Steam' and 'Steam engine'. In *Encyclopaedia Britannica*, vol. xvii, pp. 733–43 (Steam) and 743–72 (Steam engine). Third edition. 18 vols. Edinburgh: Bell and Macfarquhar.

——. 1822. *A System of Mechanical Philosophy*. Ed. David Brewster. 4 vols. Edinburgh: John Murray.

——. 1839. 'Notes on Daguerre's photography', *Edinburgh New Philosophical Journal* 27: 155–57.

Rolt, L.T.C. 1957. *Isambard Kingdom Brunel*. London: Longmans.

——. 1970. *Victorian Engineering*. London: Allen Lane.

—— and John Scott Allen. 1977. *The Steam Engine of Thomas Newcomen*. Hartington: Moorland and New York: Science History Publications.

Ronalds, Francis. 1823. *Descriptions of an Electrical Telegraph, and of Some Other Electric Apparatus*. London: R. Hunter.

Rosenberg, Nathan (ed.). 1969. *The American System of Manufactures. The Report of the Committee on the Machinery of the United States 1855, and the Special Reports of George Wallis and Joseph Whitworth 1854*. Edinburgh: Edinburgh University Press.

Ross, Ian S. 1995. *The Life of Adam Smith*. Oxford: Clarendon Press.

Ross, Sydney. 1962. 'Scientist: the story of a word', *Annals of Science* 16: 65–85.

Rudwick, Martin J.S. 1976. 'The emergence of a visual language for geological science, 1760–1840', *History of Science* 14: 148–95.

——. 1985. *The Great Devonian Controversy. The Shaping of Scientific Knowledge among Gentlemanly Specialists*. Chicago: University of Chicago Press.

——. 1992. *Scenes from Deep Time. Early Pictorial Representations of the Prehistoric World*. Chicago: University of Chicago Press.

Ruskin, John. 1903–08. *The Works of John Ruskin*. Eds E.T. Cook and A. Wedderburn. 39 vols. London: George Allen.

Russell, John Scott. 1854. 'On the progress of naval architecture and steam naviga-tion, including a notice of a large ship of the Eastern Steam Navigation Company', *Report of the British Association for the Advancement of Science* 24: 160–61.

Russell, William Howard. 1865. *The Atlantic Telegraph*. Reprinted Newton Abbot: David & Charles, 1972.

Sanderson, Michael. 1972. *The Universities and British Industry, 1850–1970*. London: Routledge.

Satow, Michael and Ray Desmond. 1980. *Railways of the Raj*. London: Scolar.

Schaffer, Simon. 1983. 'Natural philosophy and public spectacle in the eighteenth century', *History of Science* 21: 1–43.

———. 1988. 'Astronomers mark time: discipline and the personal equation', *Science in Context* 2: 115–45.

———. 1991. 'The history and geography of the intellectual world: Whewell's politics of language'. In Menachem Fisch and Simon Schaffer, eds, *William Whewell. A Composite Portrait*, pp. 201–31. Oxford and New York: Clarendon Press.

———. 1992. 'Late Victorian metrology and its instrumentation: a manufactory of ohms'. In Robert Bud and S.E. Cozzens, eds, *Invisible Connections. Instruments, Institutions, and Science*, pp. 23–56. Bellingham, Washington: Spie Optical Engineering Press.

———. 1994. 'Babbage's intelligence: calculating engines and the factory system', *Critical Inquiry* 21: 203–27.

———. 1995. 'The show that never ends: perpetual motion in the early eighteenth century', *British Journal for the History of Science* 28: 157–90.

———. 1996. 'Babbage's dancer and the impressarios of mechanism'. In Francis Spufford and Jenny Uglow, eds, *Cultural Babbage. Technology, Time and Invention*, pp. 53–80. London and Boston: Faber & Faber.

Schivelbusch, Wolfgang. 1980. *The Railway Journey. Trains and Travel in the Nineteenth Century*. Trans. Anselm Hollo. Oxford: Blackwell.

———. 1995. *Disenchanted Night. The Industrialization of Light in the Nineteenth Century*. Berkeley: University of California Press.

Schofield, Robert E. 1963. *The Lunar Society of Birmingham*. Oxford: Clarendon Press.

———. 1966. 'The Lunar Society of Birmingham: a bicentenary appraisal', *Notes and Records of the Royal Society of London* 21: 144–61.

Scoresby, William. 1854a. 'On the loss of the "Tayleur", and the changes in the action of compasses in iron ships', *Report of the British Association for the Advancement of Science* 33: 49–53.

———. 1854b. 'An inquiry into the principles and measures on which safety in the navigation of iron ships may be reasonably looked for', *Report of the British Associa-tion for the Advancement of Science* 33: 53–54.

———. 1854c. 'An inquiry into the principles and measures on which safety in the navigation of iron ships may be reasonably looked for', *Report of the British Associa-tion for the Advancement of Science* 33: 161–62.

———. 1859. *Journal of a Voyage to Australia and round the World for Magnetical Research*. Ed. A. Smith. London: Longman, Green, Longman & Roberts.

[Scotts]. 1906. *Two Centuries of Shipbuilding by the Scotts at Greenock*. London: Offices of 'Engineering'.

Scowen, F. 1976–77. 'Transoceanic submarine telegraphy', *Newcomen Society Transactions* 48: 1–10.

Secord, Anne. 1994. 'Science in the pub: artisan botanists in early 19th-century Lancashire', *History of Science* 32: 269–315.

———. 1996. 'Artisan botany'. In Nick Jardine, J.A. Secord and E.C. Spary, eds, *Cultures of Natural History*, pp. 378–93. Cambridge: Cambridge University Press.

Secord, J.A. 1981–82. 'King of Siluria: Roderick Murchison and the Imperial theme in nineteenth-century British geology', *Victorian Studies* 25: 413–42.
——. 1986a. *Controversy in Victorian Geology. The Cambrian–Silurian Dispute*. Princeton: Princeton University Press.
——. 1986b. 'The Geological Survey of Great Britain as a research school: 1839–1855', *History of Science* 24: 223–75.
——. 2000. *Victorian Sensation. The Extraordinary Publication, Reception, and Secret Authorship of Vestiges of the Natural History of Creation*. Chicago: University of Chicago Press.
——. 2003. 'From Miller to the Millennium'. In L. Borley, ed., *Celebrating the Life and Times of Hugh Miller*, pp. 328–37. Cromarty: Cromarty Arts Trust.
Sedgwick, Adam. 1833. 'Presidential address', *Report of the British Association for the Advancement of Science* 3: xxvii–xxxii.
Sekon, G.A. 1895. *A History of the Great Western Railway. Being the Story of the Broad Gauge*. London: Digby, Long.
Serpell, James. 1986. *In the Company of Animals. A Study of Human–Animal Relationships*. Oxford: Blackwell.
Shapin, Steven. 1981. 'Of gods and kings: natural philosophy and politics in the Leibniz–Clarke disputes', *Isis* 72: 187–215.
——. 1984. 'Pump and circumstance: Robert Boyle's literary technology', *Social Studies of Science* 14: 481–520.
——. 1988. 'The house of experiment in seventeenth-century England', *Isis* 79: 373–404.
——. 1991. '"A scholar and a gentleman": the problematic identity of the scientific practitioner in early modern England', *History of Science* 29: 279–327.
——. 1994. *A Social History of Truth. Civility and Science in Seventeenth-Century England*. Chicago and London: University of Chicago Press.
——. 1996. *The Scientific Revolution*. Chicago and London: University of Chicago Press.
—— and Simon Schaffer. 1985. *Leviathan and the Air-Pump. Hobbes, Boyle, and the Experimental Life*. Princeton: Princeton University Press.
Shaw, Isaac. 1970 [1831]. *Views of the Most Interesting Scenery on the Liverpool and Manchester Railway*. Newcastle: Graham.
Shields, John. 1949. *Clyde Built*. Glasgow: Maclellan.
Shinn, Terry. 1980. *L'École polytechnique, 1794–1914*. Paris: Presses Universitaires de la Fondation Nationale des Sciences Politiques.
Shortland, Michael (ed.). 1996. *Hugh Miller and the Controversies of Victorian Science*. Oxford: Clarendon Press.
Siemens, Werner von. 1845. 'Ueber die Unwendung der erhitzten Luft als Triebkraft', *Dingler's Polytechnisches Journal* 97: 324–29.
——. 1966. *Inventor and Entrepreneur. Recollections of Werner von Siemens*. Second English edition. London: Lund Humphries.
Silliman, Benjamin. 1838. 'Notice of the electro-magnetic machine of Mr Thomas Davenport, of Brandon, near Rutland, Vermont, U.S.', *Annals of Electricity* 2: 257–64.
Simmons, Jack. 1991. *The Victorian Railway*. New York: Thames and Hudson.
—— and Gordon Biddle (eds). 1997. *The Oxford Companion to British Railway History from 1603 to the 1990s*. Oxford: Oxford University Press.
Singer, Charles (ed.). 1954–58. *A History of Technology*. 5 vols. Oxford: Clarendon Press.
Skelton, Oscar. 1986. 'The evolution of Canada's railway network'. In B. Yenne, ed., *The History of North American Railroads*, pp. 123–61. London: Bison Books.

Slaven, Anthony. 1975. *The Development of the West of Scotland: 1750–1960*. London: Routledge & Kegan Paul.

Slotton, Hugh Richard. 1994. *Patronage, Practice, and the Culture of American Science. Alexander Dallas Bache and the U.S. Coast Survey*. Cambridge: Cambridge University Press, 1994.

Smiles, Samuel. 1860. *Self-Help; with Illustrations of Character and Conduct*. London: John Murray.

——. 1861–62. *Lives of the Engineers, with an Account of their Principal Works, Comprising also a History of Inland Communication in Britain*. 3 vols. London: Murray.

——. 1865. *Boulton and Watt. Principally from the Original Soho MSS; Comprising also a History of the Invention and Introduction of the Steam-Engine*. London: John Murray.

——. 1878. *Lives of the Engineers. The Steam-Engine. Boulton and Watt*. Revised edition. London: John Murray.

——. 2002 [1866]. *Self-Help; with Illustrations of Character, Conduct, and Perseverance*. Ed. P.W. Sinnema. Oxford: Oxford University Press. Second edition. First published 1866.

Smith, Adam. 1976 [1776]. *An Inquiry into the Nature and Causes of the Wealth of Nations*. Eds R.H. Campbell and A.S. Skinner. 2 vols. Oxford: Clarendon Press.

Smith, Alan. 1977–78. 'Steam and the city: the Committee of Proprietors of the Invention for Raising Water by Fire, 1715–1735', *Transactions of the Newcomen Society* 49: 5–20.

Smith, Barbara M.D. and J.L. Moilliet. 1967. 'James Keir of the Lunar Society', *Notes and Records of the Royal Society of London* 22: 144–54.

Smith, Crosbie. 1985. 'Geologists and mathematicians: the rise of physical geology'. In P.M. Harman, ed., *Wranglers and Physicists. Studies in Cambridge Physics in the Nineteenth Century*, pp. 49–83. Manchester: Manchester University Press.

——. 1989. 'William Hopkins and the shaping of dynamical geology: 1830–1860', *British Journal for the History of Science* 22: 27–52.

——. 1998a. *The Science of Energy. A Cultural History of Energy Physics in Victorian Britain*. Chicago and London: University of Chicago Press.

——. 1998b. '"Nowhere but in a Great Town": William Thomson's spiral of classroom credibility'. In Crosbie Smith and Jon Agar, eds, *Making Space for Science. Territorial Themes in the Shaping of Knowledge*, pp. 118–46. Basingstoke: Macmillan.

—— and Ian Higginson. 2001. 'Consuming energies: Henry Adams and the "tyranny of thermodynamics"', *Interdisciplinary Science Reviews* 26: 103–11.

—— and Ian Higginson. 2004. '"Improvised Europeans": science and reform in the *North American Review*'. In Geoffrey Cantor and Sally Shuttleworth, eds, *Science Serialized. Representation of the Sciences in Nineteenth-Century Periodicals*, pp. 149–79. Cambridge, MA and London: MIT Press.

——, Ian Higginson and Phillip Wolstenholme. 2003a. '"Avoiding equally extravagance and parsimony": the moral economy of the ocean steamship', *Technology and Culture* 44: 443–69.

——, Ian Higginson and Phillip Wolstenholme. 2003b. '"Imitations of God's own works": making trustworthy the ocean steamship', *History of Science* 41: 379–426.

—— and M. Norton Wise. 1989. *Energy and Empire. A Biographical Study of Lord Kelvin*. Cambridge: Cambridge University Press.

Snelders, H.A.M. 1990. 'Oersted's discovery of electromagnetism'. In Andrew Cunningham and N. Jardine, eds, *Romanticism and the Sciences*, pp. 228–40. Cambridge: Cambridge University Press.

Sobel, Dava. 1996. *Longitude. The True Story of a Lone Genius who Solved the Greatest Scientific Problem of his Time*. London: Fourth Estate.

Sorrenson, Richard. 1996. 'The ship as a scientific instrument in the eighteenth century', *Osiris* 11: 221–36.

Stafford, R.A. 1989. *Scientist of Empire. Sir Roderick Murchison, Scientific Exploration and Victorian Imperialism*. Cambridge: Cambridge University Press.

Staley, Richard. 1993. *Empires of Physics. A Guide to the Exhibition*. Cambridge: Whipple Museum of the History of Science.

Standage, Tom. 1998. *The Victorian Internet. The Remarkable Story of the Telegraph and the Nineteenth Century's Online Pioneers*. London: Weidenfeld and Nicolson.

Staudenmaier, John M. 1989a. *Technology's Storytellers. Reweaving the Human Fabric*. Cambridge, MA and London: MIT Press.

——. 1989b. 'The politics of successful technologies'. In Stephen H. Cutcliffe and Robert C. Post, eds, *In Context: History and the History of Technology. Essays in Honor of Melvin Kranzberg*, pp. 150–71. Bethlehem, London and Toronto: Lehigh University Press/Associated Universities Press.

Stephens, C. 1989. 'The most reliable time: William Bond, the New England railroads and time awareness in 19th-century America', *Technology and Culture* 30: 1–24.

Stevenson, Robert Louis. 1907. *Memoir of Fleeming Jenkin, F.R.S., LL.D.* London: Cassell.

Stewart, Larry. 1992. *The Rise of Public Science. Rhetoric, Technology, and Natural Philosophy in Newtonian Britain, 1660–1750*. Cambridge: Cambridge University Press.

Sturgeon, William. 1836–37. 'Description of an electro-magnetic engine for turning machinery', *Annals of Electricity* 1: 75–78.

Swade, Doran. 2001. *The Cogwheel Brain. Charles Babbage and the Quest to Build the First Computer*. London: Abacus.

Swinbank, Peter. 1969. 'James Watt and his shop', *Glasgow University Gazette* 59: 5–8.

Tann, Jennifer. 1974. 'Suppliers of parts: the relationship between Boulton and Watt and the suppliers of engine components, 1775–1795', *Birmingham and Warwickshire Archaeological Society Transactions* 86: 167–77.

——. 1978. 'Marketing methods in the international steam engine market: the case of Boulton and Watt', *Journal of Economic History* 38: 363–91.

——. 1979–80. 'Mr Hornblower and his crew: Watt engine pirates at the end of the 18th century', *Transactions of the Newcomen Society* 51: 95–109.

Taylor, Dr. 1841. *Steam Superseded: An Account of the Newly-Invented Electro-Magnetic Engine, for the Propulsion of Locomotives, Ships, Mills, &c. as also to the Processes of Spinning, Turning, Grinding, Sawing, Polishing &c. and Every Species of Mechanical Movement*. London: Sherwood, Gilbert and Piper.

Thackray, Arnold. 1974. 'Natural knowledge in cultural context: the Manchester model', *American Historical Review* 79: 672–709.

Thomas, R.H.G. 1980. *The Liverpool & Manchester Railway*. London: Batsford.

Thompson, E.P. 1967. 'Time, work-discipline and industrial capitalism', *Past & Present* 38: 56–97.

Thompson, Robert Luther. 1947. *Wiring a Continent. The History of the Telegraph Industry in the United States 1832–1866*. Princeton: Princeton University Press.

Thompson, Silvanus P. 1910. *The Life of William Thomson. Baron Kelvin of Largs*. 2 vols. London: Macmillan.

Tolstoy, L.N. 1957 [1869]. *War and Peace*. 2 vols. Trans. R. Edmonds. Harmondsworth: Penguin. First published 1865–69.

Topham, Jonathan. 2000. 'Scientific publishing and the reading of science in early nineteenth-century Britain: a historiographical survey and guide to sources', *Studies in History and Philosophy of Science* 31A: 559–612.

Torrens, Hugh. 1994. 'Jonathan Hornblower (1753–1815) and the steam engine: a historiographic analysis'. In Denis Smith, ed., *Perceptions of Great Engineers. Fact and Fantasy*, pp. 23–34. London: Science Museum.

——. 1995. 'Mary Anning (1799–1847) of Lyme: "the greatest fossilist the world ever knew"', *British Journal for the History of Science* 28: 257–84.

——. 2002. *The Practice of British Geology, 1750–1850*. Aldershot: Ashgate Variorum.

Trevithick, Francis. 1872. *Life of Richard Trevithick, with an Account of His Inventions*. London: Spon.

Tucker, Jennifer. 1997. 'Photography as witness, detective, and impostor: visual representation in Victorian science'. In Bernard Lightman, ed., *Victorian Science in Context*, pp. 378–408. Chicago: University of Chicago Press.

Tunbridge, Paul. 1992. *Lord Kelvin. His Influence on Electrical Measurements and Units*. London: Peter Peregrinus for Institution of Electrical Engineers.

Turner, James. 1980. *Reckoning with the Beast. Animals, Pain and Humanity in the Victorian Mind*. Baltimore: Johns Hopkins University Press.

Uglow, Jenny. 2002. *The Lunar Men. The Friends who Made the Future*. London: Faber & Faber.

Ure, Andrew. 1835. *The Philosophy of Manufactures; or an Exposition of the Scientific, Moral and Commercial Economy of the Factory System of Great Britain*. London: Charles Knight.

van der Pols, K. 1973–74. 'Early steam pumping engines in the Netherlands', *Transactions of the Newcomen Society* 46: 13–16.

Vaughan, Adrian. 1991. *Isambard Kingdom Brunel. Engineering Knight-Errant*. London: John Murray.

Vignoles, O.J. 1889. *Life of Charles Blacker Vignoles*. London: Longmans, Green.

Wachorst, Wyn. 1981. *Thomas Alva Edison. An American Myth*. Cambridge, MA: MIT Press.

Walker, Charles. 1969. *Thomas Brassey. Railway Builder*. London: Muller.

Walker, James. 1839. 'Address of the President to the annual general meeting, January 1839', *Minutes of Proceedings of the Institution of Civil Engineers* 1: 17–18.

——. 1841. 'Address of the President to the annual general meeting, February 1841', *Minutes of Proceedings of the Institution of Civil Engineers* 3: 25–26.

Walker, James Scott. 1830. *An Accurate Description of the Liverpool and Manchester Railway, the Tunnel, the Bridges, and Other Works throughout the Line, with a Sketch of the Objects which it Presents Interesting to the Traveller or Tourist*. Manchester: Liverpool and Cheshire Antiquarian Society.

Warwick, Andrew. 2003. *Masters of Theory. Cambridge and the Rise of Mathematical Physics*. Chicago and London: University of Chicago Press.

Watson, Garth. 1989. *The Smeatonians. The Society of Civil Engineers*. London: Thomas Telford.

Watters, Brian. 1998. *Where Iron Runs Like Water! A New History of Carron Iron Works, 1759–1982*. Edinburgh: John Donald.

Watts, Isaac, F.K. Barnes, J.R. Napier and W.J.M. Rankine. 1866. *Shipbuilding, Theoretical and Practical*. London: William MacKenzie.

Weiss, John Hubbel. 1982. *Making of Technological Man. The Social Origins of French Engineering Education*. London and Cambridge, MA: MIT Press.

Whipple, A.B.C. 1980. *The Clipper Ships*. Amsterdam: Time-Life Books.

Whiting, Charles Edwin. 1932. *The University of Durham 1832–1932*. London: Sheldon Press.

Whitworth, Joseph. 1854. *New York Industrial Exhibition. Special Report of Mr Joseph Whitworth*. London: Harrison and Sons.

Wiener, Martin. *English Culture and the Decline of the Industrial Spirit, 1850–1980*. Cambridge: Cambridge University Press.

Williams, Archibald. 1904. *The Romance of Modern Locomotion. Containing Interesting Descriptions (in Non-technical Language) of the Rise and Development of the Railroad Systems in all Parts of the World*. London: C. Arthur Pearson.

Williamson, G. 1856. *Memorials of the Lineage, Early Life, Education, and Development of the Genius of James Watt*. Edinburgh: Constable

Willis, Robert. 1841. *Principles of Mechanism, Designed for the Use of Students in the Universities, and for Engineering Students Generally*. London: J.W. Parker.

——. 1851. *A System of Apparatus for the Use of Lecturers and Experimenters in Mechanical Philosophy*. London: John Weale.

Wilson, George. 1852. *Electricity and the Electric Telegraph*. London: Longman.

——. 1855. *What is Technology? An Inaugural Lecture Delivered in the University of Edinburgh on November 7, 1855*. Edinburgh: Sutherland and Knox and London: Simpkin, Marshall.

Winter, Alison. 1994. ' "Compasses all awry": the iron ship and the ambiguity of cultural authority in Victorian Britain', *Victorian Studies* 38: 69–98.

——. 1998. *Mesmerized. Powers of Mind in Victorian Britain*. Chicago: University of Chicago Press.

Wise, M. Norton (ed.). 1995. *The Values of Precision*. Princeton: Princeton University Press.

—— (with the collaboration of Crosbie Smith). 1989–90. 'Work and waste: political economy and natural philosophy in nineteenth-century Britain (I)–(III)', *History of Science* 27: 263–301, 391–449; 28: 221–61.

Wishaw, Francis. 1842. *The Railways of Great Britain and Ireland Practically Described and Illustrated*. Second edition. London: John Weale. Reprinted Newton Abbot: David & Charles, 1969.

Wood, Nicholas. 1825. *A Practical Treatise on Rail-Roads, and Interior Communication in General*. London: Knight & Lacey.

Wood, Paul (ed.). 2000. *The Scottish Enlightenment. Essays in Reinterpretation*. Rochester, NY and Woodbridge: University of Rochester Press.

Yates, Joanne. 1996. 'The telegraph's effects on nineteenth-century markets and firms', *Business and Economic History* 15: 149–93.

Yenne, Bill (ed.). 1986a. *The History of North American Railroads*. London: Bison Books.

Yenne, Bill. 1986b. 'The big four and the transcontinental railroad'. In B. Yenne, ed., *The History of North American Railroads*, pp. 55–79. London: Bison Books.

Yeo, Richard R. 2001. *Encyclopaedic Visions. Scientific Dictionaries and Enlightenment Culture*. Cambridge: Cambridge University Press.

Youatt, William. 1831. *The Horse; with a Treatise on Draught*. London: Society for the Diffusion of Useful Knowledge.

Zola, Émile. 1890. *La bête humaine*. Paris: Charpentier.

Index

Note: The page numbers in italics refer to figures.

unity of, 128
and visual telegraph, 186–7
and Watt's engine, 235
empire(s),
-builders as engineers, 1, 11
and Chappe's visual telegraph, 180,
181–7, 188
cultural construction of, 41, 226
and electric telegraphy, 198–9
geo-political, 1, 226
of New World railroads, 162–71
personal and business, 1, 64, 226,
254–8
the Ottoman, 246
of railways, 130, 136, 145–56,
164, 258
shipping, 71, 257
of telegraphy, 207, 213–24
see also Empire, the British
Encyclopaedias
Cabinet Cyclopaedia, 239
Edinburgh Encyclopaedia, 240
Encyclopaedia Britannica, 83, 240
Encyclopedie, 239, 244–5
Rees' Cyclopaedia, 239
energy
dissipation of, 164
science of, 75, 272n107
engine
carbonic acid gas, *see* gaz engine
compound steam, *see* compound
steam engine
Cornish, *see* steam engine, Cornish
electro-magnetic, 45, 75–81
flame, 70
fire- , 43, 44, 46, 72
'perfect', *see* Watt, James
pumping, *see* steam engine, stationary
see also air engine; power; Newcomen
engine; steam engine; watermills;
windmill
Engineer, The, 207, 241
engineering science, 3, 74, 119, 242,
243, 250, 259n9, 287n67
Engineering, 241
engineers
as 'artists', 44, 75
canal, 160
education for, 226, 235–8, 240,
242, 252–3

education of telegraph, 216–17,
219–20
electrical, 219–20
as empire-builders, 1, 11
and experiments, 10, 11, 32, 156,
235, 145–6, 235
as gentlemen, 8, 62, 138–9, 193,
219, 226, 236, 237, 254–8
images of, 242–5
institutions of, 237–8
journals for, 240–2
as masculine, 10, 66, 235
models of, 230, 232–4, 237
and precision, 233, 234, 236
as publicists, 7, 9, 70, 71, 72, 73, 75,
76, 77, 83, 267n137
and religion, 226, 231–2,
245–54
as saboteurs, 56
as system builders, 9, 61
textbooks for, 237, 242
training of, 51, 52, 57, 60, 62, 76,
160–2
and training of naval architects,
287n47
and trust, 226, 233, 255–6
as 'visionaries', 2
Whewell's textbook for, *140*
see also systems
English and Irish Magnetic Telegraph
Co., 197, 205
Enlightenment, 13, 22
cultures of 'improvement', 131, 132
Kant and, 13–14
love of clubs, 237
and mapping, 14–16
Scottish, 41
and steamships, 88
values of rationality and independent
judgement, 165
Episcopal Church, *see* religion, and
Scottish Episcopal Church
Ericsson, 71, 73, 74, 112
Ericsson, John
and caloric engine, 71–5, 112,
138–9
and Rainhill trials, 138–9
'regenerator', 70, 72
and skill, 139
and spectacle, 139

Rastrick, John, 137
Rathbone, Samuel, 121
reading, *see* cultures, print
Red Sea and Indian Telegraph Co.,
    211–12
Reed, E.J., 287n47
Rees, Abraham, 239
*Reflections on the Decline of Science in
    England*, 21, 18
Reflexions on the Motive Power of Fire,
    see Carnot, Sadi
Reformation, 239
refrigerator, 6
reliability
    of instruments, 9
    of steam engines, 52, 82
    of technology, 7, 8, 10
    see also failure, technological
religion
    Brunel and, 253
    and Calvinist doctrines, 165, 249–50
    and Church of Scotland, 247, 249
    and Church of England, 13, 22, 34,
        248, 251–2, 253
    and Empire, 231–2, 245–6, 252–3
    and engineering, 226, 231–2, 245–54
    and engineers, 226, 231–2, 245–54
    of evangelicals, 29, 30–1, 96, 163–4,
        239, 243, 246, 247, 250–1, 257
    and Free Church of Scotland, 163–4, 247
    and 'imitation of Christ', 250
    of liberal Anglicans, 22, 34
    and natural philosophy, 243
    and natural theology, 249, 250–1, 252
    and popular beliefs, 247
    of Presbyterians, 30, 108, 109, 123
    and providence, 163–4, 245–54
    and Puritanism, 164, 165, 169–70
    and Quakers, 22, 195, 256
    of Roman Catholic Church, 165,
        167–8, 252
    and science conflicts, 245
    and Scottish Episcopal Church, 109
    and sin, 164
    and steamships, 103, 109
    of Unitarians, 30, 34, 120–1, 123, 126,
        141, 165–6, 244, 246, 262n64
    and Whig Revolution, 13
    see also Presbyterian culture;
        Unitarians

Rennie, John, 58, 59, 60, 68, 83
Rennie, John, Jr., 68
Renshaw Street Chapel, *see* Unitarians,
    in Liverpool
Repertory of Arts, 241–2
reputation
    of Airy, 19, 32–3
    of Charles Bright, 208
    of Isambard Kingdom Brunel, 66,
        150, 154–5, 156, 256
    of caloric engine, 70, 71, 72, 75
    of Clyde shipbuilders, 33, 111, 124
    confirmed by 'doing', 60
    of Cooke as telegraph engineer, 193,
        196, 197
    of Cunard, 110–12
    of Thomas Davenport, 76
    of Robert Davidson, 80
    of Elder, 117
    of electro-magnetic engine, 77
    of engineers, 241
    of John Ericsson, 70, 71, 75
    of gaz engine, 66
    of Alfred Holt, 121, 126–7
    at Institution of Civil Engineers,
        262n80
    of Robert Napier, 108–11
    of railway engineers, 136, 151, 156
    and showmanship, 59
    of steam engine, 52, 53, 56, 61
    of the Stephensons, 141, 152
    of telegraph and telegraph engineers,
        179–80, 195–7, 214, 215
    of Thomson's patent compass, 33
    of Watt, 52, 55, 56, 61, 62, 63, 83,
        86, 240
    see also character; trust
Restoration Court, 13
Reuter, Julius, 204, 209
Revel, Jacques, 14–15
Riddle, Edward, 233
roads, 131, 134, 135
Robertson, Joseph, 241
Robinson, Eric, 62
Robison, John
    On engineering education, 229
    and genesis of James Watt's engine, 49
    Glasgow College student, 44, 46, 48
    and international brain drain, 65
    on steam engine, 83

New England, 165–6
Stephensons as, 141
United States
civil war in, 199
Coastal Survey, 234
decline of railroads in, 163
electric telegraphy and, 197–9
faith in progress of, 162–3
and the 'frontier', 198–9
invention in, 162–3
railroad systems of, 149, 162–71, 181
railroads and power in, 162
early railways in, 162
rising power of, 40, 234
and river steamboats in, 165
superior sailing ships of, 97
units, *see* standards
Universal Private Telegraph Co., 197
University College London, 161,
    189, 251
University of Glasgow, 92, 117
Ure, Andrew, 230
utilitarianism, 108
Utrecht Observatory, 189

Vail, Alfred, 198
Valparaiso, 117, *118*
Van Horne, William, 174, 175
Vanderbilt, Cornelius, 165
Varley, C.F., 209, 212, 214, 217
Vestiges of Creation, 149
Victoria, Queen, 194, 195, 208, 209,
    214, 221–2
Vignoles, Charles Blacker, 138–9, 150,
    161, 233
visual culture, *see* cultures, visual
voltaic pile, 188
von Buch, Leopold, 39
von Soemmering, S.T., 188–9

Wajcman, Judy, x
Walker, Joshua, 190, 191
Walker, James, 236, 251
War and Peace, 93
waste (of power)
and I.K. Brunel's time, 68, 69
avoided in John Ericsson's caloric
    engine, 70
Alfred Holt on, 122–7
as moral issue, 3, 42

in Newcomen engines, 52
Rankine on, 114–16
avoided in James Watt's perfect
    engine, 48, 49, 51
see also Presbyterian culture
watermills, 42, 46
waterwheels, 115
Watt, James
and air engine, 69
and Albion Mills, 135
apprenticeship, 46, 54
Arago's *éloge*, 85
arrival in Birmingham, 54, 55
and Joseph Black, 235
and Boulton, 227, 255
business 'empire' of, 45, 49, 52, 56
and canal work, 54, 61
Thomas Carlyle on, 84, 85
as chemist, 55
as craftsman, 4, 82
and 'credit', 61
death of, 83
and Delftfield Pottery, 45
and Desaguliers, 144
and education of son, 62
and Enlightenment culture, 132
as 'genius', 85
Glasgow connections, 44, 45, 46, 233
'governor' of, 58, *59*, 72,
on high-pressure steam, 81, 134–5
house of, 257
as icon, 82–7
images of, 82, 244, 246
indicator diagram of, 236
as instrument maker, 44, 45, 46, 54
as inventor, 44
kettle of, 48, 82, 85, 86
memorial of, 83, 84
and miracles, 63
as merchant, 45
and moral values, 249–50
and 'parallel motion', 58, *59*
partnership with John Craig, 45
and patents, 45, 53, 56, 62, 63, 81
'perfect' engine, 42, 48, 49, 74;
    see also standards
as philosopher, 244
and Presbyterian culture, 45
re-invention as engineer, 132
reputation of, 240

CPSIA information can be obtained at www.ICGtesting.com
Printed in the USA
LVOW080400221211

260575LV00003B/1/A